S0-BLL-997

QUANTITATIVE ANALYSIS OF ORGANIC MIXTURES

Part One: General Principles

QUANTITATIVE ANALYSIS OF ORGANIC MIXTURES

Part One: General Principles

T. S. MA, Ph.D.

Professor of Chemistry
City University of New York

ROBERT E. LANG, Ph.D.

Assistant Area Director, New York Seaport, U.S. Customs Service
Formerly Director, U.S. Customs Laboratory, New York

A Wiley-Interscience Publication

JOHN WILEY & SONS
New York . Chichester . Brisbane . Toronto

Library of Congress Cataloging in Publication Data
Ma, Tsu Sheng, 1911-
Quantitative analysis of organic mixtures.

"A Wiley-Interscience publication."
Includes bibliographical references and indexes.
CONTENTS: pt. 1. General principles.
1. Chemistry, Analytic – Quantitative. 2. Chemistry, Organic. I. Lang, Robert E., joint author. II. Title.
QD271.7.M3 547′.35 78-23202
ISBN 0-471-55800-1

Printed in the United States of America

10 9 8 7 6 5 4 3 2 1

Preface

It is generally known that most organic samples submitted to quantitative analysis are either impure materials or mixtures containing several major components. On the other hand, in undergraduate and advanced courses in analytical chemistry the student learns primarily how to deal with pure compounds, while treatises on organic analysis touch upon only a few cases of mixtures. In an attempt to fill the needs of my students and those of practicing analysts whom I met during several lecture tours around the world, I have prepared this book which highlights the analysis of mixtures and impure samples.

For convenience, this work is divided into two volumes. Part One presents the general principles in handling organic materials intended for quantitative analysis. It deals with the problems of sample preparation, whether or not separation is necessary, how to choose the most suitable finishing technique, and so on. I have written this part to serve as a textbook for advanced students, a guidebook in analytical laboratories, and a reference volume for the research organic chemist who wishes to determine the constituents in a reaction mixture. Experimental details are not described, but adequate references are given so that the reader can easily locate the source and obtain the laboratory procedures.

Part Two deals with organic mixtures in fields such as pharmaceutical formulations and natural products and is intended primarily for use by the practicing analyst. It includes discussions of the type of compounds commonly encountered in the particular field and current methods for determining them followed by an extensive survey of the literature on mixtures for which analyses have been reported. Dr. Robert E. Lang, who was my assistant at New York University, collaborated in writing this part. Dr. Lang has served in supervisory capacities for three decades in the analysis of foods, drugs, industrial chemicals, and commercial products; his experience in working with a wide range of actual samples was most valuable.

Although only a few names are mentioned in the chapters, I am in debt to the many experts whom I consulted in various areas. My daughter Juliana Mei-Mei helped me in compiling the references and typing parts of the manuscript.

New York, New York
February 1979

T. S. Ma

Contents

Contents
Part Two

Tables

CHAPTER 1

Introduction: Problems in the Analysis of Organic Mixtures

1.1 GENERAL REMARKS

Whereas many treatises are available for organic analysis, there is no book that deals with the general methods for determining one or more organic compounds in complex organic mixtures. The present volume is an attempt to fill this gap.

Quantitative analysis of organic mixtures is much more complicated than quantitative inorganic analysis. For instance, if a procedure has been worked out for the determination of chloride ion, it can be utilized to analyze all kinds of samples; possible interferences can be anticipated and, if present, they can be removed without impeding the analysis. In contrast, estimation of the content of a given organic compound, say salicylic acid, in a complex and unknown mixture frequently cannot follow a fixed method, because the analytical results may be vitiated due to the nature of the sample and the presence of extraneous substances.

The purpose of this monograph is fourfold:

1. To guide the reader in selecting a suitable method to analyze a new mixture.
2. To suggest ways and means for checking a procedure already in use, when a discrepancy arises.
3. To show how to improve the existing methods and develop new methods.
4. To review the literature on the analysis of organic mixtures.

These topics are discussed fully in the various chapters. A brief resume of the first three subjects and some examples are given below.

1.1.1 Selecting an Analytical Method

When it is desired to ascertain the percentage of a known compound in a given sample, the first step usually involves a literature search to determine the availability of official, standard, and recommended methods. It should be noted, however, that the published methods may not be applicable to the new problem or may not suit the particular sample. Some procedures are liable to give erroneous results. For example, Herry and Janmot[1] found that the IUPAC method for the saponification value of alkyl resins,[2] in which a solution of the resin in toluene is heated for 1 hr with 0.5 *N* ethanolic KOH, gives low values. In the recommended method for chemical assay of sennoside in senna fruit and leaf,[3] after successive extractions with HCl in ethanol, $CHCl_3$, ether, and 1 *N* NaOH, the aqueous extract is measured at 515 and 440 nm; if the ratio of E_{515} to E_{440} is <1.25, the result is rejected. Crompton[4] reported that the isoquinoline conductometric titration procedure of Bonitz is unsuitable for determining the individual components of mixtures of triethyl- and hydrodiethylaluminum, although it does give an approximation of the total activity in samples containing up to 60% of the latter component. In the comparison of analytical values for commercial heparin, Kavanagh and Jaques[5] observed a wide variation in composition and no relationship could be established between biological activity and other analytical data.

One should bear in mind that the chemical reaction on which an analytical method is based is not restricted to the organic compound to be determined. Thus the colorimetric determination of thalidomide[6] utilizing the hydroxamic acid–Fe^{3+} complex is vitiated by the presence of foreign substance that forms hydroxamic acids.[7] The determination of quinol by oxidation with cupric tartrate[8] demands the absence of other reducing substances in the sample. In the determination of aliphatic nitriles by reaction with alkaline hydrogen peroxide followed by titration of the remaining alkali, all compounds that are oxidized under the experimental conditions interfere.[9]

1.1.2 Checking a Method

1.1.2.1 Inconsistency of Analytical Results

A method for quantitative analysis of organic compounds in a mixture should be checked and compared with other methods to establish its reliability and merit. Wilson[10] has discussed the problem of adequately describing analytical methods so that their advantages and disadvantages may be compared unambiguosly. Standard samples are frequently used to test a chosen method; these samples should be prepared by specified procedures, such as those recommended by the Division of Chemical Standards, National Physical Laboratory, England.[11]

It should be recognized that uncertainty exists in any method. Swietoslawska[12] has presented theoretical calculations to show the relationship between uncertainty in the composition of a standard substance and error in a determination in which the standard is used, directly or indirectly, for comparison. It is not uncommon to obtain erratic results when a known procedure is utilized to analyze certain samples. The causes vary. For example, Dandoy et al.[13] found that mono- and diesters of diols with perfluorinated acids gave inexact results of poor reproducibility when subjected to conventional analysis for ester and hydroxyl groups and for residual acidity. Legradi[14] found that his earlier method for determining methanol and formaldehyde in mixtures was satisfactory only if a blank determination was carried out. Anderson and Zaidi[15] studied conditions for the Zeisel reaction and found that many thioalkyl compounds do not yield alkyl iodide. In a method for the determination of carotene, Quackenbush[16] reported that variations in results obtained could not be attributed mainly to deterioration. In the fluorometric determination of corticosteroids in plasma, de Langen and Whittlestone[17] found that the unreliable results obtained were due to varying uptake of water by the solvent. In the determination of amino acid composition of food proteins, Knipfel et al.[18] ascribed inter- and intralaboratory variation in results to sample manipulation rather than to the subsequent analysis.

1.1.2.2 Effect of the Environment

The environment in which a compound is determined may affect the analytical result tremendously. Temperature and pH are two factors frequently studied. For example, Meakin et al.[19] investigated the effects of heat, pH, and buffer materials on the determination of sulfacetamide and sulfanilamide. Matrka and Sagner[20] studied the influence of temperature on the nitrite titration of aromatic amines. Hwang[21] reported that the conventional method for determining reducing sugars in samples of high sucrose content is unsatisfactory because of the high alkalinity of the Fehling's solution used. Suzuki and Ishida[22] studied the influence of pH on the iodine azide reaction for determining various sulfur compounds. For the colorimetric determination of fatty acids using copper nitrate–triethanolamine–acetic acid, Weenink[23] found that this reagent can vary during storage, and, if the pH falls below 7.1, the sensitivity of the method is greatly affected. Reiss[24] described a procedure to determine adrenaline and noradrenaline in the presence of one another using the same $NaNO_2$ reagent and measuring at the same wavelength, but at pH 1.9 and 1.1, respectively.

The influence of extraneous substances on an analytical method may be beneficial, or it may be undesirable. Studying the effect of neutral salts on potentiometric acid–base titration of weak amines, Kreshkov et al.[25] found that the potential changes at the end point are more distinct in the presence of salts of alkali or alkaline earth metals than in their absence. For the amperometric titration of carbonyl compounds with 2,4-dinitrophenylhydrazine,[26] Zobov and Lyalikov[27] recommended adding 0.1 g of activated carbon to ensure complete coagulation of the precipitate; however, high results were obtained in the presence of 0.5 g of activated carbon. In the determination of morphine, Porsius[28] traced the cause of error to the presence of codeine, while Hundley and Castillo[29] recommended removing oxidation products of morphine by elution with ethyl ether. When an electron-capture detector is used for pesticide residue analysis by gas chromatography, Bevenue et al.[30] found that distilled water collected from an apparatus with plastic fittings contains contaminants yielding a response.

When extraction is involved in the analytical method, the results may be influenced in a variety of ways. For instance, Zak et al.[31] reported that the spectrophotometric determination of barbiturates in blood serum may give incorrect results because of spectral distortion by the $CHCl_3$ used for extraction. During studies on natural fluorescent compounds, Crosby and Aharonson[31a] observed that paper and thin-layer chromatograms of extracts exhibited fluorescent zones attributed to constituents of the solvents used for extracting the samples. For the extraction–photometric determination of alkaloids using acid dyes, Babko and Konyushko[32] observed that the extent of extraction decreases with increasing pH of the solution, but neither the pH nor the color of the dye affects the color of the extract. Petrashkavich and Starobinets[33] studied the influence of halide or thiocyanate ions on the extraction of alkaloids and found that the salting-out effect increases in the order Cl^-, Br^-, I^-, SCN^-, F^-. For the determination of amines and quaternary bases as complexes with picrate, Gustavii[34] studied the relationship between extraction constants and the nature of the cation and the organic phase; side effects were sometimes observed. For the determination of dicoumarol in plasma, which involves extraction at $pH < 1$ into an organic phase, Nagashima et al.[35] found it to be unsuitable at concentration of > 50 μg of the compound per ml of plasma; this is due, not to limited solubility in the organic phase, but to a concentration- and pH-dependent interaction of dicoumarol with plasma protein.

1.1.2.3 *Interferences and Need for Their Removal*

It is unusual that an organic compound can be determined in a mixture without consideration of the possible interference due to the other substances present in the sample. The need to remove the interferences depends on the specific problem. The following cases may serve as illustrations. In the analysis of certain amino acids and carbohydrates, Brummel et al.[36] observed that low levels of some metals (e.g., copper, iron, manganese, cobalt) prevent accurate determination; the extent of interference varies with the method used and the nature of the sample. For the determination of carbohydrates with anthrone, Tong et al.[37] found that NaN_3 (added as bacteriostatic agent) seriously interfered and the interference could be reduced by increasing the concentration of the anthrone. Hydroxy indoles interfere with the determination of glucose by methods involving glucose oxidase–peroxidase–chromogen; according to Nelson and Huggins,[38] this interference results from competition as H-donors by the indoles with the chromogenic donor (e.g., guaicol, *o*-dianisidine); dilution before analysis can eliminate the interference. Bancher et al.[39] studied the influence of glucose, fructose, sucrose, and maltose on the determination of ascorbic acid; the results obtained were unaffected by up to 20 mg of glucose or maltose, or 5 mg of fructose, but preliminary purification by thin-layer chromatography is necessary with higher sugar concentrations. In fluorometric methods for determining estrogens in urine, the fluorescence is quenched by a level of glucose often encountered in patients with glycosuria. Worth [40] recommended removal of the glucose by adding 100 mg of $NaBH_4$ to 5 ml of urine and incubating the mixture at 37°C for 15 min. Certain carbonyl compounds interfere with the 2,4,6-trinitrobenzene sulfonic acid method for determining amino groups; Burger[41] reported that the interference was greatly reduced when the $NaHCO_3$ buffer of pH 8.5 was replaced by Na_2HPO_4 of pH 8.7.

Terminal epoxy groups are determined by nucleophilic attack, with titration of the liberated OH^-; however, OH^- ions can interfere by attacking the epoxide themselves. Such interference is prevented in the proposed method of Paal[42] by (*a*) buffering the solution and (*b*) using the most energetic conditions in the nucleophilic reaction so that the interfering reaction is minimized.

Lipton and Bodwell[43] found that traces of dimethyl sulfoxide cause oxidative losses of methionine, cystine, and tyrosine during hydrolysis of the proteins with 6 *M* HCl to yield methionine sulfoxide, cysteic acid, and chlorotyrosine, respectively. Addition of a crystal of phenol before hydrolysis protects tyrosine from oxidation by dimethyl sulfoxide, but oxidation of the sulfur-containing amino acids is not prevented. Furthermore, methionine sulfoxide also causes oxidative losses of cystine and tyrosine.

1.1.3 Improving Existing Methods; Devising New Methods

While an existing method is being used routinely to determine a certain compound, improvements may be feasible as the situation arises. Modification of the existing method becomes necessary if it is not exactly applicable to the particular sample. New methods are developed from time to time when unforseen problems present themselves or when new instruments or techniques become available. The following examples may serve as illustrations.

For the determination of unsaturation in high-molecular-weight fatty nitrogen derivatives by the Wijs iodine-value method,[44] Milun[45] recommended acetylation of primary and secondary amines before adding the ICl and mercuric acetate reagents, dissolution of tertiary amines in acetic acid instead of chloroform, and treatment of quaternary ammonium chlorides with sodium lauryl sulfate before the titration. For the determination of 2-methylpent-2-enal in the presence of 2-methylvaleraldehyde, Tokar and Simonyi[46] used BrCl and reported that refractometric determination of the approximate composition of the sample must be made to permit selection of a suitable size of sample for analysis.

For the spectrophotometric determination of salicylic acid in aspirin based on the ferric salicylate color, Strode et al.[47] found it necessary to control the variables that affect the color and the hydrolysis of the aspirin. Jokl and Knizek[48] observed the dependence of extinction values on concentration in the u.v. spectrophotometric determination of *o*-, *m*-, and *p*-xylenes and ethylbenzene in mixtures; the samples should be diluted with cyclohexane to a concentration of 0.02% before spectrophotometry. For the determination of flavones, Mentzer and Jouanneteau[49] found that the difficulty of differentiation of two closely related substances can be overcome by increasing the specific differences by adding suitably chosen neutral salts. Belikov et al.[50] used statistical methods to plan experiments for establishing the optimum conditions of difference photometric determination of phenazone with iron(III); the error was least when the phenazone concentrations in the final solution were 6.04, 10.33, and 4.29 m*M*. Winefordner et al.[51] put forward a mathematical expression for calculating the minimum detectable concentration of a compound with respect to minimum detectable extinction, molar extinction coefficient, and cell thickness in i.r. spectrophotometry.

Sometimes an analytical method can be improved by changing the reagents or liquid medium. Thus Harper et al.[52] recommended dimethyl sulfoxide as a solvent in the pyromellitic dianhydride method for determining alcohols and amines; it has the advantages that neither phenols nor aldehydes interfere, and the reaction time is 30 min compared with 2 hr using phthalic anhydride. Linder and Persson[53] improved their own method for determining rosin acids in the presence of fatty acids by replacing benzenesulfonic acid with sulfuric acid, whereby the esterification time is reduced from 1.5 hr to 20 min. Sasuga[54] determined free acetone in acetone cyanhydrin using hydroxylamine hydrochloride and titrating the liberated HCl with 0.1 *N* Na_2CO_3; the presence of HCN does not vitiate the determination. Stenmark and Weiss[55] determined hydroxyl groups, even in the presence of epoxy groups, by measuring the hydrogen evolved with $LiAlH_4$ in tetrahydrofuran solution.

Occasionally the conventional method is disregarded because it does not suit the situation. For instance, for the determination of acetone in the presence of formaldehyde, the carbonyl group methods are not applicable; therefore Sedivec[56] recommended conversion of the acetone into bromoform, which yields with pyridine a red color suitable for photometric purposes. When the methoxyl group is present together with other alkoxyl groups, it can be hydrolyzed to form methanol, which, after oxidation to formaldehyde, is determined spectrophotometrically.[57] Hogea et al.[58] found that acetophenone and 1-phenylethanol, by-products in the oxidation of ethylbenzene, cannot be determined by conventional methods for ketones and alcohols because ethylbenzene hydroperoxide is also present; hence a method was proposed that is based on the conversion of these two compounds to the corresponding alkyl nitrites, followed by u.v. spectrophotometry. For the determination of diphenylamine in explosives, the classical bromination method does not differentiate between diphenylamine and its nitro and nitroso derivatives; therefore Marvillet and Tranchant[59] determined diphenylamine by titration in aqueous media.

The recent advance in instrumentation has stimulated the development of numerous new methods for the analysis of organic mixtures. Many mixtures that were previously very difficult to analyze can now be conveniently determined by means of these instruments. The gas chromatograph is a good example. For many years the differential determination of alkoxyl groups was a tedious and sometimes impossible task. Kirsten and Nilsson[60] undertook a thorough investigation of the Zeisel method for propoxy, butoxy, and other compounds and concluded that it would be difficult to devise an analytical method of general applicability. With the aid of the gas chromatograph, however, mixtures containing several alkoxyl groups can be easily determined.[61]

Hancock and Laws[62] described a method to determine traces of benzene and toluene by controlled nitration so that the products are *m*-dinitrobenzene and 2,4,6-trinitrotoluene, respectively; Etlis and Artyukhina[63] determined trichloroethane in dichloroethane by selectively splitting off HCl from trichloroethane without affecting dichloroethane; the colorimetric method used by Hildebrecht[64] to determine chloroform in carbon tetrachloride is based on the red color developed by the reaction of pyridine and NaOH with $CHCl_3$ in the presence of CCl_4. Nowadays the above three binary mixtures can be analyzed by gas chromatography without going through the laborious chemical reactions. It should be mentioned, however, that a vast majority of the analytical methods for determining an organic compound in mixtures are dependent on the chemical properties of the compound (see Section 1.2).

1.2 ANALYSIS OF MIXTURES BASED ON THE FUNCTIONAL-GROUP CHARACTERISTICS OF ITS COMPONENTS

1.2.1 Relative Amounts of the Functionl Group to be Determined

1.2.1.1 Major, Minor, or Trace Components

When the analytical sample consists of a simple organic compound (molecular weight below 600), one can easily find a method of determination based on the functional group present in the molecule. [7,65] However, if the sample is a mixture or a polymeric substance, the problem may become very involved and sometimes extremely difficult to solve. The first consideration is the relative amount of the particular functional group in the sample. The compound containing this functional group may constitute the major component of the mixture. On the other hand, the functional group may be present only in very small proportions. For example, in the analysis of terminal groups in polymeric substances, the functional group to be determined is usually less than 1%.

Some methods for functional group analysis specify large samples. For instance, Gertsev and Makarov-Zemlyanskii[66] described the determination of hydroxy groups in carbohydrate compounds by transesterification with trimethyl borate, which requires 5 g of the compound. Other methods, however, stipulate that the concentration of the functional group in the sample should be below certain level. Thus, for the determination of phenol by monobromination, Kozak and Fernando[67] found that the maximum amount of phenol applicable is 100 μg per 50 ml of solution, because dibromination occurs above this concentration. Beyrich and Pohlouderk-Fabini[68] described colorimetric methods for methanol up to 4 mg per 5 ml, and for ethanol up to 2 mg per 10 ml of solution. Robins et al.[69] determined α-keto acids colorimetrically at the nanomole level. Evstratova[70] described the analysis of glucose in the presence of lactose in which glucose can be determined in concentrations of 0.01 to 5% in the presence of 1.75% of lactose.

Binary mixtures comprising the same two compounds may require different methods of analysis depending on which component is present in the predominant proportion. For example, Mitra et al.[71] recommended a colorimetric method to determine trace amounts of acetic acid in acetic anhydride, while Belskii and Vinnik[72] used acylation to determine $<1\%$ of acetic anhydride in acetic acid. Dahmen[73] determined formic acid in the presence of acetic acid by oxidation with lead tetraacetate, but recommended mass spectrometry to determine small quantities of acetic acid in the presence of formic acid.

1.2.1.2 Increasing the Sensitivity of a Method

When the concentration of the functional group in the analytical sample is small, it becomes necessary to enhance the sensitivity of the method of determination to attain the desired precision and accuracy. Various ways to accomplish this are illustrated below.

Schole[74] determined microgram amounts of the methoxy group by measuring the methyl iodide photometrically after it was converted into a colored complex. Rozental and Tomaszewski[75] determined xylose in blood at the ppm level colorimetrically. Jordan and Veatch[76] determined trace concentrations of carbonyl compounds by spectrophotometric measurements of the corresponding 2,4-dinitrophenylhydrazones. Ponomarenko and Amelina[77] studied the determination of phenols based on their quenching effect on the fluorescence of luminol; the concentrations (micrograms per milliliter of final solution) for which the method is most suitable are: phenol, 20 to 600 μg; catechol, 3 to 70 μg; quinol, 3 to 100 μg. For the determination of carbohydrates at concentrations in the range of 3 to 15 ng per 100 ml, Rogers et al.[78] described a method based on the conversion to 2-furaldehydes, which condense with resorcinol to yield the intensely fluorescing xanthenone derivatives. Indirect fluorescence was utilized by Ermolenko and Chirkova[79] to determine carboxyl group concentrations of 0.001 to 7% in cellulosic material.

Besada and Gawargious[80] determined 50 to 150 μg of α-amino alcohols by an amplification reaction based on iodimetric titration. Seiler and Wiechmann[81] determined amines, phenols, and amino acids as 5-dimethylaminonaphthalene-1-sulfonyl derivatives; using fluorometry, the concentration can be as low as 10^{-12} *M*. Konitzer et al.[82] described a method to increase by twenty-fold the sensitivity of the ninhydrin reaction to determine α-amino acids by extracting the colored product into isoamyl alcohol–isopropyl ether. For the determination of bases by nonaqueous titration with $HClO_4$, Kiciak and Minczewski[83] reported that adding an acetic acid–formic acid (1 :1) mixed solvent increases the sensitivity to permit titration with 0.001 *N* solution.

The iodine azide reaction was used to determine thioureas (2 to 30 μg) and xanthates (0.04 to 0.8 ppm) by Kurzawa et al.[89,85] Lee and Samuels[86] determined low concentrations of mercapto groups in proteins by treating the sample with ^{14}C-labeled *N*-ethyl maleimide; subsequent hydrolysis yielded the ^{14}C-labeled cysteine-*N*-ethyl maleimide adduct to be measured. Hood and Winefordner[87] used iodoethane as a heavy-atom-solvent diluent in the determination of polycyclic aromatic hydrocarbons to increase sensitivity of measurement in phosphorimetry. Flockhart and Pink[88] observed that certain polycyclic hydrocarbons are readily converted quantitatively into free radicals on the surface of a silica–alumina catalyst, thus permitting them to be determined at the microgram level by electron spin resonance.

1.2.2 Influence of Neighboring Groups

While a functional group confers certain chemical properties on an organic compound, the reactivity of this functional group may be influenced by the neighboring groups in the molecule. Thus primary, secondary, and tertiary alcohols react differently. The reactivity of an amino group attached to the benzene ring is dependent on its position (*o*rtho, *m*eta, or *p*ara) relative to another substituent. For the purpose of quantitative analysis, it is best when the chemical reaction goes to completion with respect to the compound in the sample. This does not always happen, however. Therefore numerous studies have been undertaken to test the applicability of a particular reaction. Typical examples are given in Table 1.1.

Table 1.1 The influence of Neighboring Groups on Chemical Reactivity

Functional group[a]	Synopsis	Reference[b]
	Oxygen Functions	
Alkoxyl	Resonance-stabilized —OCH_3 gives low results by the Zeisel method	Pietrogrande[89]
Carbonyl	Oximation: single or double branching in the α-position in aliphatic aldehydes or in the 2- and 6-positions in aromatic aldehydes reduces the accuracy because of steric hindrance	Veibel[90]
	Oximation: different conditions for active and less active oxo compounds	Omböly[91]
	Oximation: conditions for ketones with α-hydrogen	Klimova[92]
	Oximation: competing rates between vanillin and acetovanilone	Fowler[93]
	Oximation: some compounds 10 min, others up to 3 hr	Terentev[94]

Table 1.1 (*Contd.*)

Functional group[a]	Synopsis	Reference[b]
	2,4-Dinitrophenylhydrazone: hindering action of —OH for phenolic carbonyl compounds	Wildenhain[95]
	2,4-Dinitrophenylhydrazone: nonquantitative nature for ketones	Lawrence[96]
	Using unsymmetrical dimethylhydrazine, aromatic but not aliphatic aldehydes can be determined in the presence of ketones	Siggia[97]
	Schiff base with phenethylamine: applicable to salicylaldehyde but not to *p*-dimethylamino-benzaldehyde because the *p*-dimethylamino group hinders the reaction	Potapov[98]
Carbohydrate	Using guaiacol, pink color to fructose, not glucose	Karacsony[99]
	Using phenol + H_2SO_4, green color to fructose only	Livingston[100]
	Using ethyl malonate, intense colors to mono-saccharides, but faint colors to disaccharides	Harada[101]
Carboxyl	Time required for hydroxamic acid formation varies from 2 to 120 min, depending on the substituents	Goldenberg[102]
	Using $KMnO_4$, tartaric, fumaric, maleic, and malic acids give quantitative results, but citric acid does not	Polak[103]
	Using $KMnO_4$, Cu greatly accelerates oxidation of tartrate, but has no effect on citrate	Zolotukhin[104]
	Using $Ce(SO_4)_2$, extent of oxidation varies	Ma[105]
	Using ditelluratoargentate, tartaric acid is oxidized at 80°C, but malonic acid is not affected; however, both acids are oxidised at 100°C	Jaiswal[106]
Ester	Saponification: for hindered esters; dependent on the size, number, and symmetry of the ester groups, and the positions occupied by them	Jordan[107]
Hydroxyl	Acylation with cyclic anhydrides: unfavorable equilibrium for alcohols with pK_a 13.5	Cohen[108]
	Using pyromellitic dianhydride, nonquantitative for tertiary alcohols	Siggia[109]
	Acylation: conditions for mixtures of primary and secondary alcohols	Siggia[110], Ilina[111]
	Acetylation: 5 min for cyclohexanol, 65 min for tertiary alcohols	Schenk[112]

Table 1.1 (*Contd.*)

Functional group[a]	Synopsis	Reference[b]
	Acetylation: for spatially hindered phenols and cyclohexanols	Zelenetskaya[113]
	Acetylation in the presence of ester groups	Vioque[114]
	Using 3-nitrophthalic anhydride, secondary alcohols react only partially	Floria[115]
	Nitration: quantitative for primary alcohols, incomplete for most secondary alcohols	Schenk[116]
	BrCl oxidation: for primary and secondary alcohols, but inapplicable to CH_3OH	Konishi[117]
	Using chlorotriphenylmethane for primary and secondary hydroxyl end groups, widely different rates are obtained	Fijolka[118]
	Dehydration: for tertiary and easily dehydrated primary and secondary alcohols	Petrova[119]
Lactone	Conversion to hydroxamic acid: for glucurono-lactone in the presence of glucuronamide, different reactivity	Yamanaka[120]
	Nitrogen Functions	
Amino	Using salicylaldehyde for primary aliphatic amines, aliphatic secondary and tertiary amines do not interfere, but ammonia and amine salts do	Johnson[121]
	Conditions for using salicylaldehyde to determine primary arylalkylamines	McCoubrey[122]
	Vanillin for primary arylamines	Deeb[123]
	Reaction with CS_2, for secondary amines	Umbreit[124]
	$HClO_4$ titration of tertiary amine, after acetylation of primary and secondary amines	Ruch[125]
	HCl titration of tertiary amine, after reacting primary and secondary amines with phenyl isothiocyanate	Miller[126]
	Complex formation for tertiary aromatic amine, after acetylation of primary and secondary amines	Schenk[127]
	Diazotization of primary amine: anomalies	Kainz,[128] Hori[129]
Hydrazino	$NaNO_2$ titration of 2,4-dinitrophenylhydrazones of ketones: not for ketones containing a C=O group, a $C_6H_5CH_2$ group, or a single H atom adjacent to C=N	Baldinus[130]

Table 1.1 (*Contd.*)

Functional group[a]	Synopsis	Reference[b]
Hydrazido	$NaNO_2$ titration of hydrazide: not for secondary hydrazides with strongly electronegative substituents	Schill[131]
	Sulfur Functions	
Mercapto	$AgNO_3$ titration: not for cysteine derivatives and thioglycolic acid; influence of nearby amino or carboxyl groups	Nedic,[132] Sluyterman[133]
Sulfonic	Amine salts of *o*-, *m*-, and *p*-benzenedisulfonic acids: conditions for precipitation	Spryskov[134]
Sulfoxide	Using HI, linear dialkylsulfoxides are reduced in a few minutes, whereas *ter*-butyl methyl sulfoxide requires 17 hr	Hogeveen[135]
Thiourea	Color reaction with sodium nitroprusside: for thiosemicarbazide and aliphatic, but not aromatic thiosemicarbazones	Kitaev[136]
	Unsaturated Functions	
Alkene	Mixtures of alkenes with different double bonds, reactivity toward mercuric acetate	Kreshkov[137]
	Use of $HgSO_4$: pent-1-ene, pent-2-ene, and 3-methylbut-1-ene do not react	Buzlanova[138]
	Hydrogenation: for chlorostyrene, after the vinyl group has reacted, the ring double bond starts	Bezborodko[139]
	Miscellaneous Functions	
Acidic	Nonaqueous titration: effects of substituents	Mathews[140]
	Using $Ba(OH)_2$ for humic acids, carboxy groups react; some phenolic hydroxy groups react partially	Avgushevich[141]
Basic	Use of sodium lauryl sulfate: influence of other functional groups	Pellerin[142]
	$HClO_4$ titration for amides: electrophilic groups typically decrease the basic strength to an extent such that the end point cannot be detected	Chatten[143]
Hydrocarbon	CrO_3 oxidation of ethylnitrobenzene: the *p*-compound quantitative, but only minute amounts of the *o*-isomer react	Neumann[144]
Phenolic	Bromination: 8-hydroxyquinoline in 1 min, but 8-hydroxy-4-methylquinoline requires 30 min	Corsina[145]

Table 1.1 (*Contd.*)

Functional group[a]	Synopsis	Reference[b]
	Color with amidopyrine: the extinction values at the same concentration decrease in the order phenol, *m*-cresol, *o*-cresol, *p*-cresol	Voloskovets[146]
Organic halogen	Reaction with thiosulfate: quantitative for halogen-substituted carboxylic acid salts, esters, and ketones and some alkyl and arylalkyl halides, but not for methyl bromide or allyl chloride and bromide	Ashworth[147]
	Reaction with morpholine: for benzoyl chloride and palmitoyl chloride, but not for short-chain aliphatic acid chlorides	Omböly[148]

[a]The functional groups are arranged according to the writer's classification scheme.[7,149,150]
[b]Only the first author of joint authors is given here and in all other tables throughout the book; see complete citation at the end of the chapters.

1.2.3 Influence of Reactive Impurities

In an analytical sample, any substance that does not constitute the compound to be determined can be considered as an impurity or extraneous matter. Reactive impurities are those that react either with the test compound or with the reagent used.

By necessity, one is always confronted with impurities when dealing with the analysis of mixtures. In the case where the impurity reacts with the compound to be determined, the sample received should be kept under such conditions that no change of composition occurs during storage.[151] If this is not possible, the extent of the reaction during that period should be ascertained. In the case where the impurity is known to react with the reagents used, it may or may not vitiate the analytical result, depending on the experimental conditions. Sometimes the impurity that reacts with the test compound originates from the reagent used. For instance, Smith and Wu[152] observed that phytolaccagenin, which has one double bond and an α-diol group, consumed 3-chloroperoxybenzoic acid equivalent to two double bonds; it was established that 3-chlorobenzoic acid (an impurity in the peroxy acid reagent) catalyzes the dehydration of the α-diol to ketone, which reacts with the peroxy acid. In the analysis of oleic acid by reaction with perbenzoic acid with I^-, Schmalz and Geiseler[153] found that errors are caused by the reaction of the free iodine so formed with the olefin.

Table 1.2 gives a number of references to reports on the influence of reactive extraneous substances on functional group analysis. When it is known

Table 1.2 Influence of Reactive Impurities

Functional group	Synopsis	Reference
	Oxygen Functions	
Acyl	Hydroxamic acid–Fe(III) complexes: conditions	Cheronis,[7] Bayer[154]
Carbonyl	In autoxidezed fats, peroxides interfere	Muzuno[155]
	Anthrone reaction for aliphatic aldehydes: influence of carbohydrates	Kwon[156], Helbert[157]
	Indirect polarography for aliphatic aldehydes: ketones and aromatic aldehydes react much more slowly	Fedoronko[158]
Carbohydrate	Aldoses in presence of ketoses: selective oxidation by $HClO_2$	Stitt[159]
	Resorcinol method for fructose: interference by tryptophan	Sheth[160]
Epoxy	Pyridinium chloride method for β-epoxide: phenols and alcohols interfere slightly, α-epoxides interfere seriously	Keen[161]
Hydroxyl	Indirect polarography for alcohols: peroxides do not interfere	Sudnik[162]
	Complex with $Ce(NO_3)_4 \cdot 2NH_4NO_3$: carbonyl groups do not interfere, but hydroxy acids cannot be determined	Obtemperanskaya[163]
Peroxy	Iodimetry: formic acid and furaldehyde do not interfere	Badovskaya[164]
Salts	Partial oxidation with IO_4^- for tartrate: citrate does not interfere	Nisli[165]
	Hydroxyflavone complex fluorescence quenching for citrate: tartrate, oxalate interfere	Guyon[166]
	Nitrogen Functions	
Amino	Oxidation with $HAuCl_4$ for amino acids: hydroxyl and carbonyl compounds interfere	Masood[167]
Cyano	Hydrogenation in the presence of acid amides	Huber[168]

Table 1.2 (*Contd.*)

Functional group	Synopsis	Reference
Nitro	Color with iminobispropylamine: influence of other functional groups	Schrier[169]
Urea	Biacetyl monoxime reaction: serious interference by NaN_3	Keyser[170]
	Sulfur Functions	
Mercapto	$AgNO_3$ titration: low results when the ratio of elementary sulfur to thiol is >1 :1.	Karchmer[171]
	Unsaturated Functions	
Alkene	Bromine method: for mixture of but-2-enylidene-acetone and diacetone alcohol, both addition to double bond and substitution of methylene group are quantitative	Polyanskii[172]
Alkyne	$AgClO_4$ method: no interference from aldehydes, but acids of K_A 10^{-4} to 10^{-1} and amines of K_B 10^{-9} to 10^{-12} interfere	Barnes[173], Gutterson[174]
	Miscellaneous Functions	
Acidic	Acidimetry: complexes formed between ethoxylated nonylphenol and phenol, resorcinol, catechol, quinol, or phlorglucinol drastically increase the K_A of the original phenols from 10^{-10} to 10^{-6}	Mohr[175]
	Acidimetry: for volatile acids in water, esters interfere, as they are partially hydrolyzed when the sample is evaporated	Stradomskaya[176]
Basic	$HClO_4$ titration: interference from other acids decreases in the order HI, HBr, HCl, HNO_3, H_2SO_4, H_3PO_4, oxalic acid, picric acid, and maleic acid	Kashima[177]
	Using $HClO_4$ in acetic acid for primary and secondary amines, the acetic anhydride (impurity in acetic acid) interferes	Posgay[178]
Organic halogen	Using *m*-nitroaniline for acid halides, some anhydrides (e.g., acetic, maleic) also react	Litvinenko[179]
	Hexamine titration of acid halides and sulfonyl halides: organic acids and alkyl halides do not interfere	Terentev[180]

that the impurities present interfere with the determination, steps must be taken to remove the interference (see Section 1.2.5).

1.2.4 Influence of Other Extraneous Matter and the Environment

Besides the reactive impurities, other extraneous matter in the millieu may affect the analysis. Thus the amount of water present in the system sometimes is of much concern. In the saponification of esters with methanolic KOH, Greive et al.[181] observed low and variable results because of the absence of water. For the determination of fatty acid esters and chloesteryl esters by the hydroxamic acid method, Skidmore and Entenman[182] reported that the most important single factor is the amount of water present during the reaction. For the titration of isomeric phthalic acids in acetone with ethanolic $(C_2H_5)_4NOH$ or KOH, Kreshkov et al.[183] found that water diminishes the differentiating action of acetone; the best medium for this titration is acetone–water (25 :2). Similarly, Lin[184] confirmed that water has a detrimental effect on the differential titration of a combination of salicyclic acid and acetylsalicylic acid. In the differential thermometric titration of very weak acids, Keily and Hume[185] found that even traces of water interfere. Hall and Harris [186] studied the effect of water on the polarographic determination of chloroacetone and observed that tetrachloroacetone was so readily hydrolyzed that it gave no reduction wave. Hsiao and Chiou[187] investigated the effect of water vapor on the electron-capture detector during gas-chromatographic analysis and reported that the presence of 1 μl of H_2O in a sample for a g.l.c. column may reduce the peak height of a fluorocarbon by 40%. In contrast, for the extraction of the insecticides aldrin and heptachlor from soil with hexane–acetone, Williams[188] found that the presence of 5% of H_2O is essential. Burke and Porter[189] reported that the decrease in the amount of pp′-TDE extracted by acetonitrile from kale is directly proportional to the decrease in sample moisture content. For the determination of methanol by the Widmark method, Dynakowski and Kubalski[189a] found that diluted H_2SO_4 should be used in place of conc. H_2SO_4.

The liquids that serve as solvents in analysis are not supposed to be reactive. Nevertheless, they may influence the determination to a certain extent. For instance, Feuge et al.[190] observed that highly purified cyclopropenoid acid esters appeared to have a purity of only 83 to 86% by titration with hydrobromic acid in anhydrous acetic acid; but if toluene is used instead of acetic acid, correct cyclopropenoid content and sharp end points are obtained. For mixtures containing H_2SO_4 and water-soluble fatty acids, Mankovskaya and Udovenko[191] found that, in the presence of 90% acetone, the dissociation of fatty acids is so much reduced that these acids are not titrated with 0.5 N alkali. Morales[192] studied the relative acidities of phenols and benzoic acids in dimethyl sulfoxide and water, and the results show that phenols display a greater increase in relative acidity in dimethyl sulfoxide than do benzoic acids.

Colorimetric determinations may be influenced by extraneous liquids in the system. For example, Still et al.[193] found that, contrary to earlier reports, ethanol interferes in the determination of formaldehyde or glycolic acid with chromotoropic acid or naphthalene-2,7-diol in H_2SO_4 medium; molar extinction coefficients are lowered up to 65 and 87%, respectively. Kazitsyna and Mishchenko[194] studied the effect of 32 liquids on the absorption intensity of α-isobutylimino-*o*-cresol at 400 nm and found that the extinction coefficients ranged from 0.8 to 3.78. Dimroth and Reichardt[195] investigated colorimetric analysis in binary solvent mixtures; when pyridinium-*N*-phenol

Table 1.3 Influence of Nonreactive Impurities

Functional group	Synopsis	Reference
	Oxygen Functions	
Carbohydrate	Colorimetric method for invert sugar in the presence of sucrose: certain salts (e.g., phosphate) inhibit the reaction	Bernaerts[196]
	Phenol–H_2SO_4 method: change of slope depends on the borate concentration, large effect with pentoses and less with hexoses	Lin[197]
	$NaIO_4$ method for glucose: when twice the stoichiometric amount is used, or when pH is increased, more $NaIO_4$ is consumed	Dusic[198]
	$NaIO_4$ method for pentoses: the reactant ratios depend on the amount of $NaIO_4$ reagent present and are independent of pH and time of reaction	Dusic[199]
Ester	Colorimetric method for citrate: effect of EDTA	Geisler[200]
Hydroxyl	For inositol by precipitation as ferric inositol hexaphosphate: affected by the ratio of Fe to P, the acid concentration, and the nature of other ions present	Anderson[201]
	For vicinal diols through complex formation with styrene-4-boronic acid: the complexing reaction is highly pH dependent	Elliger[202]
Salt	Hydrolysis of ketosteroid glucuronide and sulfate conjugates: presence of formaldehyde decreases the recovery	Goldzieher[203]

Table 1.3 (*Contd.*)

Functional group	Synopsis	Reference
	Nitrogen Functions	
Amino	For tertiary amines through polarography of amine oxides: the slope of the current–potential curves is affected by the ionic strength and the surface-active materials in the solution	Hoffmann[204]
	For amino acids from protein: yield of tryptophan is decreased in the presence of carbohydrates	Isobe[205]
Heterocyclic	For pyrrolizidine alkaloids with 4-dimethyl-aminobenzaldehyde: solvent and temperature effects	Bingley[206]
	Thiochrome method for vitamin B_1: low results ($\sim 50\%$) in the presence of 4-aminosalicylic acid	Sen[207]
Nitro	Polarographic method: surface-active compounds (e.g., dodecyltrimethylammonium chloride) shift the two waves to more cathodic potentials under acidic conditions, whereas under basic conditions large shifts towards less cathodic potentials are found in the second wave	Pietrzyk[208]
	Unsaturated Functions	
Alkene	Bromine addition: the rate of bromination is related to the dielectric constant of the solvent	Hanna[209]
	Miscellaneous Functions	
Acidic	Nonaqueous titration of phenols: effect of association with the solvent	Bruss[210]
	Indicator titration of petroleum products: misleading results when additives are present	British Standards Institution[211]
Basic	For weak bases: strong aqueous solutions of neutral salts enhance the potentiometric break	Critchfield[212]
	$HClO_4$ titration of alkaline carboxylates: effect of phenol	Vasiliev[213]
Organic halogen	For trichloroacetic acid with tris(1,10-phenanthroline)Fe(II): chloroacetic acid does not interfere, but dichloroacetic acid does	Yamamoto[214]

betaines were used as test compounds, the absorption maximum was displaced from 453 nm in pure H_2O toward longer wavelengths as the organic-solvent content increased (e.g., 805 nm in pure piperidine). Further discussion on the influence of solvent in spectrophotometric methods is given in Section 1.3.

Miscellaneous factors that may influence the analytical results obtained by chemical methods based on functional group reactions are given in Table 1.3. It should be mentioned that the reports cited in this table are only for purposes of illustration. One factor common to all media but frequently overlooked is the effect of storage of the analytical sample. The compound to be determined or its relative proportion in the mixture may undergo significant change during the storage period. McCarthy[215] studied the effects of storage in polyethylene and poly(vinyl chloride) containers on some aqueous preservative solutions; storage in glass or poly(vinyl chloride) is satisfactory for all but sorbic acid, whereas polyethylene is unsuitable for aromatic alcohol and phenolic preservatives. Adamski and Pawelczyk[216] reported on the storage of amidopyrine in aqueous solutions; at 20°C, a 10% decomposition in 1% aqueous solution takes place in 60 days. In contrast, Hanok and Kuo[217] investigated the stability of a reconstituted serum for the assay of 15 chemical constituents and found that storage up to 3 weeks at −15°C does not affect the assay of any of the constituents. Working on chlortetracycline in feedingstuffs, Katz and Fassbender[218] observed that epimerization occurs rapidly; the temperature of storage appears to be significant. For the evaluation of the stability of a compound upon storage, the analytical procedure chosen should discriminate between the pure compound and its decomposition products. For example, the *N*-bromosuccinimide method is suitable for determining penicillins, but cannot be used for the determination of stability, as pointed out by Alicino,[219] because the degradation products interfere.

1.2.5 Removal of Interferences

When the analytical sample is known to contain impurities that interfere with the determination of a specific compound, it is self-evident that such interference should be removed. The first task, however, is to identify the nature of the impurities and the extent of interference that may occur in the analysis. For instance, the influence of 27 excipients on the analysis of tablets and capsules by nonaqueous titrimetry in 11 solvents was studied by Chatten and Mainville[220]; of these, the excipients that consume titrant when dissolved alone do not interfere in the presence of stronger acids or bases. Soliman et al.[221] developed a nonaqueous titration procedure to determine diphenylhydramine and naphazoline in eye drops to prevent the interference from sodium acetate and other ingredients that occurs in spectrophotometric determinations. Braverman et al.[222] observed an apparent large increase in the urinary steroid content, as determined colorimetrically, after the administration of hexamine mandelate; it is recommended that this and other formaldehyde-forming drugs should not be given before the determination of steroids and catecholamines. Chrzanowski et al.[223] found that high results for the determination of theophylline in blood by gas chromatography were due to contact of the sample with the butyl rubber stopper of the evacuated glass tube used for its collection; an apparent theophylline content of up to 5.5 μg per ml was obtained when 10 ml of water remained in contact with the stopper for 60 sec and up to 50 μg per mole was obtained when the contact was for 60 min. Pileggi and Kessler[224] reported that the determination of iodine compounds in serum by the Ce(IV)–As(III) redox method is inhibited by an unidentified substance; however, this inhibition can be prevented by adding bromide or chloride before the colorimetric determination.

Understandably, the ways and means to remove interferences vary widely, depending on the nature of the problem. A few examples are cited below. Kanter[225] found that acetone was a source of error in the colorimetric determination of meprobamate; this difficulty was avoided by using a mixed solvent prepared by washing 25 ml of $CHCl_3$–CCl_4 (1 :1) with H_2O (1 ml), conc. aq. NH_3 (2 drops), and saturated aq. KCl (2 drops). For the nonaqueous titration of chlortetracycline hydrochloride, Kelemen and Kerenyi[226] reported that cellulose acetate phthalate, used in making pellets, interferes with the complete extraction by the conventional procedure; this interference is prevented by adding ethyl acetate and waiting for 30 min to dissolve the coating before titration. The presence of pectins and Fe(II) interferes with oxalate determination by the permanganate method; Abaza et al.[227] solved the problem by increasing the acidity of the water-soluble oxalate medium and by separately determining the oxalate in the Fe(II) complex. Iron and plant pigments interfere with analysis of plants for fluoroacetic acids; Vickery et al.[228] recommended steam distilling the fluoro compounds from the acidified aqueous extract. The presence of DDT interferes with the determination of camphechlor in crop residues; Kawano et al.[229] treated the mixture at 0°C with H_2SO_4–fuming HNO_3(1 :1) to remove the DDT.

Chromatography[151] is often used to remove the interfering substances in analytical procedures. Thus Brandt et al.[230] described a simple method to prevent vitamin C interference with urinary glucose determinations by passing the sample through a small column (10 mm $\times$ 5 mm) prepared in a Pasteur pipette packed with Bio-Rad AGl-X4 (Cl^- form, 200 to 400 mesh). Cerda and Eek[231] removed acids, anhydrides, and phenols from organic compounds by gas-chromatographic separation using a precolumn containing Chromosorb W and lead oxide. Some other methods for removal of interferences are given in Table 1.4.

Table 1.4 Some Methods for the Removal of Interferences

Functional group	Synopsis	Reference
	Oxygen Functions	
Carbonyl	Acetaldehyde in crotonaldehyde: heat with $NaHSO_3$ and distill off acetaldehyde in $NaHCO_3$ solution	Sjostrom[232]
	iso-Butyraldehyde in *n*-butyraldehyde: $NaBH_4$ reduction, then react *iso*-butanol with H_2SO_4+salicylaldehyde	Primavesi[233]
	Cyclohexanone in cyclohexanol: colorimetric for ketone, add hexamine and acetic anhydride to prevent oxidation of alcohol	Maslennikov[234]
Carboxyl	Carboxylic acids in carboxylates: direct oximation using conditions under which esters do not react	Pesez[235]
Peroxy	Hydroperoxide in hydrogen peroxide: polarography and precipitation of H_2O_2–metal complexes	MacNevin,[236] Bruschweiler[237]
	Nitrogen Functions	
Amino	Primary in secondary and tertiary: Schiffs base with salicylaldehyde in acetic acid	Milun[238]
	Secondary in primary: coulometric titration with Hg(II); remove primary by reaction with salicylaldehyde	Przybylowicz[239]
	Aliphatic amines in ammonia: colorimetric with chloro-2,4-dinitrobenzene	Eklandius[240]
Heterocyclic	Barbiturates in colored impurities: titrate with 0.1 *N*[110] $AgNO_3$	Tologyessy[241]
	Creatinine in ketone bodies by alkaline picrate method: keep the sample for several days before determination	Watkins[242]
	Papaverine in narcotine by reineckate precipitation: use excess of $CHCl_3$	Lee[243]
Nitro	Nitropropanes in nitrite: extract with benzene	Estes[244]
	Sulfur Functions	
Mercapto	Thiols in hydrogen sulfide: pass gas mixture through 0.1 *N* NaOH at 90°C to absorb H_2S selectively	Hammer[245]
	Glutathione in nitrite: remove nitrite by sulfamic acid	Mortensen[246]

Table 1.4 (*Contd.*)

Functional group	Synopsis	Reference
	Miscellaneous Functions	
Active hydrogen	Alcohols in aldehydes, esters, amines: titrate with lithiumaluminumdibutylamide	Jordan[247]
Phenolic	*m*-Aminophenol in 4-aminosalicylic acid by 4-aminophenazone: suppress the reaction of 4-aminosalicylic acid by using 25% aq. NH_3	Franc[248]

1.3 ANALYSIS OF MIXTURES BASED ON THE MOLECULAR PROPERTIES OF THE CONSTITUENTS; INFLUENCE OF THE ENVIRONMENT

Besides using the chemical reactions of the functional groups for determinations (see Section 1.2), quantitative analysis of an organic compound may be performed by methods that are dependent on the properties of its whole molecular structure. These properties include spectral characteristics (e.g., i.r. absorption), physical constants (e.g., density), nuclear magnetic resonance, mass fragmentation, biochemical behavior (e.g., enzymatic response), and so on. Detailed classification of the methods is given in Chapter 2, Section 2.7. In the following paragraphs, we discuss briefly the features of these methods with emphasis on the influence of the environment on the results of analysis. Two points should be noted: (*a*) With the exception of the enzymatic methods, no reagent is used to react with the compound. (*b*) The property that is quantitatively measured is attributed to the entire molecule. Thus, while the i.r. determination of acetone is related to the carbonyl group, the absorbance measured must be standardized by using acetone itself, and not any other ketonic compound.

1.3.1 Infrared Absorption

Since the infrared spectrometer is now a common instrument in chemical laboratories, analytical procedures based on i.r. absorption have gained much popularity. Infrared spectroscopy can perform determinations that are unique. Thus Higuchi et al.[249] described a method to determine the concentration of conformational isomers of *meta*-disubstituted benzenes in CS_2. Beckering et al.[250] studied the spectra of phenols, catechols, and guaicols in the range 2800 to 3000 cm^{-1} and found that no two compounds have the same absorption pattern. Katon et al.[251] recorded the spectra of 78 aliphatic amides; most exhibit three characteristic bands in the 700 to 250 cm^{-1} region, and the wavelengths at which these occur depend on the structure in the vicinity of the amide group, but appear to be specific for a given class. Yamaguchi et al.[252] proposed a simple equation for calculating the wave number for the C=O group of carbonyl compounds; for quantitative analysis, however, this information is usually obtained experimentally using the pure compound.

Powers et al.[253] determined aromatic aldehydes in the presence of aromatic ketones using the absorption band in the 2.21 μ region because it is relatively free from interference; the only interference is that from terminal epoxides and cyclopropyl groups. Saier et al.[254] showed that aromatic and aliphatic aldehydes all give two intense bands in the 2600 to 2900 cm^{-1} region; aromatic compounds are determined by integrating both bands, whereas aliphatic compounds are determined by integrating only the lower frequency. Anderson and Zaidi[255] determined small amounts of alcohols in aqueous solutions; acetals, esters, and ethers may be sufficiently water soluble to interfere. Lippmaa[256] determined phenolic hydroxy groups by measuring at 3420 and 3540 cm^{-1}. The ratio of molar extinction coefficients for the two maxima is constant; other hydroxyl compounds (alcohols and acids) interfere.

The occurrence of polymorphism should be considered carefully in i.r. spectroscopy.[257] For instance, in the i.r. determination of chloroamphenicol palmitate, Borka and Backe-Hansen[258] observed that the process of grinding with KBr converts the physiologically inactive β-polymorph (843 cm^{-1}) into the α-form (868 cm^{-1}). The pressure used in making the discs appears to have no effect.

1.3.2 Ultraviolet Absorption

The environment may have significant influence in u.v. absorption of the compound to be determined. Pellerin et al.[259] investigated the effect of pH and buffers on phenolic compounds at the 230 to 350 nm region; for monohydric phenols the spectra at a fixed pH remain unchanged regardless of whether carbonate, phosphate, or borate buffer solutions are used, but catechol at pH 8.2 (maximum at 275 nm in carbonate or phosphate) forms a complex with borate that causes a peak shift to 282 nm with a higher absorption, indicating higher acidity. Yarborough et al.[260] studied temperature dependence between 5 and 30°C of 19 compounds in methanol or *iso*-octane. The variation ranged from 0% per °C for acetone to 0.74% per °C for toluene; the absorbance generally decreased with increasing temperature. Baczyk[261] reported on the change of absorption maxima for ascorbic acid when it is dissolved in a 1% solution of other organic acids: acetic acid and trichloroacetic acid cause slight shifts; citric acid causes a moderate shift to shorter wavelength and a decrease in extinction, while oxalic acid causes a shift to longer wavelength and a very large decrease in extinction.

Englis et al.[262] determined benzoic and salicylic acids in food products using the absorption at 227 and 236 nm, respectively; careful consideration of the solvent and pH is emphasized. Doyle and Burgan[263] recorded the absorption maxima and molar extinction coefficients of monosaccharides; these workers pointed out that the method is useful for determining individual sugars when the identities are known, but it is not suitable for unknown mixtures. Kracmar and Kracmarova[264] discussed the influence of substituents and solvent on the absorption of naphthalene and 39 of its derivatives. Santavy et al.[265] studied the effect of methoxy and methylenedioxy groups on the spectra of aromatic compounds with a conjugated carbonyl chromophore. For the differentiation of steroids (77 spectra recorded), Brandstätter-Kuhnert et al.[266] found that the refractive indexes of the melt are more suitable than u.v. absorption. Recently Chiang[266a] investigated the quantitative relationship between molecular structure and physicochemical properties such as the electronic spectra of polyenes.

1.3.3 Fluorometry

Goldman and Wehry[267] studied the influence of solvent and temperature on the fluorescence of hydroxyquinolines and discussed the implication on fluorometric determination of *N*-heteroaromatic compounds. Fine and Koehler[268] studied the pH effects on fluorescence of umbellifererone; the band at 460 nm is independent of pH for all values >2.2, while the maximum shifts to 480 nm for pH<2.2. For the determination of anthracene, fluorene, and phenanthrene in mixtures, Thommes and Leininger[269] found that only the fluorescence of anthracene at 400 nm is free from interference, and readings at 316 and 350 nm are dependent on the concentrations of all three compounds. For mixtures containing *o*-, *m*-, and *p*-hydroxybenzoic acids, only the *o*-isomer fluoresces upon excitation at 314 nm at pH 5.5, while both the *o*- and *m*-isomers fluoresce at pH 12; the *p*-isomer does not fluoresce under these conditions.[270] According to Grant and Patel,[271] for mixtures containing a phenol and a phenolic acid, under optimum conditions of pH and wavelength, the fluorescence due to the phenol can be determined without interference from the phenolic acid and vice versa. For the determination of serotonin in biological materials, Thompson et al.[272] observed that the presence of ascorbic acid causes a diminution in the intensity of fluorescence, but the use of ascorbic acid is essential to stabilize the serotonin. The effect of external pressure on the fluorescence of carbonyl derivatives of anthracene has been investigated by Mitchell et al.[272a]

1.3.4 Polarography

For the polarographic determination of ethyl nitrate and ethyl nitrite in aqueous solutions, Blyumberg and Pikaeva[273] found that the potentials do not vary with pH, but are greatly affected by the presence of ethanol or ether. Kemula et al.[274] observed the following effect of oxygen on the polarographic behavior of hydrazo compounds: When a deaerated solution of hydrazobenzene in ethanol and aq. NH_3–NH_4Cl buffer are mixed in the absence of air, only one wave due to the hydrazo–azo redox couple appears. If the solutions are mixed before deaerating, or if oxygen is passed through the deaerated solution for up to 60 min and then removed, the anodic branch of the wave increases at the expense of the cathodic branch, and a new cathodic wave due to H_2O_2 appears. In the polarography of fructose in alkaline media, Swann et al.[275] observed a well-defined wave that was depressed by addition of borate; in 0.1 *M* LiCl–0.01 *M* LiOH, however, the waves for fructose and for fructose plus borate have the same characteristics. Usami[276] studied the polarography of aldehydes in dimethylformamide. Crotonaldehyde and acraldehyde give one-step waves at −1.48 and −1.29 V, respectively, while saturated aldehydes give a one-step wave at −2.25 V. The waves of unsaturated aldehydes are not affected by vinyl acetate, but the waves of saturated aldehydes are covered by that of vinyl acetate. In the presence of water in dimethylformamide, the wave of acetaldehyde is shifted to −1.89 V.

Kelly[277] determined compounds containing dichloroacetamido groups using the first reduction wave of the chlorine from the dichloroacetamido group; the stability of these compounds at various pH values and the effects of various elements or groups in the molecule on the half-wave potentials were investigated. Ma et al.[278] studied the influence of neighboring groups on heterocyclic *N*-oxides. Tsuji and Elving[279] made polarographic measurements of the relative strengths of Brönsted acids in pyridine and concluded that individual acids may be determined in a mixture or in a sample containing other polarographically reducible compounds. Brüschweiler and Minkoff[280] and Kuta and Quackenbush[281] studied peroxides; the limiting diffusion currents are proportional to the concentration of the individual peroxides and are additive in mixtures.

1.3.5 Other Methods

Methods using biological reagents (enzymes, bacteria cultures) are usually specific and require careful control of the environment. The preparation and storage of the reagents sometimes must follow strict procedures. For instance, von Korff[282] described a specific enzymatic method to determine microgram amounts of acetate using an enzyme preparation obtained from rabbit-heart ventricle; pyruvate and sodium ions must be removed, while formate, propionate, butyrate, acetoacetate, and fluoroacetate do not interfere. Blecher and Glassmann[283] studied the effect of pH on the determination of glucose in the presence of sucrose using glucose oxidase; they found that all commercial preparations of the enzyme were contaminated with sucrase, which yielded glucose from sucrose during the assay. Later Schiweck[284] reported that the hydrolysis of sucrose by sucrase is completely inhibited by the use of 0.5 *M* Tris buffer at pH 8.0. Goetz[285] discussed the specificity and interferences in the determination of glucose in biological fluids by the glucose dehydrogenase method.

Buisman et al.[286] studied the effect of isomerization on the chemical and biological assay of vitamin D. Biological assays give the potential vitamin D content (i.e., the sum total of calciferol and precalciferol), provided that solutions of the sample and a reference have been equilibrated by simultaneous heating to ensure that both have the same ratio of the isomers. The actual vitamin D content (i.e., calciferol only) can be determined after separation or by an appropriate physicochemical procedure.

Nuclear magnetic resonance can be influenced by the environment. Thus Cavalli and Cancellieri[287] studied the effect of solvents on methyl groups and reported that benzene is most suitable for differentiation of the methyl-group protons of alcohols from those of hydroperoxides. Gehring and Reddy[288] examined the effect of nitro-group substitution on the chemical shift of the methylene protons of diphenylmethane and found it to be additive.

Raven et al.[289] investigated the effect of cyanide, serum, and other factors on the determination of vitamin B_{12} by a radioisotope method. Unless cyanide ion was present, low values were obtained both for hydroxocobalamin solution and for sera with a vitamin B_{12} concentration of >450 pg per ml. It was also shown that serum increases the vitamin B_{12} binding capacity of intrinsic factor solution and thus improves the accuracy of assay of crystalline vitamin B_{12} solution when an intrinsic factor control is used.

Space does not permit full discussion of the influence of environment in various analytical methods. Suffice it to say that careful consideration of the pertinent factors is warranted when one devises or adapts a procedure to determine an organic compound in a sample that contains other ingredients.

1.4 THE QUESTION OF SAMPLE SIZE: AVERAGE SAMPLE VERSUS SINGLE SAMPLE ANALYSIS

In the author's opinion, the aim of performing quantitative analysis of a mixture is to obtain the analytical data within the accuracy required and time available using the simplest and most economical method. When two or more methods are equally acceptable, the one that uses the smaller sample size is generally preferred, provided that it does not entail expensive equipment and special manipulative skill. For the analysis of pure organic compounds, milligram quantities (0.1 millimole or less) are usually sufficient for titrimetry, and microgram amounts (1 μmole or less) are suitable for colorimetry. The case of mixtures is more complicated as it depends on the proportion and nature of the constituents. Sometimes the whole sample is taken for analysis. Thus Lowry et al.[290] discussed the analysis of single cells (0.005 to 0.03 μg dry weight). Jeffs et al.[291] described the determination of insecticides on single seeds.

Among the various categories of organic mixtures, pharmaceutical preparations are probably the most important and numerous. These mixtures are processed in bulk quantities and are then divided into small dosages as tablets or in ampuls. The medicinal content of each dose must be within certain limits as specified by governmental regulations or guaranteed by the manufacturer. To check the contents, it must be decided whether to analyze single units or to combine a number of units to be analyzed and obtain an average value. Since the medicine is administered to the patient in separate dosages, analysis using individual tablets or ampuls seems to be the logical approach. For example, de Meijer[292] used single tablets to determine content uniformity of diazepam and oxazepam, while Indemans[293] determined reserpine in admixture with dihydrallazine sulfate and hydrochlorothiazide, digoxin, and digitoxin preparations. Roberts and Siino[294] and Gänshirt and Polderman[295] measured the stability of steroids. Duda and Elste[296] described the procedures for the single-tablet assay of aspirin, ascorbic acid, isoniazid, phenobarbitone, and other compounds, including samples containing two or three active ingredients. Shane and Kowglansky[297] determined aspirin, salicylamide, paracetamol, and caffeine by independent methods with one tablet, the procedure being amenable to automation.

The development of microchemical methods and sensitive measurement techniques has made possible the accurate analysis of individual doses of medicine. For instance, Sennello and Argoudelis[298] utilized gas chromatography for the simultaneous determination of pyridoxine, ascorbic acid, and nicotinamide in vitamin capsules. Kadin et al.[299] determined estrogens in anabolic vitamin tablets by quantitative paper chromatography. Knizhnik and Senov[300] applied thin-layer chromatography to determine sulfonamide preparations. Jacobsen and Thorgersen[301] recommended pulse polarography for determining nicotinamide in multivitamin tablets. Soeterboek and van Thiel[302] discussed the methods to study the stability of glyceryl trinitrate in tablets, citing gas chromatography, colorimetry, polarography, and i.r. spectrophotometry. Table 1.5 gives some examples of spectraphotometric methods for the determination of drugs in dosage forms.

Table 1.5 Examples of Spectrophotometric Determination of Pharmaceutical Preparations

Method and material analyzed	Reference
Colorimetric, for phenacetin in tablet mixtures	Lee[303]
Colorimetric, for isopropamide	Santoro[304]
Colorimetric, for panthenol in injections	Bukowska[305]
Colorimetric, for atropine	Gupta[306]
Colorimetric, for glycosides in tablets, drops, ampuls	Khafagy[307]
U.v., for aspirin, caffeine, and phenacetin in tablets and capsules	Machek[308]
U.v., for santonin	Biesemeyer[309]
U.v., for 4-acetamidobenzaldehyde, isonicotinoylhydrazone, and thiacetazone	Nino[310]
U.v., for pemoline	Kos[311]
Fluorometric, for atropine and hyoscyamine in tablets and injections	Roberts[312]
Internal reflectance spectroscopy, for tablet coatings	Warren[313]
Difference spectrophotometry, for drugs in dosage forms	Doyle[314]

Table 1.6 gives a number of methods in which the contents of only a single tablet, capsule or ampul are determined. A wide range of medicinals can be analyzed in this way. On the other hand, there are many published methods for the analysis of pharmaceutical preparations that call for an unnecessarily large amount of working material. As shown in Table 1.7, some procedures stipulate the preparation of 100 ml to 1 liter of solution while using 10 μl to 1 ml for the subsequent operations; other methods demand quantities of a decigram or more of the compound to be determined when milligram amounts are sufficient; still other methods specify the analytical sample to be a combination of 10 dosage units. By applying the microchemical approach (i.e., using the minimum quantity of working material to obtain the desired chemical information), these methods can be easily modified so that a single tablet, capsule, or ampul suffices. It should be mentioned that random checking of individual dosage units is the best way to detect adultera-

Table 1.6 Some Methods Using a Single Dosage for Analysis of Pharmaceutical Mixtures

Compounds determined	Quantity	Reference
Alkaloids	2–5 mg	Sobiezewska[315]
Ascorbic acid	4–20 mg	Bark[316]
Barbiturates	80 mg	Gupta[317]
Caffeine, codeine phosphate	30 mg	Stevens[318]
Codeine, caffeine	1–65 mg	James[319]
2,2′-Dehydromorphine	$>1\ \mu g$	Hammerstingl[320]
Dextropropoxyphene HCl (a), paramethasone acetate (b)	32 mg a, 250 μg b	Stevenson[321]
Digoxin	0.25 mg	Myrick[322]
Ethinylestradiol	0.1 mg	Becker[323]
Ethanol		Motz[324]
Hyoscyamine (a), atropine (b), hyoscine (c), phenobarbitone (d)	100 μg a+b, 6 μg c, 16 mg d	Zimmerer[325]
Isoxsuprine HCl	10 mg	Bryant[326]
Mestranol	80 μg	Templeton[327]
Nitroglycerin	5 μg	Bell[328]
Estrogen in megestrol acetate, norethisterone	30–100 mg	Miller[329]
Prednisolone		Meister[330]
Reserpine		Kabadi[331]
Salicylic acid in acetylsalicylic acid		Reed[332]
Steroids	5 mg	Beyer[333]
Tetracene (a), lead styphnate (b)	1–10 mg a, 1–40 mg b	Wild[334]
Theopaverin, theophenal, theoserpin		Grabowska[335]
Theophylline (a), sodium pentobarbitone (b), papaverine HCl (c)	76 mg a, 15 mg b 14 mg c	Helgren[336]
Tinidazole		Slamnik[337]

tion or criminal negligence. The following case is cited for illustration. During the war there were large shipments of morphine ampuls. The manufacturer normally put in each ampul 30% over the required quantity. An unscrupulous dealer removed 25% of the ampuls in every batch and replaced them with blanks. The lot of supply passed the assay using the average value method, but the wounded in the field who were administered the blank ampuls certainly suffered.

Table 1.7 Some Methods That Can Be Adapted to Single-Dosage Analysis (see text)

Compounds determined	Synopsis of published method	Reference
	Methods That Use Large Samples	
Acetylsalicylic acid, codeine, phenacetin	0.3 g sample, 0.1 *N* titrant	Jeske[338]
Acetylsalicylic acid, phenacetin, caffeine	0.5–4 g sample	Wirth[339]
Amidopyrine, *o*-hydroxy-quinoline-*m*-sulfonic acid	0.4–0.5 g sample, 0.1 *N* titrant	Vasiliev[340]
Amidopyrine, papaverine HCl, phenobarbitone, atropine	0.15–0.45 g sample, 0.1 *N* titrants	Vasiliev[341]
Barbiturates	0.6–0.9 g sample	Beguin[342]
Carbutamide	0.1 *N* titrant	Wartmann-Hafner[343]
Cyanoacetic acid hydrazide	0.12 g sample, 0.1 *N* titrant	Grabowicz[344]
Cyclamates	0.15 g sample	Meadows[345]
Diphenylhydantoin	0.25 g sample	Biedebach[346]
p-Hydroxypropiophenone	0.15 g sample, 0.1 *N* titrant	Beral[347]
Isoniazid	0.2 g sample, 0.1 *N* titrant	Ionescu[348]
Isoniazid	0.1 g sample, coulometry	Stoicescu[349]
4-(Isopropylamino)phenazone	0.2 g sample, 0.1 *N* titrant	Grabowicz[350]
Nicotinamide	0.1 g sample, 0.1 *N* titrant	Devyatnin,[351] Kavarana[352]
Pentobarbitone	0.25 g sample, 0.1 *N* titrant	Tomaskova[353]
Pentylenetetrazole	0.1 g sample, 0.1 *N* titrant	Throop[354]
Salicylic acid, acetylsalicylic acid	0.1 g sample	Ali[355]
Sulfonamides	0.5 g sample, 0.1 *N* titrant	Faber[356]
Sulthiame	0.2 g sample, 0.1 *N* titrant	Popescu[357]
Tetracene	0.1–0.2 g sample, 0.1 *N* titrant	Ballreich[358]
Theophyllinate	0.4–1 g sample, 0.1 *N* titrant	Bukowska[359]
Thiamine HCl, amitriptyline HCl, cyanocobalamin, ascorbic acid	0.1 *N* titrant	Dobrecky[360]

Table 1.7 (*Contd.*)

Compounds determined	Synopsis of published method	Reference
	Methods Involving Large Volumes of Solutions	
Aspirin, phenacetin, caffeine	Extract 0.6 g sample into 50 ml, use only 15 μl	Dertinger[361]
Iodoamino acids	Prepare 100 ml soln; use 0.05 ml	Wachholz[362]
Salicylic acid, acetylsalicylic acid, salicylamide, caffeine, phenacetin	Take 1 g sample to make 1 liter soln; use 1 ml	Clayton[363]
Santonin, phenolphthalein	100 ml. soln.; use 1 ml	Ozsöz[364]
Sodium 4-aminosalicylate, *m*-aminophenol	Take 0.1 g to prepare 1 liter; use only 10 ml	Kalinowski[365]
	Methods that Specify ***10*** *Dosage Units*	
Atropine	10 suppositories, colorimetric	Czlonkowska[366]
Bellergot alkaloids	10 tablets, colorimetric	Czyczewska[367]
Chlorzoxazone	10 tablets, 0.1 N titrant	Beral[368]
2-Dimethylaminoethyl-*p*-chlorophenoxyacetate HCl	10 tablets, 0.1 N titrant	Beral[369]

As to the other types of analytical samples, the need to take the whole sample for quantitative analysis seldom arises. For most cases of organic mixtures, the problem is how to obtain representative samples by taking small fractions of the material. This subject is discussed in Chapter 2, Section 2.2.

REFERENCES

1. F. Herry and J. L. Janmot, *Double Liaison*, **20**, 288 (1973).
2. *Pure Appl. Chem.*, **33**, 417 (1973).
3. Pharmaceutical Society of Great Britain and Society for Analytical Chemistry, Joint Committee on Methods for the Evaluation of Drugs, *Analyst*, **90**, 582 (1965).
4. T. R. Crompton, *Anal. Chem.*, **39**, 268 (1967).
5. L. W. Kavanagh and L. B. Jaques, *Arzneim.-Forsch.*, **23**, 605 (1973).
6. T. Yamanka and T. Sato, *Arch. Pract. Pharm., Jap.*, **22**, 189 (1962).

7. N. D. Cheronis and T. S. Ma, *Organic Functional Group Analysis*, Wiley, New York, 1964.
8. U. Shankar, *Chim. Anal. (Paris)*, **51**, 376 (1969).
9. D. H. Whitehurst and J. B. Johnson, *Anal. Chem.*, **30**, 1332 (1958).
10. A. L. Wilson, *Talanta*, **17**, 21 (1970).
11. Division of Chemical Standards, National Physical Laboratory, Teddington, England, *Rep. Natl. Phys. Lab., Chem.*, **26**, 1973.
12. J. Swietoslawska, *Chem. Anal. (Warsaw)*, **17**, 1039 (1972).
13. J. Dandoy, A. Alloing-Bernard, and C. Renson-Deneubourg, *Ind. Chim. Belge*, **36**, 689 (1971).
14. L. Legradi, *Magy. Kem. Foly.*, **71**, 17 (1965).
15. D. M. W. Anderson and S. S. H. Zaidi, *Talanta*, **9**, 611 (1962).
16. F. W. Quackenbush, *J. Assoc. Off. Agric. Chem.*, **36**, 857 (1953).
17. H. de Langen and W. G. Whittlestone, *N. Z. J. Sci.*, **13**, 337 (1970).
18. J. E. Knipfel, J. R. Aitken, D. C. Hill, B. E. McDonald, and B. D. Owen, *J. Assoc. Off. Anal. Chem.*, **54**, 777 (1971).
19. B. J. Meakin, I. P. Tansey, and D. J. G. Danes, *J. Pharm. Pharmac.*, **23**, 252 (1971).
20. J. P. Matrka and Z. Sagner, *Chem. Prumy.*, **12**, 549 (1962).
21. M. W. Hwang, *J. Chin. Chem. Soc., Ser. II*, **9**, 298 (1962).
22. S. Suzuki and T. Ishida, *Jap. Analyst*, **11**, 299, 377 (1962).
23. R. O. Weenink, *Clin. Chim. Acta*, **24**, 186 (1969).
24. R. Reiss, *Z. Anal. Chem.*, **145**, 265 (1955).
25. A. P. Kreshkov, L. N. Shvetsova, G. P. Svistunova, and E. A. Emelin, *Zh. Anal. Khim.*, **26**, 369 (1971).
26. N. D. Cheronis and T. S. Ma, *Organic Functional Group Analysis*, Wiley, New York, 1964, p. 144.
27. E. V. Zobov and Y. S. Lyalikov, *Izv. Akad Nauk Turkm. SSR.*, **1958**, 93
28. A. J. Porsius, *Pharm. Weekbl. Ned.* **106**, 885 (1971).
29. H. K. Hundley and G. D. Castillo, *J. Assoc. Off. Anal. Chem.*, **57**, 738 (1974).
30. A. Bevenue, J. N. Ogata, Y. Kawano, and J. W. Hylin, *J. Chromatogr.*, **60**, 45 (1971).
31. B. Zak, C. C. Kepisty, and E. S. Baginski, *Microchem. J.*, **16**, 488 (1971).
31a. D. G. Crosby and N. Aharonson, *J. Chromatogr.*, **25**, 330 (1966).
32. A. K. Babko and V. S. Konyushko, *Zh. Anal. Khim.*, **21**, 486 (1966).
33. S. F. Petrashkevich and G. L. Starobinets, *Izv. Akad. Nauk Beloruss. SSR, Ser. Khim. Nauk*, **1970** (5), 35.
34. K. Gustavii, *Acta Pharm. Suec.*, **4**, 233 (1967).
35. R. Nagashima, G. Levy, and E. Nelson, *J. Pharm. Sci.*, **57**, 58 (1968).
36. M. Brummel, C. M. Gerbeck, and R. Montgomery, *Anal. Biochem.*, **31**, 331 (1969).

37. H. K. Tong, K. H. Lee, and H. A. Wong, *Anal. Biochem.*, **51**, 390 (1973).
38. D. R. Nelson and A. K. Huggins, *Anal. Biochem.*, **59**, 46 (1974).
39. E. Bancher, P. Riederer, and F. Wurst, *Ernachr.-Umsch.*, **18**, 417 (1971).
40. H. G. J. Worth, *Clin. Chim. Acta*, **49**, 53 (1973).
41. W. C. Burger, *Anal. Biochem.*, **57**, 306 (1974).
42. T. Paal, *Magy. Kem. Foly.*, **79**, 455 (1973).
43. S. H. Lipton and C. E. Bodwell, *J. Agric. Food Chem.*, **21**, 235 (1973).
44. N. D. Cheronis and T. S. Ma, *Organic Functional Group Analysis*, Wiley, New York, 1964, p. 363.
45. A. J. Milun, *Anal. Chem.*, **33**, 123 (1961).
46. G. Tokar and I. Simonyi, *Magy. Kem. Foly.*, **68**, 333 (1962).
47. C. W. Strode, Jr., F. N. Stewart, H. O. Schott, and O. J. Coleman, *Anal. Chem.*, **29**, 1184 (1957).
48. J. Jokl and J. Knizek, *Acta Chim. Acad. Sci. Hung.*, **33**, 17 (1962).
49. C. Mentzer and J. Jouanneteau, *Bull. Soc. Chim. Biol.*, **37**, 887 (1955).
50. V. G. Belikov, N. I. Kokovkin-Shcherbak, and S. K. Mutsueva, *Zavod. Lab.*, **33**, 1049 (1967).
51. J. D. Winefordner, J. J. Cetorelli, and W. J. McCarthy, *Talanta*, **15**, 207 (1968).
52. R. Harper, S. Siggia, and J. G. Hanna, *Anal. Chem.*, **37**, 600 (1965).
53. A. Linder and V. Persson, *J. Am. Oil. Chem. Soc.*, **34**, 24 (1957).
54. H. Sasuga, *J. Chem. Soc. Jap., Ind. Chem. Sect.*, **59**, 1117 (1956).
55. G. A. Stenmark and F. T. Weiss, *Anal. Chem.*, **28**, 1784 (1956).
56. V. Sedivec, *Chem. Listy*, **51**, 63 (1957).
57. A. P. Mathers and M. J. Pro, *Anal. Chem.*, **27**, 1662 (1955).
58. I. Hogea, V. Parausanu, and C. Scodigor, *Rev. Chim. Rom.*, **20**, 10 (1969).
59. L. Marvillet and J. Tranchant, *Chim. Anal. (Paris)*, **40**, 293 (1958).
60. W. J. Kirsten and S. K. Nilsson, *Mikrochim. Acta*, **1960**, 983.
61. T. S. Ma and A. S. Ladas, *Organic Functional Group Analysis by Gas Chromatography*, Academic, London, 1976, p. 45.
62. W. Hancock and E. Q. Laws, *Analyst*, **81**, 37 (1956).
63. V. S. Etlis and L. M. Artyukhina, *Zavod. Lab.*, **21**, 919 (1955).
64. C. D. Hildebrecht, *Anal. Chem.*, **29**, 1037 (1957).
65. I. M. Kolthoff and P. J. Elving, *Treatise on Analytical Chemistry*, Part II, Vol. 14, Wiley, New York, 1971.
66. V. V. Gerstev and Y. Y. Makarov-Zemlyanskii, *Zh. Anal. khim.*, **23**, 417 (1968).
67. G. S. Kozak and Q. Fernando, *Anal. Chim. Acta*, **26**, 541 (1962).
68. T. Beyrich and R. Pohlouder-Fabini, *Ernahrungsforschung*, **5**, 441 (1960).
69. E. Robins, N. R. Roberts, K. M. Eydt, O. H. Lowry, and D. E. Smith, *J. Biol. Chem.*, **218**, 897 (1956).
70. K. I. Evstratova, *Biokhimiya*, **23**, 181 (1958).

71. B. C. Mitra, P. Ghosh, and S. R. Palit, *Anal. Chem.*, **36**, 673 (1964).
72. V. E. Beiskii and M. I. Vinnik, *Zh. Anal. Khim.*, **19**, 375 (1964).
73. E. A. M. F. Dahmen, *Chim. Anal. (Paris)*, **40**, 430 (1958).
74. J. Schole, *Z. Anal. Chem.*, **193**, 321 (1963).
75. M. Rozental and L. Tomaszewski, *Clin. Chim. Acta*, **50**, 311 (1974).
76. D. E. Jordan and F. C. Veatch, *Anal. Chem.*, **36**, 120 (1964).
77. A. A. Ponomarenko and L. M. Amelina, *Zh. Anal. Khim.*, **18**, 1244 (1963).
78. C. J. Rogers, C. W. Chambers, and N. A. Clarke, *Anal. Chem.*, **38**, 1851 (1966).
79. I. N. Ermolenko and G. N. Chirkova, *Zh. Anal. Khim.*, **18**, 994 (1963).
80. A. Besada and Y. A. Gawargious, *Talanta*, **21**, 1247 (1974).
81. N. Seiler and M. Wiechmann, *Z. Anal. Chem.*, **220**, 109 (1966).
82. K. Konitzer, I. Körner, and S. Yoigt, *Z. Anal. Chem.*, **223**, 436 (1966).
83. S. Kiciak and J. Minczewski, *Chem. Anal. (Warsaw)*, **8**, 425 (1963).
84. Z. Kurzawa and M. Krzymien, *Chem. Anal. (Warsaw)*, **13**, 1047 (1968).
85. Z. Kurzawa, W. Wojciak, and R. Solecki, *Chem. Anal. (Warsaw)*, **12**, 1007 (1967).
86. C. C. Lee and E. R. Samuels, *Can. J. Chem.*, **42**, 164 (1964).
87. L. V. S. Hood and J. D. Winefordner, *Anal. Chem.*, **38**, 1922 (1966).
88. B. D. Flockhart and R. C. Pink, *Talanta*, **9**, 931 (1962).
89. A. Pietrogrande, F. Bordin, and G. D. Fini, *Mikrochim. Acta*, **1966**, 1156.
90. S. Veibel and I. G. K. Anderson, *Anal. Chim. Acta*, **14**, 320 (1956).
91. C. Omböly, *Acta Pharm. Hung.*, **33**, 15 (1963).
92. V. A. Klimova and K. S. Zabrodina, *Izv. Akad. Nauk SSSR*, **1959**, 175.
93. L. Fowler, H. R. Kline, and R. S. Mitchell, *Anal. Chem.*, **27**, 1688 (1955).
94. A. P. Terentev, S. I. Obtemperanskaya, and N. Dyk-Hoe, *Zh. Anal. Khim.*, **19**, 902 (1964).
95. W. Wildenhain and G. Henseke, *Chimia*, **20**, 357 (1966).
96. R. C. Lawrence, *Nature*, **205**, 1313 (1965).
97. S. Siggia and C. R. Stahl, *Anal. Chem.*, **27**, 1975 (1955).
98. V. M. Potapov, G. P. Moiseeva, and A. P. Terentev, *Zh. Anal. Khim.*, **18**, 275 (1963).
99. D. Karacsony, *Elelmez. Ipur*, **8**, 309 (1954).
100. E. M. Livingston, R. K. Maurmeyer, and A. Worthman, *Microchem. J.*, **1**, 261 (1957).
101. T. Harada, *Biochim. Biophys. Acta*, **63**, 334 (1962).
102. V. Goldenberg and P. E. Spoerri, *Anal. Chem.*, **30**, 1327 (1958).
103. H. L. Polak, H. F. Pronk, and G. den Boef, *Z. Anal. Chem.*, **190**, 377 (1962).
104. V. K. Zolutukhin, and A. S. Molotkova, *Tr. Kom. Anal. Khim. Akad. Nauk SSSR*, **5** (8), 179 (1954).
105. T. S. Ma and W. L. Nazimowitz, *Mikrochim. Acta*, **1969**, 345, 821.
106. P. K. Jaiswal and K. L. Yadava, *J. Indian Chem. Soc.*, **51**, 750 (1974).

107. D. E. Jordan, *Anal. Chem.*, **36**, 2134 (1964).
108. J. L. Cohen and G. P. Fong, *Anal. Chem.*, **47**, 313 (1975).
109. S. Siggia, J. G. Hanna, and R. Culmo, *Anal. Chem.*, **33**, 900 (1961).
110. S. Siggia and J. G. Hanna, *Anal. Chem.*, **33**, 896 (1961).
111. A. I. Ilina and V. G. Nedavnyana, *Maslob-Zhir. Prom.*, **1961** (10), 31.
112. G. H. Schenk, P. Wines, and C. Mojzis, *Anal. Chem.*, **36**, 914 (1964).
113. A. A. Zelenetskaya, I. V. Levina, and G. P. Borodina, *Tr. Vses. Nauchno-Issled. Inst. Sint. Nat. Dushistykh Veshchestr.*, **1965** (7), 180.
114. E. Vioque and M. P. Maza, *Grasas Aceites*, **13**, 207 (1962).
115. J. A. Floria, I. W. Dobratz, and J. H. McClure, *Anal. Chem.*, **36**, 2053 (1964).
116. G. H. Schenk and M. Santiago, *Anal. Chem.*, **39**, 1795 (1967).
117. K. Konishi, Y. Mori, H. Inoue, and M. Mozoe, *Anal. Chem.*, **40**, 2198 (1968).
118. P. Fijolka, *Plaste Kautsch.*, **18**, 431 (1971).
119. L. N. Petrova and E. N. Novikova, *Zh. Anal. Khim.*, **12**, 411 (1957).
120. T. Yamanaka, S. Asai, and J. Aoki, *Arch. Pract. Pharm. Jap.*, **22**, 60 (1962).
121. J. B. Johnson and G. L. Funk. *Anal. Chem.*, **28**, 1977 (1956).
122. A. McCoubrey, *J. Pharm. Pharmacol.*, **8**, 442 (1956).
123. E. N. Deeb, *Drug Stand.*, **26**, 175 (1958).
124. G. R. Umbreit, *Anal. Chem.*, **33**, 1572 (1961).
125. J. E. Ruch and F. E. Critchfield, *Anal. Chem.*, **33**, 1569 (1961).
126. M. Miller and D. A. Keyworth, *Talanta*, **10**, 1131 (1963).
127. G. H. Schenk, P. Warner, and W. Bazzelle, *Anal. Chem.*, **38**, 907 (1966).
128. G. Kainz and F. Schöller, *Naturwissenschaften*, **42**, 209 (1955); G. Kainz and F. Kasler, *Mikrochim. Acta*, **1960**, 62.
129. M. Hori, I. Aoki, H. Kashiwagi, and S. Kusumoto, *Ann. Rep. Takeda Res. Lab.*, **22**, 59, 64 (1962).
130. J. G. Baldinus and I. Rothberg, *Anal. Chem.*, **34**, 924 (1962).
131. G. Schill, C. Ericson, and K. Platzak, *Acta Pharm. Suec.*, **1**, 37 (1964).
132. M. Nedic and I. Berkes, *Acta Pharm. Jugosl.*, **13**, 13 (1963).
133. L. A. A. Sluyterman, *Biochem. Biophys. Acta*, **25**, 402 (1957).
134. A. A. Spryskov and S. P. Starkov, *Zh. Obsch. Khim.*, **26**, 2607 (1956).
135. H. Hogeveen and F. Montanari, *Gazz. Chim. Ital.*, **94**, 176 (1964).
136. Y. P. Kitaev and G. K. Budnikov, *Zavod. Lab.*, **28**, 806 (1962).
137. A. P. Kreshkov, L. N. Balayatinskaya, and S. M. Chesnokova, *Zh. Anal. Khim.*, **28**, 1571 (1973).
138. M. M. Buzlanova, N. A. Kozhikhova, and N. G. Polyanskii, *Zh. Anal. Khim.*, **18**, 1125 (1963).
139. G. L. Bezborodko, *Plast. Massy*, **1961** (1), 59.
140. D. H. Mathews and T. R. Welch, *J. Appl. Chem.*, **8**, 701 (1958).
141. I. V. Avgushevich and N. M. Karauaev, *Pochvovedenie*, **1965** (4), 97.

142. F. Pellerin, J. A. Gautier, and D. Demay, *Talanta*, **12**, 847 (1965).
143. L. G. Chatten and C. K. Orbeck, *J. Pharm. Sci.*, **53**, 1306 (1964).
144. J. Neumann, V. Stepankova, and Z. Aunicky, *Chem. Prum.*, **8**, 244 (1958).
145. A. Corsini and R. P. Graham, *Anal. Chim. Acta*, **8**, 583 (1963).
146. A. L. Voloskovets, *Ukr. Khim. Zh.*, **30**, 1347 (1964).
147. M. R. F. Ashworth and M. Winter, *Anal. Chim. Acta*, **29**, 75 (1963).
148. C. Omböly and E. Dersi, *Acta Pharm. Hung.*, **32**, 88 (1962).
149. T. S. Ma, in *Proceedings of the International Symposium on Microchemistry 1958*, Pergamon, London, 1959, p. 151.
150. A. Dodson, C. H. Hughes, and G. Ingram, in *Annual Reports on the Progress of Chemistry*, Vol. LX, The Chemical Society, London, 1964, p. 542.
151. T. S. Ma and V. Horak, *Microscale Manipulations in Chemistry*, Wiley, New York, 1976, p. 377.
152. W. T. Smith and A. Wu, *Anal. Lett.*, **3**, 393 (1970).
153. E. O. Schmalz and G. Geiseler, *Z. Anal. Chem.*, **190**, 233 (1962).
154. E. Bayer and K. H. Reuther, *Chem. Ber.*, **89**, 2541 (1956).
155. G. R. Mizuno and J. R. Chipault, *J. Am. Oil Chem. Soc.*, **42**, 839 (1965).
156. T. W. Kwon and B. M. Watts, *Anal. Chem.*, **35**, 733 (1963).
157. J. R. Helbert and K. D. Brown, *Anal. Chem.*, **28**, 1098 (1956).
158. M. Fedoronko, J. Königstein, and M. Bullova, *Chem. Zvesti*, **22**, 25 (1968).
159. F. Stitt, S. Friedlander, H. J. Lewis, and F. E. Young, *Sugar Ind. Abstr.*, **15**, 727 (1953).
160. A. R. Sheth and S. S. Rao, *Experientia*, **19**, 362 (1963).
161. R. T. Keen, *Anal. Chem.*, **29**, 1041 (1957).
162. M. V. Sudnik and M. F. Romantsev, *Zh. Anal. Khim.*, **29**, 1031 (1974).
163. S. I. Obtemperanskaya and N. Dyk-Hoe, *Vestn. Mosk. Univ., Ser. Khim.*, **1964** (3), 83.
164. L. A. Badovskaya, R. R. Vysyukova, and V. G. Kulnevich, *Zh. Anal. Khim.*, **22**, 1268 (1967).
165. G. Nisli and A. Townshend, *Talanta*, **15**, 1480 (1968).
166. J. C. Guyon and J. Y. Marks, *Mikrochim. Acta*, **1969**, 731.
167. A. Masood and O. C. Saxena, *Microchem. J.*, **13**, 178, 316, 321 (1968).
168. W. Huber, *Z. Anal. Chem.*, **197**, 236 (1963).
169. M. Schrier, A. Fono, and T. S. Ma, *Mikrochim. Acta*, **1965**, 1091; **1967**, 218.
170. J. W. Keyser and C. C. Entwistle, *Lancet*, **1971-I**, 111.
171. J. H. Karchmer, *Anal. Chem.*, **29**, 425 (1957).
172. N. G. Polyanskii, V. S. Markevich, and N. E. Shtivel, *Zh. Anal. Khim.*, **19**, 1132 (1964).
173. L. Barnes, *Anal. Chem.*, **31**, 405 (1959).
174. M. Gutterson and T. S. Ma, *Microchem. J.*, **5**, 601 (1961).

175. K. H. Mohr and F. Wolf, *Z. Anal. Chem.*, **233**, 269 (1968).
176. A. G. Stradomskaya and I. A. Goncharova, *Gidrokhim. Mater.*, **41**, 78 (1966).
177. T. Kashima and K. Kano, *J. Pharm. Soc. Jap.*, **76**, 50 (1956).
178. E. Posgay, *Acta Pharm. Hung.*, **35**, 266 (1965).
179. L. M. Litvinenko and A. P. Grekov, *Uch. Zap. Khark. Univ.*, **76**, 59 (1956).
180. A. P. Terentev, S. I. Obtemperanskaya, M. M. Buzlanova, and T. E. Vlasova, *Zh. Anal. Khim.*, **17**, 900 (1962).
181. W. H. Greive, K. F. Sporek, and M. K. Stinson, *Anal. Chem.*, **38**, 1264 (1966).
182. W. D. Skidmore and C. Entenman, *J. Lipid Res.*, **3**, 356 (1962).
183. A. P. Kreshkov, I. Y. Guretskii, N. T. Smolova, and A. I. Ryaguzov, *Zh. Anal. Khim.*, **25**, 451 (1970).
184. S. L. Lin, *J. Pharm. Sci.*, **56**, 1130 (1967).
185. H. J. Keily and D. N. Hume, *Anal. Chem.*, **36**, 543 (1964).
186. M. E. Hall and E. H. Harris, *Anal. Chem.*, **41**, 1130 (1969).
187. J. H. Hsiao and W. L. Chiou, *J. Pharm. Sci.*, **63**, 1776 (1974).
188. I. H. Williams, *J. Assoc. Off. Anal. Chem.*, **51**, 715 (1968).
189. J. A. Burke and M. L. Porter, *J. Assoc. Off. Anal. Chem.*, **50**, 1260 (1967).
189a. R. Dynakowski and J. Kubalski, *Acta Pol. Pharm.*, **18**, 21 (1961).
190. R. O. Feuge, Z. Zarins, J. L. White, and R. L. Holmes, *J. Am. Oil Chem. Soc.*, **46**, 185 (1969).
191. N. K. Mankovskaya and S. A. Udovenko, *Maslab. Zhir. Prom.*, **1956** (3), 30.
192. R. Morales, *Anal. Chim. Acta*, **48**, 309 (1969).
193. R. H. Still, K. Wilson, and B. W. J. Lynch, *Analyst*, **93**, 805 (1968).
194. L. A. Kazitsyna, V. V. Mishchenko, *Vestn. Mosk. Gos. Univ., Ser. Khim.*, **1970**, 600.
195. K. Dimroth and C. Reichardt, *Z. Anal. Chem.*, **215**, 344 (1966).
196. J. Bernaerts, *Sucr. Belge.*, **74**, 167 (1955).
197. F. M. Lin and Y. Pomeranz, *Anal. Biochem.*, **24**, 128 (1968).
198. Z. Dusic and A. Berka, *Acta Pharm. Jugosl.*, **25**, 43 (1975).
199. Z. Dusic, *Acta Pharm. Jugosl.*, **5**, 35 (1975).
200. J. Geisler and Z. Szot, *Chem. Anal. (Warsaw)*, **12**, 427 (1967).
201. G. Anderson, *J. Sci. Food Agric.*, **14**, 352 (1963).
202. C. A. Elliger, B. G. Chan, and W. L. Stanley, *J. Chromatogr.*, **104**, 57 (1975).
203. J. W. Goldzieher and A. de la Pena, *Clin. Chem.*, **14**, 1125 (1968).
204. H. Hoffmann, *Arch. Pharm. (Berl.)*, **304**, 849 (1971).
205. S. Isobe, *Ann. Rep. Natl. Inst. Nutr. Tokyo*, **1959**, 45.
206. J. B. Bingley, *Anal. Chem.*, **40**, 1166 (1968).
207. S. P. Sen and A. K. Sengupta, *Analyst*, **89**, 558 (1964).
208. D. J. Pietrzyk and L. B. Rogers, *Anal. Chem.*, **34**, 936 (1962).
209. J. G. Hanna and S. Siggia, *Anal. Chem.*, **37**, 690 (1965).

210. D. B. Bruss and G. A. Harlow, *Anal. Chem.*, **30**, 1836 (1958).
211. British Standards Institution, B.S. 4839 (1969).
212. F. E. Critchfield and J. B. Johnson, *Anal. Chem.*, **30**, 1247 (1958).
213. R. Vasiliev, E. Sisman, M. Jecu, and I. Chialda, *Rev. Chim. Buchar.*, **11**, 347 (1960).
214. Y. Yamamoto, T. Kumamaru, and Y. Uemura, *Anal. Chim. Acta*, **39**, 51 (1967).
215. T. J. McCarthy, *Pharm. Weekbl. Ned.*, **105**, 557 (1970).
216. R. Adamski and K. Pawelczyk, *Farm. Pol.*, **25**, 979 (1969).
217. A. Hanok and J. Kuo, *Clin. Chem.*, **14**, 58 (1968).
218. S. E. Katz and C. A. Fassbender, *J. Assoc. Off. Anal. Chem.*, **50**, 821 (1967).
219. J. Alicino, *J. Pharm. Sci.*, **65**, 300 (1976).
220. L. G. Chatten and C. A. Mainville, *J. Pharm. Sci.*, **52**, 146 (1963).
221. S. A. Soliman, H. Abdine, and M. G. Morcos, *Can. J. Pharm. Sci.*, **11**, 63 (1976).
222. L. E. Braverman, H. Reinstein, H. Loayza, and D. Pomfret, *Clin. Chem.*, **14**, 374 (1968).
223. F. Chrzanowski, P. J. Niebergall, R. Mayock, J. Taubin, and E. Sugita, *J. Pharm. Sci.*, **65**, 735 (1976).
224. V. J. Pileggi and G. Kessler, *Clin. Chem.*, **14**, 339 (1968).
225. S. L. Kanter, *Clin. Chim. Acta*, **17**, 147 (1967).
226. I. Kelemen and I. Kerenyi, *Acta Pharm. Hung.*, **37**, 60 (1967).
227. R. H. Abaza, J. T. Blake, and E. J. Fisher, *J. Assoc. Off. Anal. Chem.*, **51**, 963 (1968).
228. B. Vickery, M. L. Vickery, and J. T. Ashu, *Phytochemistry*, **12**, 145 (1973).
229. Y. Kawano, A. Bevenue, H. F. Beckman, and F. Erro, *J. Assoc. Off. Anal. Chem.*, **52**, 167 (1969).
230. R. Brand, K. E. Guyer, and W. L. Banks, Jr., *Clin. Chim. Acta*, **51**, 103 (1974).
231. V. Cerda and L. Eek, *Quim. Anal.*, **29**, 22 (1975).
232. E. Sjostrom, *Acta Chem. Scand.*, **7**, 1392 (1953).
233. C. R. Primavesi, *Analyst*, **78**, 647 (1953).
234. A. S. Maslennikov, *Zh. Anal. Khim.*, **13**, 599 (1958).
235. M. Pesez and J. Bartos, *Talanta*, **21**, 1306 (1974).
236. W. M. MacNevin and P. F. Urone, *Anal. Chem.*, **25**, 1760 (1953).
237. H. Bruschweiler, G. J. Minkoff, and K. C. Salooja, *Nature*, **172**, 909 (1953).
238. A. J. Milun, *Anal. Chem.*, **29**, 1502 (1957).
239. E. P. Przybylowicz and L. B. Rogers, *Anal. Chim. Acta*, **18**, 596 (1958).
240. L. Eklandius and H. K. King, *Biochem. J.*, **65**, 128 (1957).
241. J. Tologyessy and M. Sarsunova, *Z. Anal. Chem.*, **195**, 429 (1963).
242. P. J. Watkins, *Clin. Chem. Acta*, **18**, 191 (1967).
243. K. T. Lee and C. G. Farmilo, *J. Pharm. Pharmacol.*, **10**, 427 (1958).
244. F. L. Estes and P. K. Baughman, *Anal. Chem.*, **33**, 473 (1961).
245. C. G. B. Hammer, *Sven. Kem. Tidskr.*, **67**, 307 (1955).

246. R. A. Mortensen, *J. Biol. Chem.*, **203**, 855 (1953).
247. D. E. Jordan, *Anal. Chim. Acta*, **30**, 297 (1964).
248. J. Franc, *Cesk. Farm.*, **4**, 4 (1955).
249. S. Higuchi, S. Tanaka, and H. Ishikawa, *Jap. Analyst*, **25**, 328 (1976).
250. W. Beckering, C. M. Frost, and W. W. Fowkes, *Anal. Chem.*, **36**, 2412 (1964).
251. J. E. Katon, W. R. Feairheller, Jr., and J. V. Pustinger, Jr., *Anal. Chem.*, **36**, 2126 (1964).
252. M. Yamaguchi, Y. Hayashi, and S. Matsukawa, *Jap. Analyst*, **10**, 1106 (1961).
253. R. M. Powers, J. L. Harper, and H. Tai, *Anal. Chem.*, **32**, 1287 (1960).
254. E. L. Saier, L. R. Cousins, and M. R. Basila, *Anal. Chem.*, **34**, 824 (1962).
255. D. M. W. Anderson and S. S. H. Zaidi, *Anal. Chim. Acta*, **30**, 303 (1964).
256. E. T. Lippmaa, *Tr. Tallin. Politekh. Inst.*, **A195**, 35 (1962).
257. (a) R. J. Mesley and C. A. Johnson, *J. Pharm. Pharmacol.*, **17**, 329 (1965); (b) B. L. van Duuren, New York Univ. Medical Center, private communication, 1974.
258. L. Borka and K. Backe-Hansen, *Acta Pharm. Suec.*, **5**, 271 (1968).
259. F. Pellerin, R. Chasset, and M. F. Le Baron, *Ann. Pharm. Fr.*, **26**, 421 (1968); **27**, 719 (1969).
260. V. A. Yarborough, J. F. Haskin, and W. J. Lambdin, *Anal. Chem.*, **26**, 1576 (1954).
261. S. Baczyk, *Z. Lebensmittelunters.*, **131**, 215 (1966).
262. D. T. Englis, B. B. Burnett, R. A. Schreiber, and J. W. Miles, *J. Agric. Food Chem.*, **3**, 964 (1955).
263. R. J. Doyle and H. M. Burgan, *Anal. Biochem.*, **17**, 171 (1966).
264. J. Kracmar and J. Kracmarova, *Pharmazie*, **26**, 157 (1971).
265. F. Santavy, L. Hruban, V. Simaner, and D. Walterova, *Collect. Czech. Chem. Commun.*, **35**, 2418 (1970).
266. M. Brandstätter-Kuhnert, E. Junger, and A. Kofler, *Microchem. J.*, **9**, 105 (1965).
266a. M. C. Chiang, *Scientia Sinica*, **1**, 207 (1978).
267. M. Goldman and E. L. Wehry, *Anal. Chem.*, **42**, 1178 (1970).
268. D. W. Fink and W. R. Koehler, *Anal. Chem.*, **4** , 990 (1970).
269. G. A. Thommes and E. Leininger, *Talanta*, **7**, 181 (1961).
270. G. A. Thommes and E. Leininger, *Anal. Chem.*, **30**, 1361 (1958).
271. D. J. W. Grant and J. C. Patel, *Anal. Biochem.*, **28**, 139 (1969).
272. J. H. Thompson, C. A. Spezia, and M. Angulo, *Anal. Biochem.*, **31**, 321 (1969).
272a. D. J. Mitchell, G. B. Schuster, and H. G. Drickamer, *J. Am. Chem. Soc.*, **99**, 1145 (1977).
273. E. A. Blyumberg and V. L. Pikaeva, *Zh. Anal. Khim.*, **10**, 310 (1955).
274. W. Kemula, E. Najdeker, and Z. Kublik, *J. Electroanal. Chem.*, **5**, 211 (1963).
275. W. B. Swann, W. M. McNabb, and J. F. Hazel, *Anal. Chim. Acta*, **28**, 441 (1963).
276. S. Usami, *Jap. Analyst*, **9**, 216 (1960).
277. C. A. Kelly, *Talanta*, **11**, 175 (1964).

278. T. S. Ma, M. R. Hackman, and M. A. Brooks, *Mikrochim. Acta.* **1975-II,** 617.
279. K. Tsuji and P. J. Elving, *Anal. Chem.*, **41**, 286 (1969).
280. H. Brüschweiler and G. J. Minkoff, *Anal. Chem. Acta*, **12**, 186 (1955).
281. E. J. Kuta and F. W. Quackenbush, *Anal. Chem.*, **32**, 1069 (1960).
282. R. W. von Korff, *J. Biol. Chem.*, **210**, 539 (1954).
283. M. Blecher and A. B. Glassmann, *Anal. Biochem.*, **3**, 343 (1962).
284. H. Schiweck, *Zucker*, **16**, 170 (1963).
285. W. Goetz, *Glas.-Instr. Tech. Fachz. Lab.*, **20**, **330** (1976).
286. J. A. K. Buisman, K. H. Hanewald, F. J. Mulder, J. R. Roborgh, and K. J. Keuning, *J. Pharm. Sci.*, **57**, 1326 (1968).
287. L. Cavalli and G. Cancellieri, *Analyst*, **100**, 46 (1975).
288. D. G. Gehring and G. S. Reddy, *Anal. Chem.*, **37**, 868 (1965).
289. J. L. Raven, M. B. Robson, P. L. Walker, and P. Barkhan, *Guy's Hosp. Rep.*, **117**, 89 (1968).
290. O. H. Lowry, N. R. Roberts, and M. L. W. Chang, *J. Biol. Chem.*, **222**, 97 (1956).
291. K. A. Jeffs, K. A. Lord, and R. J. Tuppen, *J. Sci. Food Agric.*, **19**, 195 (1968).
292. P. J. J. de Meijer, *Pharm. Weekbl. Ned.*, **108**, 849 (1973).
293. A. W. M. Indemans, *Pharm. Weekbl. Ned.*, **108**, 785 (1973); **109**, 161 (1974).
294. H. R. Roberts and M. R. Siino, *J. Pharm. Sci.*, **52**, 370 (1963).
295. H. G. Gänshirt and J. Polderman, *J. Chromatogr.*, **16**, 510 (1964).
296. H. Duda and U. Elste, *Dtsch. Apoth. Ztg.*, **110**, 593 (1970).
297. N. Shane and M. Kowglansky, *J. Pharm. Sci.*, **57**, 1218 (1968).
298. L. T. Sennello and C. J. Argoudelis. *Anal. Chem.*, **41**, 171 (1969).
299. H. Kadin, M. S. Ugolini, and H. R. Roberts, *J. Pharm. Sci.*, **53**, 1313 (1964).
300. A. Z. Knizhnik and P. L. Senov, *Izv. Vysch. Uchebn. Zaved. Khim.*, **12** (3), 289 (1969).
301. E. Jacobsen and K. B. Thorgersen, *Anal. Chim. Acta*, **71**, 175 (1974).
302. A. M. Soeterboek and M. van Thiel, *Pharm. Weekbl.*, **110**, 169 (1975).
303. K. T. Lee and C. S. Chan, *J. Pharm. Pharmacol.*, **12**, 624 (1960).
304. R. S. Santoro, *J. Am. Pharm. Assoc. Sci. Ed.*, **49**, 666 (1960).
305. H. Bukowska, W. Grzegorzewicz, and E. Pawlak, *Acta Pol. Pharm.*, **21**, 169 (1964).
306. V. D. Gopta and N. M. Ferguson, *Am. J. Hosp. Pharm.*, **23**, 168 (1969).
307. S. M. Khafagy and A. N. Girgis, *Planta Med.*, **25**, 350 (1974).
308. G. Machek, *Sci. Pharm.*, **29**, 73 (1961).
309. M. E. Biesemeyer, *J. Assoc. Off. Agric. Chem.*, **45**, 593 (1962).
310. N. Nino and N. Taneva, *Farm., Sofia*, **18** (5), 1 (1968).
311. J. Kos and V. Vukcevic-Kovacevic, *Bull. Sci. Cons. Acad. RSF Yugosl.*, **14**, 387 (1969).
312. L. A. Roberts, *J. Pharm. Sci.*, **58**, 1015 (1969).

313. R. J. Warren, I. B. Eisdorfer, W. E. Thompson, and J. E. Zarembo, *Microchem. J.*, **12**, 555 (1967).
314. T. D. Doyle and F. R. Fazzari, *J. Pharm. Sci.*, **63**, 1921 (1974).
315. M. Sobiczewska, *Farm. Pol.*, **20**, 35 (1964).
316. L. S. Bark and L. Kershaw, *Analyst*, **100**, 873 (1975).
317. V. D. Gupta, *Indian J. Pharm.*, **25**, 161 (1963).
318. M. R. Stevens, *J. Pharm. Sci.*, **64**, 1686 (1975).
319. T. James, *J. Pharm. Sci.*, **62**, 1500 (1973).
320. H. Hammerstingl and G. Reich, *J. Chromatogr.*, **101**, 408 (1974).
321. C. E. Stevenson and I. Comer, *J. Pharm. Sci.*, **57**, 1227 (1968).
322. J. W. Myrick, *J. Pharm. Sci.*, **58**, 1018 (1969).
323. A. Becker and F. Ehinger, *Z. Anal. Chem.*, **198**, 162 (1963).
324. R. J. Motz, *Int. Choc. Rev.*, **23**, 218 (1968).
325. R. O. Zimmerer, Jr. and L. T. Grady, *J. Pharm. Sci.*, **59**, 87 (1970).
326. R. Bryant, D. E. Mantle, D. L. Timma, and D. S. Yoder, *J. Pharm. Sci.*, **57**, 658 (1968).
327. R. J. Templeton, W. A. Arnett, and I. M. Jakovljevic, *J. Pharm. Sci.*, **57**, 1168 (1968).
328. F. K. Bell, *J. Pharm. Sci.*, **53**, 752 (1964).
329. J. H. McB. Miller and P. Duguid, *Proc. Anal. Div. Chem. Soc.*, **13**, 9 (1976).
330. P. D. Meister, C. A. Schlagel, J. E. Stafford, and J. L. Johnson, *J. Am. Pharm. Assoc., Sci. Ed.*, **47**, 576 (1958)
331. B. N. Kabadi, A. T. Warren, and C. H. Newman, *J. Pharm. Sci.*, **58**, 1127 (1969).
332. R. C. Reed and W. W. Davis, *J. Pharm. Sci.*, **54**, 1533 (1965).
333. W. F. Beyer, *J. Pharm. Sci.*, **55**, 200 (1966).
334. A. M. Wild, *Chem. Ind. (Lond.)*, **1963**, 819.
335. I. Grabowska, K. Marcinkowska, and J. Wodkiewicz, *Farm. Pol.*, **31**, 825 (1975).
336. P. F. Helgren, F. E. Chadde, and D. J. Campbell, *J. Am. Pharm. Ass., Sci. Ed.*, **46**, 644 (1957).
337. M. Slamnik, *J. Pharm. Sci.*, **65**, 736 (1976).
338. S. Jeske and Z. E. Kalinowska, *Farm. Pol.*, **20**, 819 (1964).
339. C. M. P. Wirth, *Pharm. Acta Helv.*, **34**, 283 (1959).
340. R. Vasiliev, E. Sisman, and V. Scinteie-Pazarina, *Rev. Chim. Buchar.*, **15**, 46 (1964).
341. R. Vasiliev, V. Scinteie-Pazarina, and E. Sisman, *Rev. Chim. Buchar.*, **15**, 163 (1964).
342. M. Beguin, *Pharm. Acta Helv.*, **34**, 146 (1959).
343. F. Wartmann-Hafner and J. Büchi, *Pharm. Acta Helv.*, **40**, 592 (1965).
344. W. Grabowicz, *Chem. Anal. (Warsaw)*, **9**, 697 (1964).
345. G. S. Meadows, *J. Assoc. Publ. Anal.*, **6**, 97 (1968).

346. F. Biedebach and G. Manns, *Mitt. Dtsch. Pharm. Ges.*, **30**, 1 (1960).
347. H. Beral and T. Constantinescu, *Rev. Chim. Buchar.*, **14**, 235 (1963).
348. I. Ionescu, D. Popescu, and S. Enache, *Rev. Chim. Buchar.*, **14**, 532 (1963).
349. V. Stoicescu, C. Ivan, and H. Beral, *Revta Chim.*, **19**, 484 (1968).
350. W. Grabowicz, *Chem. Anal. (Warsaw)*, **13**, 1273 (1968).
351. V. A. Devyatnin and M. Y. Moizhes, *Med. Prom. SSSR.*, **17**, (11), 50 (1963).
352. H. H. Kavarana, *Indian J. Pharm.*, **20**, 360 (1958).
353. V. Tomaskova, M. Blesova, and M. Zahradnicek, *Cesk. Farm.*, **13** (3), 93 (1964).
354. L. J. Throop, *J. Pharm. Sci.*, **54**, 308 (1965).
355. S. L. Ali, *Chromatographia*, **7**, 655 (1974).
356. J. S. Faber, *J. Pharm. Pharmacol.*, **6**, 187 (1954).
357. D. Popescu, R. Schneider, and M. Vaeni, *Revta Chim.*, **19**, 734 (1968).
358. K. Ballreich, *Z. Anal. Chem.*, **195**, 274 (1963).
359. H. Bukowska and Z. Nasierowska, *Acta Pol. Pharm.*, **22**, 31 (1965).
360. J. Dobrecky and A. M. Z. de Suffriti, *Proanalisis*, **2**, 10 (1969).
361. G. Dertinger and H. Scholz, *Pharm. Ind. Berl.*, **34**, 114 (1972).
362. E. Wachholz and S. Pfeifer, *Pharmazie*, **24**, 459 (1969).
363. A. W. Clayton and R. E. Thiers, *J. Pharm. Sci.*, **55**, 404 (1965).
364. B. Ozsöz, *Turk. Bull. Hyg. Exp. Biol.*, **23**, 172 (1963).
365. K. Klainowski and Z. Sykulska, *Acta Pol. Pharm.*, **14**, 255 (1957).
366. H. Czlonkowska and I. Gradecka, *Farm. Pol.*, **20**, 739 (1964).
367. S. Czyszewska, F. Kaczkarek, J. Lutomski, and H. Speichert, *Herb. Pol.*, **12** (2), 87 (1966).
368. H. Beral, V. Gamentzy, and I. Bucur, *Revta Chim.*, **16**, 322 (1965).
369. Beral, E. Dumitrescu, P. Grintescu, and I. Calafeteanu, *Farm. Ed. Prat.*, **18**, 544 (1963).

CHAPTER 2

Preparation of the Sample and Selection of the Analytical Method

2.1 SAMPLING TECHNIQUES

2.1.1 Samples from Bulk Materials

Bicking[1] has written an excellent article on the problems encountered in sampling chemicals in bulk form, citing the committee reports of the American Society for Testing and Materials (ASTM) and other references. Sampling has been defined in engineering terms as "the operation of removing a part convenient in size for testing from a whole which is of much greater bulk, in such a way that the proportion and distribution of interest (e.g., chemical composition) in the sample represent, within measurable limits of error, the proportion and distribution of the quality of the whole."

In the collection of samples from bulk material, the selection of suitable containers and appropriate devices depends on the nature, condition, and structure of the material. For collecting gaseous mixtures, the container may be made of steel, glass, or plastic; a sampling probe and delivery line direct the gases from the bulk source toward the container, and the flow of gases is effected by water displacement, expansion into an evacuated chamber, or purging by an innocuous gas. The sampling techniques for collecting liquid materials,[2] as recommended by ASTM, are given in Table 2.1. Various conditions in which solids may be found at the time of sampling and the type of sample device appropriate to each[3] are shown in Table 2.2.

Generally speaking, bulk material is not homogeneous. Therefore a number of samples must be drawn from various segments of the lot. This number is determined statistically and is called "sample size" (Note: this is a

Table 2.1 Techniques for Collecting Liquid Samples[2]

Technique	For bulk material in
Bottle or thief sampling	Storage tanks, tank trucks, tank cars, ships or barges
Tap sampling	Storage tanks, tank trucks, tank cars
Continuous sampling	Pipelines, filling lines, transfer lines
Tube sampling	Drums, carboys, cans, or bottles
Jar sampling	Free or open discharge streams

statistical term and does not mean the size of the sample subsequently taken for chemical analysis). If a prior estimate of the standard deviation of the lot is available, the formula for calculating sample size is as follows[4]:

$$n = \left(\frac{t\sigma'}{E}\right)^2$$

where n = the sample size, that is, number of samples required
σ' = the prior estimate of the standard deviation of the lot
E = the maximum allowable difference between the estimate to be made from the sample and the actual value
t = a probability factor to give a selected level of confidence that the difference is greater than E

Efforts should be made to minimize bias in obtaining samples from bulk materials. Stratification is a common phenomenon for solid mixtures; thus finer materials tend to settle to the bottom of a carton, leaving larger particles

Table 2.2 Sampling of Solids

Sampling device	Condition
Auger sampler	Compacted material, natural deposits
Split tube thief	Non-free-flowing material in bales, boxes, and so on
Concentric tube thief	Free-flowing material in hoppers, drums, cans, bags, and so on
Hand scoop or automatic crosscut sampler	Conveyers, chutes, free-flowing streams
Automatic gravity flow auger sampler	Conveyer pipes, spouts, hoppers
Shovel or automatic vacuum probe sampler	Pile on the ground, or in a bin, railway car, truck, or hold of a ship

on top. Even for single phase liquid materials such as vegetable oil in a drum, the composition at different levels may vary.

It is advisable to follow established standard sampling methods. For example, there are ASTM sampling standards for volatile solvents, petroleum products, soaps, detergents, rosin, pine oil, pine tars, liquid oils, fatty acids, paints, varnishes, synthetic elastomers, plasticizers, and so on. The British Standards Institution[5] has published two methods for sampling powdered detergents, including details of apparatus suitable, in particular, for physical mixtures, the constituents of which may segregate on shaking. The Tobacco Research Council[6] has recommended methods for the classification, sampling, and selection of cigarettes for the analysis of tobacco smoke. Sampling of pesticides such as 2,4-D formulations was reported by the Collaborative Pesticides Analytical Committee.[7] A review written by Stakheaev and Kuznetsov[8] discussed the error of sampling and methods of reducing such errors, citing 89 references.

It should be mentioned that sampling of bulk materials sometimes has important economic implications. For instance, the price of fertilizer is based on its nitrogen content and the customs duty on molasses is determined by sugar analysis. When the transaction involves tons of the material, discrepancies of fractions of a percent may have a large economic effect.

2.1.2 Laboratory Samples

After the samples obtained from bulk materials are brought to the analytical laboratory, they may be analyzed separately or composited and then analyzed. Usually the amount of material drawn is much more than the quantity actually used in the chemical analysis. Thus another sampling process takes place. An excellent discussion of this subject was written by Benedetti-Pichler.[9] Various techniques are available, the most common being "coning and quartering." This method involves piling the material to form a cone, which is then flattened and separated into quarters; opposite quarters are taken and the other two are rejected; the procedure is repeated until the remaining quantity is within the range required for the analytical method selected. Rowland[10] constructed a simple sample divider from a phonograph turntable and a plastic dish of six segments of equal size and of the same shape; the dish is placed on the turntable and allowed to rotate at 20 rpm while the powder material is fed by gravity into the rotating segments through a funnel firmly clamped above the turntable. Gooden[11] described the following turntable technique for the automatic sampling of granulated pesticides. A sieve receiver pan is fixed on a turntable, and a metal cup is placed within the pan. A funnel is so adjusted above the pan that its outlet just clears the cup, and the funnel and cup are set off-center with respect to the pan. The turntable is rotated at 45 rpm, and the stock sample is passed through the funnel; the portion for testing falls within the cup, and the remainder falls in the pan. The position of the cup within the pan determines the ratio of the weight of test sample to stock sample.

It should be noted that a vast majority of organic mixtures submitted to analysis do not go through the statistical sampling process. The reader is referred to another monograph[12] for the general techniques to collect gases, liquids, or solids. For the purpose of quantitative analysis, special care should be taken so that the samples obtained have the composition that is representative of the whole. If multiplicate determinations give erratic answers or continuous changing of results in one direction, it is a clear indication that the samples analyzed are not representative.

Sometimes a large amount of the laboratory sample is required to provide sufficient quantity of the compound to be determined, and special techniques are devised to meet particular situations. For instance, for the colorimetric determination of reducing sugars in natural waters, Ivleva et al.[13] described a method that involves passing 300 ml of the test water through two chromatographic columns, discarding the first 100 ml, and evaporating the remaining 200 to 20 ml, adding 0.5 ml of a fine suspension of $CaCO_3$ and evaporating to dryness before dissolution in ethanol and color development with *p*-aminohippuric acid. Scholz[14] determined alkylphenols in drinking and surface waters by a combined evaporation–extraction procedure prior to paper and gas chromatography. Farrow et al.[15] recommended sublimation of pesticide residues from plant material at 2 μ pressure and 85°C before gas chromatography.

Collection of volatile materials for analysis may require special techniques. For instance, for the collection of ethanol, acetaldehyde, and ethyl acetate in apples in different stages of maturity, Paillard[16] described a method to convey the vapors from the apples by a current of air, adsorb them on active carbon, then desorb by heating the carbon *in vacuo*, and condense the volatile matter at −80°C. For the collection of diphenyl vapor from fruit cartons, Wells et al.[17] aspirated it from the carton, adsorbed it on activated alumina, and then eluted with ethanol. Morgan and Day[18] designed a column to trap flavor volatiles for gas-chromatographic analysis. Reymond et al.[19] collected highly volatile constituents of tea or coffee by suspending the material (10 g) in water (100 ml) at 80°C, passing a stream of nitrogen at 50 ml per min, and condensing the volatile matter in a trap at −80°C. For the collection of high-boiling flavor volatiles from aqueous ethanolic solutions, Shipton and Whitfield[19a] used a special distillation flask (100 ml to 10 liter) that is connected to a U-tube oil separator and two condensers and also, by way of a liquid-air cold trap, to the vacuum pump. Novak et al.[20] described a method of sampling by conventional syringes from moderately pressurized closed systems. Lawless et al.[20a] described a special sampling device through which a weighed amount of aerosol can be injected into 10 ml of acetone, the propellant being allowed to evaporate. Sampling and analysis of aerosols was discussed at a recent symposium.[21]

2.2 DISINTEGRATING AND MIXING

With the exception of pulverized materials such as some pharmaceutical preparations, solid samples usually go through the disintegrating process (known as "sample reduction") prior to chemical analysis. Very large pieces of the material may be broken up in a crusher, while moderate-sized pieces may be reduced in the ball mill. Most organic mixtures, however, can be ground to fine powder by means of the mortar and pestle. After it is ground for some time, the material is sifted through a sieve of suitable porosity, and the particles that do not pass the sieve are returned to the mortar for further grinding and are again sifted. It should be recognized that some organic compounds undergo chemical changes during the disintegrating step; if these occur, the analytical results will not represent the composition of the original mixture. Sublimation also may cause difficulty; thus Gore et al.[21a] reported that the content of salicylic acid is not a reliable basis for judging the degradation of aspirin in tablets because of the ready sublimation of salicylic acid.

Evidently, solid samples should be thoroughly mixed, unless the whole material (e.g., single tablet) is taken for analysis. Mixing of fluid materials is also recommended, since stratification may occur while the samples are standing in the containers. If a liquid sample contains suspended matter, the sample should be homogenized.

2.3 DISSOLUTION OF THE SAMPLE

In all methods of organic analysis based on functional group reactions or spectral characteristics, it is necessary to dissolve the compound in a suitable medium. In the case of an organic mixture containing several types of compounds, the chosen solvent may bring some components into solution while leaving other components undisturbed. This is the basis of the classification scheme of organic compounds by solubility proposed by Shriner et al.[22] as shown in Table 2.3. Water, diethyl ether, 5% $NaHCO_3$, 5% NaOH, 5% HCl, conc. H_2SO_4, and 85% H_3PO_4 are used as solvents. The types of compounds that may belong to the respective solubility groups are given in Table 2.4, compiled by Cheronis et al.[23,24] While such a scheme does not provide a

Table 2.3 Classification of Organic Compounds by Solubility[22]

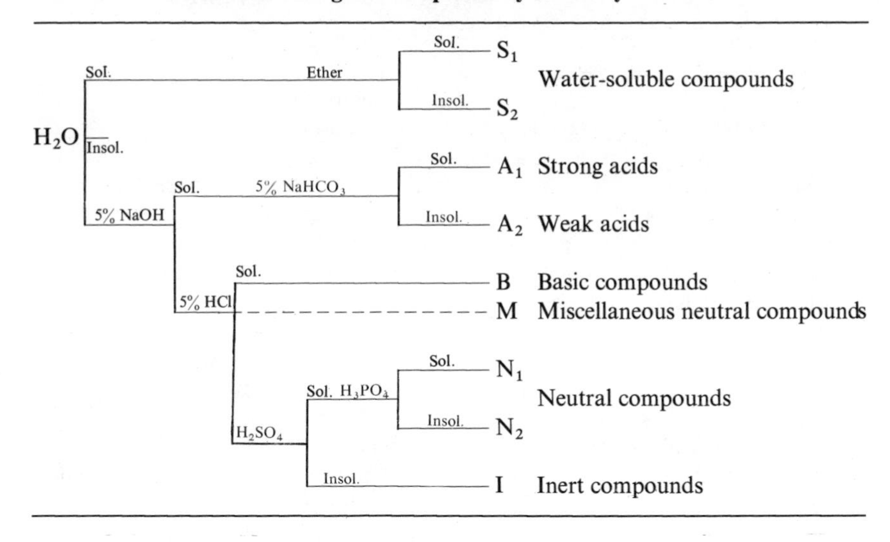

Table 2.4 Divisional Solubility Classifications[23a,b]

Division S_2[c]	*Division S_1*[d]
1. Only C, H, and O present:	1. Only C, H, and O present:
DIBASIC AND POLYBASIC ACIDS	ALCOHOLS
HYDROXY ACIDS	ALDEHYDES AND KETONES
POLYHYDROXY ALCOHOLS	CARBOXYLIC ACIDS
POLYHYDROXY PHENOLS	Acetals
Simple carbohydrates	Anhydrides
2. Metals present:	Esters
SALTS OF ACIDS AND PHENOLS	Ethers
Miscellaneous metallic compounds	Some glycols
3. Nitrogen present:	Lactones
AMINE SALTS OF ORGANIC ACIDS	Polyhydroxyphenols
AMINO ACIDS	2. Nitrogen present:
AMMONIUM SALTS	AMIDES
Amides	AMINES
Amines	Amino heterocyclics
Amino alcohols	Nitriles
Semicarbazides	Nitro paraffins
Semicarbazones	Oximes
Ureas	3. Halogen present:
4. Halogen present:	Halogen-substituted compounds of 1 above

Table 2.4 (*Contd.*)

Division S_2[c]

HALO ACIDS
Acyl halides (by hydrolysis)

Halo alcohols, aldehydes, and so on

5. Sulfur present:
   SULFONIC ACIDS
   Alkyl sulfuric acids

   Sulfinic acids
6. Nitrogen and halogen present:
   Amine salts of halogen acids
7. Nitrogen and sulfur present:
   AMINO DISULFINIC ACIDS
   BISULFATES OF WEAK BASES
   Cyano sulfonic acids
   Nitro sulfonic acids

Division S_1[d]

4. Sulfur present:
   Hydroxy heterocyclic sulfur compounds
   Mercapto acids
   Thio acids
5. Nitrogen and halogen present:
   Halogenated amines, amides, and nitriles
6. Nitrogen and sulfur present:
   Amino heterocyclic sulfur compounds

Division B

AMINES[e]
Amino acids
Amphoteric compounds (e.g., amino phenols, aminothiophenols, amino sulfonamides)
Aryl-substituted hydrazines
N-Dialkyl amides

Division A_1

1. Only C, H, and O present:
   ACIDS[f] AND ANHYDRIDES
2. Nitrogen present:
   AMINO ACIDS

   NITRO ACIDS
   Cyano acids
   Heterocyclic nitrogen carboxylic acids
   Polynitro phenols
3. Halogens present:
   HALO ACIDS
   Polyhalo phenols
4. Sulfur present:
   SULFONIC ACIDS
   Sulfinic acids
5. Nitrogen and sulfur present:
   Amino sulfonic acids
   Nitrothiophenols
   Sulfates of weak bases
6. Sulfur and halogens present:
   SULFONHALIDES

Division A_2

1. Only C, H, and O present:
   ACIDS[g]
   ANHYDRIDES
   PHENOLS, including esters of phenolic acids
   Enols
2. Nitrogen present:
   AMINO ACIDS
   NITRO PHENOLS
   Amides[h]
   Amino phenols
   Amphoteric compounds
   Cyano phenols
   Imides
   N-monoalkyl aromatic amines
   N-Substituted hydroxylamines
   Oximes
   pri- and *sec*-Nitro paraffins
   Trinitro aromatic hydrocarbons
   Ureides
3. Halogens present:
   HALO PHENOLS

Table 2.4 (*Contd.*)

Division M[i]	4. Sulfur present:
1. Nitrogen present:	Mercaptans (thiols)
ANILIDES AND TOLUIDIDES	Thiophenols
AMIDES	5. Nitrogen and halogen present:
NITRO ARYLAMINES	Polynitro halogenated aromatic hydrocarbons
NITRO HYDROCARBONS	Substituted phenols
Amino phenols	6. Nitrogen and sulfur present:
Azo, hydrazo, and azoxy compounds	Aminosulfonamides
Di- and triaryl amines	Amino sulfonic acids
Dinitrophenylhydrazines	Aminothiophenols
Nitrates	Sulfonamides
Nitriles	Thioamides
2. Sulfur present:	
Mercaptans	*Division N*
N-Dialkyl sulfonamides	ALCOHOLS
Sulfates; sulfonates	ALDEHYDES AND KETONES
Sulfides; disulfides	ESTERS
Sulfones	ETHERS
Thio esters	UNSATURATED HYDROCARBONS[j]
Thiourea derivatives	Acetals
3. Nitrogen and sulfur present:	Anhydrides
Sulfonamides	Lactones
4. Nitrogen and halogen present:	Polysaccharides[k]
Halogenated amines, amides, nitriles	
	Division I
	HYDROCARBONS[l]
	Halogen derivatives of hydrocarbons
	Diaryl ethers

[a] Nitrogen, halogens, and sulfur are absent unless specified.
[b] In this table, the more common classes are printed in small capital letters.
[c] Moderate-weight compounds with two or more polar groups, except for the sulfonic and sulfinic acids when only one polar group is necessary.
[d] Generally monofunctional compounds with five carbons or less.
[e] Amines with sufficiently strong negative substituents as well as diaryl and triaryl amines fall into Division M.
[f] Generally with 10 carbons or less; many form colloidal soap solutions.
[g] High-molecular-weight acids form colloidal soaps.
[h] Including *N*-monoalkyl amides.
[i] Only the most common classes are listed.
[j] Noncyclic unsaturated hydrocarbons and those unsaturated cyclics that are easily sulfonated, such as di- or polyalkyl-substituted benzenes.
[k] Char in the acid.
[l] Including most of the cyclic hydrocarbons and all of the saturated noncyclic hydrocarbons.

clear-cut separation of the different categories, it does indicate the presence or absence of certain compounds in the mixture, and also the solvents that may be suitable for the preparation of the solutions for quantitative analysis of the material.

Sometimes specific solvents are required for certain methods of chemical analysis. Thus, in the procedures for nonaqueous titration of acids and bases, the solvents are usually specified.[25] For the hydroxyl-group determination in high-molecular-weight alcohols and complex mixtures, Jordan[26] found that an alkane–ethyl acetate solvent can effect nearly instantaneous dissolution of the sample. For the determination of alcohols, phenols, and alkoxysilanes, Magnuson and Cerri[27] recommended 1,2-dichloroethane solvent, claiming it to be far superior to the conventional ethyl acetate. Sfiras and Demeilliers[28] reported that the use of dimethyl sulfoxide as solvent in the saponification of difficultly saponifiable esters greatly accelerates the reaction. In the nonaqueous cyanometric determination of substituted isoquinoline derivatives, Gimesi and Rady[29] described a method that involves dissolving the sample in acetone, precipitating the quaternary ammonium salt by adding methyl iodide, dissolving the salt in $CHCl_3-(CH_3)_2CHOH$, and titrating with 0.1 *N* KCN in 1 :9 $H_2O-(CH_3)_2CHOH$ mixed solvent.

Certain insoluble substances can be solubilized by forming molecular complexes with suitable reagents. For instance, Brock et al.[30] found that aqueous solutions of purines dissolve aromatic hydrocarbons to a high degree. Eisenbrand and Becker[31] observed the increase in water solubility of benzopyrene by the addition of caffeine, but this is limited by the 2% solubility of caffeine itself. However, by adding 17% of ascorbic acid, the solubility of caffeine is increased to 8%, and in this solution the solubility of benzopyrene is increased to 440 mg per liter. According to Bellen and Bellen,[32] sparingly soluble acids (e.g., toluic, anthranilic) can be brought into solution by treating them with hexamine and a little water to form a paste, which is then dissolved in water and titrated with an aqueous alkali solution.

Since finely divided material is more easy to dissolve than large particles, the analytical sample is usually pulverized prior to dissolution. It should be remembered, however, that the pressure being applied during grinding may cause changes in the organic compounds. Furthermore, if the substance is susceptible to oxidation, exposure of fine powder to air increases the danger. Ultrasonic agitation is a useful aid to dislodge organic compounds from a surface or biological cellular material so that they can pass into the solvent medium. For the dissolution of polymers, it is often advisable to treat the sample first with a small portion of the solvent to allow swelling before the addition of the rest of the solvent.

In the preparation of sample solutions for physical methods of determination (see Section 2.7), it is important that the organic compounds do not undergo decomposition or chemical change in the selected solvent system. Obviously solvents for spectral absorptiometry must be transparent in the regions of the spectrum to be measured. The useful regions of the solvents for infrared spectroscopy were reported by Pristera[33]; carbon disulfide, carbon tetrachloride, and tetrachloroethylene are the most transparent solvents, but they are not good solvents for a large number of organic compounds. Other solvents for the infrared range include chloroform, benzene, cyclohexane, 2-nitropropane, tetranitromethane, diethyl ether, methyl formate, acetone, and dimethyl formamide. The common solvents for the ultraviolet range are water, alcohols, ethers, chloroform, acetonitrile, and saturated hydrocarbons; the same solvents, as well as aromatic hydrocarbons and ketones, are used for the visible region. Kracmar and Kracmarova[34] have studied the effect of the solvent on the positions of the characteristic ultraviolet absorption bands of compounds containing the benzene chromophore.

2.4 SELECTION OF THE ANALYTICAL METHOD

After the sample is obtained, the analyst decides how to perform the analysis. It should be recognized that the selection of the analytical method depends not only on the material to be analyzed, but also on other factors, such as the following:

1. Is the analysis to be performed for one sample only, or occasionally, or frequently with the same kind of working material?
2. What equipment and reagents are readily available?
3. What is the time allowed for the analysis to be completed?
4. What are the training and skills of the staff?
5. Is there any restriction by government regulatory agencies or legal requirement concerning the method to be used?

2.4.1 Official and Standard Methods

When it is mandatory to carry out the determination of a certain component in the mixture by a specified procedure, the analytical method is restricted and must be followed meticulously. Every independent nation may have its own official methods of chemical analysis. For example, in the United States there are the *Official Methods of Analysis of the Association of Official Analytical Chemists* (AOAC),[35] and the *U.S. Pharmacopoeia* (USP)[36] (although U.S. Pharmacopoeia Convention, Inc. is not a governmental establishment). While the analytical results obtained by these methods can be submitted to the courts or arbitration officers in other countries, the findings may not be accepted as final. On the other hand, for legal evidence in the United States, a method given in *U.S. Pharmacopoeia* is preferable to one that is considered superior by the chemist who performs the analysis. For instance, Das Gupta[37] recently described a spectrophotometric method for determining piperazine citrate in a formulation that is simpler, more sensitive, and much more rapid than the USP XIX procedure, but in a litigation, the lawyer would very likely demand that the analysis of the sample be carried out by the gravimetric method specified in the *U.S. Pharmacopoeia*.

Standard methods are generally analytical procedures recommended by professional societies. Among the well-known standard methods are those published by the American Society for Testing and Materials (ASTM).[38] These methods have been verified by members of the society and shown to give reliable results with the test samples. However, it does not guarantee that the procedure described is suitable for all kinds of mixtures. Recommended methods for certain purposes are published by interested groups of workers from time to time. For example, recommended methods for the analysis of drying oils[39] was published by a committee of the International Union of Pure and Applied Chemistry (IUPAC). Lamprecht[40] reported on the analysis of paint solvents subject to legal requirements. It should be noted that some methods are known to be nonquantitative but are accepted as long as they give consistent results, for example, the tentative method for the determination of the solvent composition of electrodeposition paint.[41]

2.4.2 Choice of Unofficial Methods

As mentioned in the previous section, official methods are restrictive. Contrastingly, if the analysis may be performed by any suitable method, then it is up to the analyst to decide which method to be used. Some guidelines are given below.

1. If the method entails chemical change of the compound to be determined, the chemical reaction should be understood unequivocally. If the published method does not include a discussion of the chemistry involved, consult the comprehensive treatise on organic functional group analysis.[25] It should be noted that some analytical procedures have been proposed even though the chemical changes have not been fully verified. For instance, Egginton and Graham[42] studied the titrimetric method for determining phenylhydrazine with *N*-bromosuccinimide and casted doubt on the assumed course of the reaction. While such methods may be

practical for analyzing pure compounds, they should be avoided in the case of mixtures.

2. The chemical reaction should be selective or specific for the compound to be determined. It may be mentioned that the commonly recommended redox reagents (e.g., titanous chloride,[43] ceric sulfate,[44] potassium permanganate[45]) react with many types of compounds. In certain cases, however, it is possible to use these reagents to analyze mixtures. For example, Flaschka and Garrett[46] determined formic acid in the presence of acetic acid by means of permanganate oxidation and subsequent EDTA titration, because acetic acid is not oxidized under the experimental conditions.
3. It is advantageous to utilize chemical changes that yield more than one measurable product. For instance, carbamates[47] can be analyzed by determining the amounts of alcohol, amine, and/or carbon dioxide. Measurement of two or more products is a convenient way to cross-check the analytical results. Another example is the determination of 1,2-diols by modified Zeisel reactions; thus Anderson and Zaidi[48] employed vapor-phase infrared spectroscopy to measure both the alkyl iodide and ethylene or propene produced by the diols.
4. The analytical results must be reproducible and accurate. For instance, in the determination of compounds containing propoxyl and butoxyl groups by measuring the alkyl halides, Anderson et al.[49,50] reported that some samples required HBr while other samples required HI for the cleavage reaction to produce quantitative results. On the other hand, it is not advisable to use the cleavage reaction to differentiate esters and acetals from ethers.[51]
5. The experimental operations should be simple and easy to perform. Furthermore, if the analytical procedure is to be carried out frequently, the cost of the reagents and maintenance of the equipment should be considered.
6. In some methods successive chemical changes are prescribed. Understandably, the method that requires fewer steps is preferred. It may not be practicable, however, to use the one-step analytical method. For instance, when a peroxide is mixed with compounds that react readily with iodine, direct iodimetry cannot be used. Therefore Dulog and Burg[52] proposed a method in which the peroxide is reacted with trialkyl- or triphenylphosphine to form substituted phosphine oxide; the resulting phosphorus-containing compound is then determined gravimetrically, colorimetrically, or by indirect cerimatric titration. For the determination of carboxylic amides, Siggia and Stahl[53] reduced the amides to the amines with lithium aluminum hydride, separated the resulting amides by steam distillation, and titrated them with standardized acid.

7. A method that consists of measuring a certain physical property of the compound without alteration has the advantage that the sample can be recovered readily. These physical methods (see Section 2.7), however, generally require more expensive apparatus than do the chemical methods. Some physical methods are also subject to more interferences. For example, Rashkes[54] determined the acetates of natural compounds (steroids, alkaloids) by infrared spectrophotometry using the areas of the C═O and C—O—C bonds; any extraneous substance containing these linkages interferes. In contrast, chemical methods involving hydrolysis and measurement of the acetic acid formed[55] do not suffer from this drawback.

2.5 SEQUENCE OF OPERATION IN ANALYZING MIXTURES

Except for certain cases (see Chapter 3), the first step in the analysis of an organic mixture is the separation process. After separation, the portion of the sample containing the compound to be determined may go through an intermediate treatment, such as:

1. Oxidation.
2. Reduction.
3. Degradation (e.g., hydrolysis, decarboxylation, dehydration, cleavage with HI).
4. Combination (e.g., derivatization, condensation).
5. Other.

Subsequently, either the end product of the chemical reaction or the original compound is determined by measurement by an appropriate technique. A wide range of methods for analytical measurement is currently available, as can be seen from the following list:

A. Chemical methods of finish
   1. Titrimetric methods
      a. Acidimetry
      b. Alkalimetry
      c. Iodometry
      d. Argentometry
      e. Oxidimetry
      f. Chelometry
      g. Aquametry

- h. Potentiometry
- i. Coulometry
- j. Enthalpimetry
2. Gravimetric methods
3. Gasometric methods
4. Other

B. Physical and instrumental methods of finish
1. Colorimetry
2. Absorption spectrophotometry
   - a. Ultraviolet
   - b. Visible
   - c. Infrared
3. Fluorometry
4. Nuclear magnetic resonance spectrometry
5. Mass spectrometry
6. Polarography
7. Polarimetry
8. Gas chromatographic response
9. Other

2.6 CHEMICAL METHODS FOR ANALYZING COMPOUNDS IN MIXTURES

Until very recently, most of the analytical procedures used in the determination of organic compounds involved chemical transformation of the compounds. These procedures are known as "wet methods," because the sample is usually dissolved in a solvent and the reagent that reacts with the compound is generally prepared in the form of a solution. Solid reagents, however, also can be used. Thus Flaschka and Weiss[56] discussed titrimetry using solid titrants and consider it preferable when the titrant is unstable in solution, or is required in very small amounts, or when titration without dilution is desirable. Dukhota and Fedoseev[57] performed titration by adding reagent paper containing a known amount of reagent per unit area.

All chemical methods are based on the reaction of certain functional groups present in the compound to be determined. To familiarize the reader with the myriad variations of analytical procedures, we present several tables giving the classification of these methods and examples of the compounds reported to be amenable to the respective procedures. Tables 2.5 and 2.6 present a number of titrimetric methods but do not include direct acid–base titration of the original compound. This subject is discussed in Chapter 3. Examples of titrimetric procedures with visual end-point indication are

Table 2.5 Examples of Titrimetric Methods Using Visual End-Point Detection

Type of compound	Chemical reaction involved	Reference
Oxygen functions	Formaldehyde, formic acid, acetic acid: cerate–chromate oxidation	Sharma[58]
	Acetone–isobutyl methyl ketone: phase titration with H_2O–acetone	Dunnery[59]
	Esters: saponification in dimethyl sulfoxide	Vinson[60]
	Ethanol in esters, aldehydes: phase titration with H_2O	Rogers[61]
	Anhydrides of *N*-carboxy-α-amino acids: titration of CO_2 produced by polymerization induced by sodium methoxide	Ballard[62]
	Reducing sugars in citric, tartaric acids: Fehling's solution and iodometric titration	Maslowska[63]
	Fats: saponification, then complexation of the fatty acids	Garcia-Villanora[64]
	Barium salts of nonvolatile acids: titration with H_2SO_4	Dernovskaya[65]
	Tocopherol, roboflavine: complexation with Pb(II), Zn picrates	Gajewska[66]
Nitrogen functions	Aliphatic primary amines: react with pentane-2,4-dione	Critchfield[67]
	Aliphatic primary and secondary amines: react with CS_2	Critchfield[68]
	Aliphatic and aromatic amines: react with tetraphenylboron	Ievinsh[69]
	Tertiary amines, quaternary ammonium compounds: react with Bi-EDTA	Budesinsky[70]
	Heterocyclics: two-phase titration with sodium lauryl sulfate	Pellerin[71]
	Heterocyclics: react with aluminum chloro-*iso*-propylate–HCl complex	Tokar[72]
	Hydrazine derivatives: react with Hg(II)–EDTA	Budesinsky[73]
	Nitro, nitroso, azo compounds: titration with $CrSO_4$	Tandon[74]
	Aliphatic nitro compounds: react with KOH	Khailov[75]
Sulfur functions	Butanethiols: titration with 4-diethylamino-phenylmercury acetate	Busev[76]

Table 2.5 (*Contd.*)

Type of compound	Chemical reaction involved	Reference
	Thiols: react with *o*-hydroxymercuribenzoid acid	Wronski[77]
	Chlorobenzenesulfonic acids: conversion to N-methylanilinesulfonic acids, bromination, and titration of the H_2SO_4 liberated	Spryskov[78]
	Sulfanilic acid: titration with NH_4NaHPO_4	Saxena[79]
Unsaturated functions	C=C compounds: react with mercuric acetate; then complexometric titration of Hg(II)	Budesinsky[80]
Miscellaneous functions	Hydrolyzable H in dimethoxyborane: react with I_2 and titration of excess with $Na_2S_2O_3$	Krol[81]
	Phenols and anilines: bromination with oxazones as indicators	Ruzicka[82]
	Chloropropionic acids: conversion into pyruvic acid, which reacts with $Hg(NO_3)_2$	Marquardt[83]

Table 2.6 Examples of Titrimetric Methods Using Instrumental End-Point Detection

Instrumental technique	Chemical reaction of the method	Reference
Potentiometric	Ascorbic acid, glutathione: iodate, *N*-bromosuccinimide, or dichlorophenolindophenol oxidation	Huber[84]
	Hydrazine derivatives: KIO_4 oxidation	Berka[85]
	Nitro compounds: $FeSO_4$ reduction	Rybacek[86]
	Nitro compounds: $CrCl_2$ reduction	Bottei[87]
	Azo and hydrazo compounds: $TiCl_3$ reduction	Nebbia[88]
	Disulfides: $NaBH_4$ reduction, titration of thiols	Stahl[89]
	C=C compounds: lead tetraacetate titration in the presence of excess Br^-	Sigler[90]
	Polynuclear hydrocarbons: form trinitrofluorenone complexes, titration with $(C_4H_9)_4NOH$	Cundiff[91]
	Organoaluminum compounds: titration with pyridine	Uhniat[92]
Amperometric	Ascorbic acid, tocopherol, and so on: titration with metal salts	Konopik[93]

Table 2.6 (*Contd.*)

Instrumental technique	Chemical reaction of the method	Reference
	Aromatic amines: titration with diazonium salt	Matrka[94]
	Heterocyclic *N*-compounds: titration with nitranilic acid	Kracmar[95]
	Thiols: titration with $AgNO_3$	Malik[96]
	SO_3H group in dyes: precipitation with benzidine C=C compounds: iodometric titration	Matrka[97] Jedlinski[98]
	Alkylaluminum compounds: titration with ethers	Mardykin[99]
Coulometric	Benzophenone, azobenzene, nitro compounds: titration with diphenyl radical anions	Maricle[100]
	N-substituted phenothiazines: Ce(IV) titration	Patriarche[101]
	Phenols, sulfonamides: titration with bromine	Patriarche[102]
	Phenols: Ce(IV) titration	Takahashi[103]
High-frequency	Carboxylates: titration with $HClO_4$	Ishidata[104]
	Salts of organic bases: titration with NaOH or $AgNO_3$	Bionda[105]
	Substituted anilines: titration with $HClO_4$	Lippincott[106]
	Xylenols: titration with HCl	Ershov[107]
	Cyclopentadienylsodium: titration with HCl	Shtifman[108]
Conductometric	Alkylsilyl phosphates and sulfates: titration with $LiOCH_3$ or $(C_2H_5)_4NOH$	Drozdov[109]
	Methylchlorosilanes: conversion to alkylthiocyanato derivatives, titration with amidopyrine	Kreshkov[110]
Spectrophotometric	Primary aliphatic amines: titration with 2-ethylhexanal	Liu[111]
	Primary aliphatic amides: titration with $Ca(OCl)_2$	Post[112]
	C=C compounds: titration with bromine	Miller[113]
	Phenobarbitone, phenytoin: titration with $(C_4H_9)(C_2H_5)_3NOH$	Agarwal[114]
Dielectrometric	Amines: titration with picric, *p*-toluenesulfonic, or trichloroacetic acid	Ishidate[115]
Pressuremetric	Hydrazine derivatives: titration with iodate, detection of nitrogen pressure	Curran[116]

given in Table 2.5; The table shows that many compounds can be determined in mixtures using simply the burette and human eye. Table 2.6 gives examples of instrumental methods for the detection of end points in titrimetry.

Some examples of gravimetry are presented in Table 2.7. Although gravimetric methods are much less frequently used today as compared with a couple of decades ago, a number of gravimetric procedures are still retained as standard methods in the latest (1975) editions of the *U.S. Pharmacopoeia*[36]

Table 2.7 Examples of Gravimetric Methods

Type of compound	Method	Reference
Oxygen functions	Vanillin: precipitation as 2,4-dinitrophenylhydrazone	Meluzin[117]
	Cyclic aldehydes: as Schiff bases with 4-aminophenazone	Manns[118]
	Musk ketone: condensation with anisaldehyde	Bogdanov[119]
	Ketosteroids: as 2,4-dinotrophenylhydrazones	Mazzella[120]
	Carboxylic acids: adducts with urea	Mankovskaya[121]
	Carboxyl groups in filter paper: react with $Pb(NO_3)_2$	Ultee[122]
	Epoxides: conversion into aldehydes then to 2,4-dinitrophenylhydrazones	Durbetaki[123]
	Sterols: as digitonides	Sperry[124]
	Naphthol isomers: as addition compounds with CrO_3	Suzuki[125]
	Aromatic hydroxy compounds: react with halo-2,4-dinitrobenzenes	Poethke[126]
Nitrogen functions	Methylamine: precipitation of nickel dithiocarbamate	Nebbia[127]
	Aliphatic amines: as tetraphenylborates	Yanson[128]
	Amino groups in resins: aminolysis, conversion into disubstituted ureas	Kappelmeier[129]
	Piperazine: as picrate	Zawadzki[130]
	Caffeine	Marine-Fort[131]
	Melamine: as picrate	Engelbrecht[132]
	Alkaloids: react with flavianic acid	Wachsmuth[133]
Sulfur functions	Saccharin: conversion to $BaSO_4$	Markus[134]
	Toluenesulfonic acid: a mine salt	Spryskov[135]

and *Official Methods of Analysis of the Association of Official Analytical Chemists*,[35] respectively. With the general availability of single-pan balances, the tedium of weighing has been greatly reduced. Digital readout and automatic recording of weights also speed up the experimental operation considerably.

Chemical methods that involve reactions forming gaseous products are particularly suited to the determination of compounds in mixtures. The evolution of a gas is a convenient means to discriminate and separate the particular compound from the other constituents in the mixture. When the quantity of the gas obtained is 0.1 millimole or larger, it can be collected in a gasometer and accurately measured. On the other hand, determination of the gas formed at the micromole level is preferably carried out in the gas chromatograph. If the gasometer and gas chromatograph are not available, the gas can be absorbed in a suitable reagent solution and determined by titrimetry or spectrophotometry. The reader is referred to reference 47, which gives the numerous functional group reactions that produce gases. Some examples are given in Table 2.8 to illustrate the variety of gases formed.

Table 2.8 Examples of Chemical Methods Involving Gas Formation

Type of compound	Reaction	Gas formed	Reference
Oxygen functions	Alkoxyl compounds: heating with HI	Alkyl iodide	Schachter,[136] Araki[137]
	Carboxylic acids: decarboxylation	CO_2	Ma[138]
	Uronic acids: heating in HCl	CO_2	Barker[139]
	Substituted aromatic acids: heating with H_3PO_4, HCl, or $ZnCl_2$	CO_2	Jaunzems[140]
	Formates: oxidation with $K_3Fe(CN)_6$	CO_2	Desmukh[141]
	Ethyl cyanoacetate derivatives: heating with H_2SO_4	CO_2	Maros[142]
	Benzyloxycarbonyl compounds: heating with H_2SO_4	CO_2	Medzihradsky[143]
	Aliphatic and aromatic acids: pyrolysis of tetramethylammonium salts	Methyl ester	Bailey[144]
	Steroids: reaction with H_2SO_4 and acetic anhydride, liberation of 3 moles of SO_2 per mole of steroid	SO_2	Brieskorn[145]
Nitrogen functions	Primary amino group in peptides: reaction with $NaNO_2$ and acetic acid	N_2	Belenkii[146]

Table 2.8 (*Contd.*)

Type of compound	Reaction	Gas formed	Reference
	Aromatic amines: reaction with KNO_2 in HCl	N_2	Yamagishi[147]
	Amino acids: reaction with ninhydrin	CO_2	Maros[148]
	Diazo compounds: heating with chromic acid	N_2	Kozak[149]
	Methylhydrazine: reaction with NaOCl	CH_4	Neumann[150]
	Barbituric acids: treatment with diazobenzenesulfonic acid; excess of latter determined with NaN_3	N_2	Yamagishi[151]
	Carbamates: heating with KOH	Alcohol	Ladas[152]
Sulfur functions	Thionyl compounds: conversion into K_2SO_3 which liberates SO_2 with H_3PO_4	SO_2	Hennart[153]
	Sulfonamides: conversion into diazonium compound, then decomposition with NaN_3	N_2	Yamagishi[154]
	Dithiocarbamates: heating with H_2SO_4	CS_2	Bighi[155]
Miscellaneous functions	Active hydrogen: reaction with diborane	H_2	Martin[156]
	Borane compounds: reaction with $LiAlH_4$	H_2	Lysyj[157]
	Alkylaluminum compounds: treatment with azomethine reagent	Alkane	Hagen[158]

Since spectrophotometers are presently common equipment in chemical laboratories, analytical methods based on absorption spectroscopy have gained much popularity. Although quantitative absorptiometry entails the preparation of the calibration graph and hence requires more preliminary operations than titrimetry, it is a big time-saver for multiple determinations. Furthermore, spectrophotometry is readily adaptable to automation. Table 2.9 gives some examples of absorptiometric methods in which chemical reactions are carried out to produce the chromophoric species to be measured (for spectrophotometric methods that do not involve chemical change see Section 2.7). It should be mentioned that absorptiometric methods also have been employed for the indirect determination of organic compounds in various ways. For instance, Lee et al.[201] described a spectrophotometric

Table 2.9 Examples of Spectrophotometric Methods Involving Chemical Reactions

Type of compound	Chemical reaction	Reference
Oxygen functions	Carbonyl compounds: conversion into 2,4-dinitrophenylhydrazones	Muck[159]
	Aldehydes: conversion into acetals	Crowell[160]
	Acetone: conversion into 2,4-dinitrophenylhydrazone	Ahmadzadeh[161]
	Formaldehyde: reaction with 7-amino-4-hydroxynaphthalene-2-sulfonic acid, and so on	Sawicki[162]
	β-Ionone: with HCl and acetic acid	Pesez[163]
	Methylpentoses: with thioglycolic acid	Gibbons[164]
	Formic acid: reduction to formaldehyde	Fabre[165]
	Aromatic fatty acids: reduction to alcohols	Eckert[166]
	Ethers of benzhydrol: conversion into chloranil	Wilczynska[167]
	Ethylene oxide: conversion into formaldehyde	Critchfield[168]
	Hydroxyl compounds: with vanadium oxinate	Amos[169]
	Alcohols: with 3,5-dinitrobenzoyl chloride	Obtemperanskaya[170]
	Ethanol: conversion into acetaldehyde	Pfeil[171]
	Aliphatic nitroalcohols: reaction with alkali to produce formaldehyde	Jones[172]
	Aliphatic alcohols: conversion into nitrites	Shchukarev,[173] Shmulyakovskii[174]
	Polyhydric compounds: $NaIO_4$ oxidation to produce formaldehyde	Okuhara[175]
	Steroids: side chain oxidized to formaldehyde	Demey-Ponsart,[176] Edwards[177]
	Hydroxymethyl compounds: periodate oxidation to produce formaldehyde	Speck[178]
	Methoxyl compounds: hydrolysis, then oxidation to produce formaldehyde	Langejan[179]
	Methylenedioxy compounds: hydrolysis to form formaldehyde	Langejan[180]
Nitrogen functions	Primary amines: conversion into *N*-substituted pyrroles, which react with 4-dimethylaminobenzaldehyde	Sawicki[181]

Table 2.9 (*Contd.*)

Type of compound	Chemical reaction	Reference
	Secondary aliphatic amines: conversion into nitrosoamines	Morgan[182]
	Secondary and tertiary arylamines: mercuration, then reaction with HNO_2	Korenman[183]
	Thebaine: conversion into phenolic substances, then reaction with HNO_2	Clair[184]
	Diaminouracil: reaction with ninhydrin	Degtyarev[185]
	Ephedrine: complex with bromothymol blue	Das Gupta[186]
	Strychnine: reduction with Zn–Hg, then treatment with $NaNO_3$	Miller[187]
Sulfur functions	Thiols: conversion into pyridinethiones	Grassetti[188]
	Tertiary thiols: conversion into thionitrites	Ashworth[189]
	Thiophen: complex with alloxan	Giovannini[190]
Unsaturated functions	Olefins: conversion into iodine complexes	Long[191]
	Acetylenes: hydration to carbonyls, then conversion into 2,4-dinitrophenylhydrazones	Scoggins[192]
	Unsaturated carbonyl compounds: reaction with sulfohydrazinoazobenzene	Hünig[193]
	Terminal methylene compounds: oxidation to give formaldehyde	Lemeiux[194]
	iso-Propylidene compounds: oxidation to give acetone, then treatment with salicylaldehyde	Von Rudloff[195]
Miscellaneous functions	Benzene and derivatives: conversion into nitro and then amino compounds, which are diazotized and coupled with sulfanilic acid	Ciuhandu[196]
	Phenolic compounds: reaction with Millon's reagent	Kartashevskii[197]
	Aminophenols: reaction with $Cu(NO_3)_2$	Rzesutko[198]
	Halogenated aliphatic compounds: conversion into substituted dianilides of glutaconaldehyde	Belyakov[199]
	Bromoacetic acid: conversion into glycolic acid	Mergenthaler[200]

method to determine water-soluble carboxylic acids that is based on the decrease in absorption of ferric 5-nitrosalicylate upon addition of the acids. Guyet-Hermann et al.[202] determined ethanol in pharmaceutical preparations by measuring its effect on decreasing the color intensity of $CoCl_2$ in acetone. Tomlinson and Sebba[203] determined surfactants by adding the sample to a solution of crystal violet; the complexes formed floated to the surface and the reduction in the color of the underlying liquid was measured. Horak and Zyka[204] used Tl_2SO_4 to precipitate alkaloids; the Tl content of the precipitate was then determined photometrically after the Tl(II) was oxidized to Tl(III). Babko et al.[205] determined cysteine by its catalytic effect on the reduction of Ag(I) by Fe(II). Almassy and Dezso[206] determined oxalate ions by means of the reaction between Cr(VI) and Mn(II), which continues as long as oxalate ions are present; the residual Cr(VI) is then measured. Kertes[207] determined organic bases by a method depending on their ability to abstract protons from hexanitrodiphenylamine, thus generating the red *aci*-form to be measured.

Fluorometric methods[208] are similar to spectrophotometric methods but require a different instrument. Methods involving direct excitation of the organic compound to be determined are discussed in Section 2.7. If the original sample cannot be so analyzed, the compound can be converted through a chemical reaction into its fluorescent derivative. Thus the dansyl-(5-dimethylamino-1-naphthalenesulfonyl) derivatives of amino and hydroxy compounds are frequently used. Pesez and Bartos[208a] have given a number of examples of chemical conversions based on the functional groups present in the compounds. Some nonfluorescent compounds can be made to fluoresce by permanganate oxidation.[208j]

Thermometric methods[259] for the determination of organic compounds have been proposed recently. These methods are dependent on the precise measurement of the heat absorbed or liberated in a chemical reaction in which the sample participates. Some examples are given in Table 2.10.

Kinetic methods involve measurement of the rate of reaction between the compound to be determined and a chosen reagent. Some examples are given in Table 2.11. These methods are very useful for the simultaneous determination of two compounds that contain the same functional group and therefore can react with the given reagent (see Chapter 3, Section 3.3).

Besides the above various categories of chemical methods, there are still other analytical methods for the determination of organic compounds that involve certain chemical transformations. For instance, the enzymatic methods[248] employ specific enzymes to react with the compounds of biological origin; many examples can be found in Chapters 5 and 6. Bergmeyer and Hogen[249] have used enzymes supported on an insoluble carrier for rapid continuous analysis; a circulating system is described for the determination of

Table 2.10 Examples of Thermometric Methods

Type of compound	Chemical reaction involved	Reference
Oxygen functions	Acid anhydride: reaction with water	Greathouse[210]
	Aldehydes: reaction with *p*-tolyhydrazine	Reynolds[211]
	Carbohydrates: reaction with $NaIO_4$ in H_2SO_4	Bark[212]
	Carboxylic acids: reaction with NaOH, $NaOCH_3$, or $(C_4H_9)_4NOH$	Harries[213]
Nitrogen functions	Amines: reaction with sodium tetraphenylborate	Bark[214]
	Amines and *N*-heterocyclic compounds: reaction with HCl in isopropyl alcohol	Vaughan[215]
	Amines, pyridine, nicotine: reaction with $NaNO_2$	Daftary[216]
	Aminophylline: reaction with $AgNO_3$	De Leo[217]
	Nicotinamide: reaction with HCl	De Leo[218]
Miscellaneous functions	Grignard reagents: reaction with isopropyl alcohol	Parker[219]
	Butyl lithium: reaction with butanol	Everson[220]

glucose in less than 1 min. Sawicki and Pfaff[250] reported on the specific reagents that quench the phosphorescence of various aromatic compounds. Indirect atomic absorption spectroscopy has been proposed to determine organic compounds by using reagents that contain heavy metals, for example, aldehydes with silver,[251] surfactants with molybdenum,[252] and phthalic acid with copper.[253] Even quantitative elemental analysis can be used to determine the composition of organic mixtures; thus Reich et al.[254] used C, H, N, Br, and S results to determine a mixture of pyridine compounds, and de Reeder[255] used C, H, N, and S values to determine sulfonamide mixtures.

2.7 PHYSICAL METHODS FOR DETERMINING ORGANIC COMPOUNDS

For convenience we group together in this section analytical methods that do not depend on the chemical reaction between the organic compound and a specific reagent (see Section 2.6). Instead, these methods are based on the spectral, molecular (e.g., polarimetry), or electrochemical (e.g., polarography) characteristics of the compound to be determined. Most of these methods are "nondestructive," meaning that the original compound can be

Table 2.11 Examples of Kinetic Methods

Type of compound	Reagent used	Reference
Oxygen functions	Carbonyl compounds: $H_2NOH \cdot HCl$	Greinke[221]
	Aldehydes: $NaBH_4$	Jensen[222]
	Ketohexoses: cysteine + H_2SO_4	Bissett[223]
	Acetylacetone: $H_2NOH \cdot HCl$	Blaedel[224]
	Ketones: $H_2NOH \cdot HCl$	Toren[225]
	Glycolic acids: 2,7-dihydroxynaphthalene	Garmon[226]
	Uronic acids: H_2SO_4	Meller[227]
	Hydroxy compounds: 3,5-dinitrobenzoyl chloride	Berezin[228]
	iso-Propyl alcohol: NaOBr	Grover[229]
	Primary hydroxy groups in polyglycols: phenyl isocyanate	Willeboordse[230]
	Cholesterol: H_2SO_4	Hewitt[231]
	Esters: KOH	Munnelly[232]
	Malonic esters: KOH	Bellen[233]
	Phenylmalonate: NaOH	Jancik[234]
	Peroxy acids: sulfonated diphenylamine	Shapilov[235]
Nitrogen functions	Amines: phenyl isothiocyanate	Hanna[236]
	Amino acids: trinitrobenzenesulfonic acid	Schwerdtfeger[237]
	Amides: KOH	Siggia[238]
	Alkanolamides: KOH	Ranny[239]
	Biuret: NaOCl + phenol	Karayannis[240]
Unsaturated functions	C=C compounds: hydrogen	Flaschka[241]
Miscellaneous functions	Cresols: diazotized sulfanilic acid	Legradi[242]
	Naphthols: diazotized 2-naphthylamine-5,7-disulfonic acid	Parsons[243]
	Phenols: $KBrO_3$–KBr	Rodziewicz[244]
	Phenolic compounds: formaldehyde	Krahl[245]
	Chloronitrobenzenes: $H_2NOH \cdot HCl$	Legradi[245]
	Organophosphorus compounds: KOH	Carter[247]

recovered unchanged after the analysis has been performed. Some physical methods (e.g., mass spectrometry), however, involve chemical decomposition of the compound.

2.7.1 Spectroscopic Methods

Colored organic compounds can be determined by means of the colorimeter or by using the spectrophotometer of the visual region. A vast number of colorimetric methods, however, utilize chemical reactions (see Table 2.9). Absorption spectrophotometry in the ultraviolet region is limited to compounds that possess highly conjugated structures. In contrast, infrared spectrometry is applicable to practically all types of compounds. While the measurement is commonly made by transmission, also used is the attenuated total reflectance technique in which the infrared beam is reflected from the sample into the slit of the spectrophotometer. In certain cases, such as in the quantitative evaluation of paint vehicles, McGowan[256] reported that the reflectance technique was superior. Some examples of infrared methods are presented in Table 2.12.

Table 2.13 gives some examples of analytical methods using fluorescence or phosphorescence. It should be recognized that these methods are extremely sensitive and therefore are rarely employed for the determination of the major component in a mixture. As for colorimetry, there are more methods that involve chemical reactions to produce the fluorescent species than those that depend on the fluorescence of the original compound to be determined. The theory and applications of diffuse reflectance spectroscopy has been reviewed by Frei.[292] Raman scattering spectroscopy has been used by some workers to determine acids and anhydrides (Mironov[293]), alkylbenzenes (Nicholson[294]), and chlorosilanes (Efremova[295]), although its quantitative aspects still await further developments, as pointed out by Ewing.[246]

Analytical methods based on atomic absorption, emission, or fluorescence can be applied directly to organometallic compounds. Using similar principles, Belcher et al.[297] recently developed a technique, known as molecular emission cavity analysis (MECA), for determining nonmetals. The sample is placed in a cavity formed at the end of a metal rod; the cavity is introduced into a hydrogen-based flame, and characteristic molecular emissions are generated inside the cavity. This technique has been used to determine nanogram amounts of compounds containing nitrogen[298] or sulfur.[299]

X-ray diffraction has been utilized for quantitative analysis of organic compounds. For example, Papariello et al.[300] described the determination of intact glutethimide tablets. Kuroda et al.[301] determined 1-(2-hydroxyethyl)-2-methyl-5-nitroimidazole. Mamedaliev et al.[302] measured the proportion of

Table 2.12 Examples of Infrared Methods

Functional group	Compounds	Reference
Oxygen functions	Carbonyl compounds	Popova[257]
	Ketones	Katon[258]
	Ketosteroids	Dvoryantseva[259]
	Essential oils	Carroll[260]
	Methyl esters of pectin	Filippov[261]
	Carboxylic acids	Nicholson[262]
	Aliphatic alcohols	Bek[263]
Nitrogen functions	Primary fatty amides	Link[264]
	Anilides and substituted anilides	Jart[265]
	Azides	Lieber[266]
	Hydrazides	Jart[267]
	Isocyanates	Zharkov[268]
	Caffeine, phenacetin	Oi[269]
	Pyrrole-, indole-, carbazole-type compounds	Pozefsky[270]
Sulfur functions	Thioamides, sulfonamides	Jart[267]
	Xanthates	Pearson[271]
Unsaturated functions	Azomethines	Suydan[272]
	Myrcene	Malone[273]
Miscellaneous functions	Hydrocarbons	Kunz[274]
	Substituted alkylbenzenes	Wexler[275]
	Diphenyls, fluorenes, indanes	Hume[276]
	Naphthenes	Luther[277]
	Terpenes	Mitzner[278]
	Phenolic compounds	Bard[279]
	Polychloroprene isomers	Ferguson[280]
	Chloroacetic acids	Ito[281]

isomeric phthalic acids formed by the oxidation of dialkylbenzenes by the intensity of the x-ray diffraction rings given by the powder.

2.7.2 Nuclear Magnetic Resonance and Mass Spectrometry

Previously nuclear magnetic resonance and mass spectrometry were used exclusively by research organic chemists to elucidate molecular structures.

Table 2.13 Examples of Fluorometric and Phosphorimetric Methods

Functional group	Compounds	Reference
Oxygen functions	Aromatic aldehydes	Crowell[282]
Nitrogen functions	Amino- and nitrobenzenes	Ponomarenko[283]
	Hydrazones	Brandt[284]
	Alkaloids	Hollifield[285]
	Catecholamines	Schwedt[286]
	Indoles	Balemans[287]
	Barbituric acids, cocaine, morphine	Harbaugh[288]
Miscellaneous functions	Aromatic hydrocarbons	Fleet[289]
	Naphthalene, phenanthrene	McGlynn[290]
	Polycyclic hydrocarbons	Lavalette[291]

Thanks to the intensive efforts of the instrument manufacturers, these two techniques became available to the analytical laboratories after the costs of the machines were considerably reduced and the operational procedures were simplified. Consequently, attempts have been made to employ the commercial instruments for quantitative analysis of a wide range of compounds. However the initial investment and maintenance expenses of these machines are still very costly in comparison with the other analytical methods. Therefore, when the nuclear magnetic resonance or mass spectrometric method is chosen to determine a particular compound in the sample, the great expenditure involved must be justified. Table 2.14 gives some examples of the methods based on nuclear magnetic resonance. Table 2.15 shows some methods that utilize mass spectrometry. Brodskii et al.[343] described a procedure for quantitative mass-spectrometric analysis of complex mixtures of hydrocarbons by direct introduction of the sample into the ion source.

2.7.3 Polarographic Methods

During the first two decades after Heyrovsky discovered polarography, this technique was employed solely in inorganic analysis. Then Zuman[344] investigated its applications to the determination of organic compounds. Theoretically, any compound that contains an electrically reducible functional group can be submitted to polarographic analysis. The actual performance, however, may become complicated. A common difficulty is the insolubility of the organic compound in the liquid medium suitable for polarography. This problem is even more troublesome in the case of mix-

Table 2.14 Examples of Methods Using Nuclear Magnetic Resonance Spectrometry

Functional group	Compounds	Reference
Oxygen functions	Ketosteroids through the methoxy-imino derivatives	Fales[303]
	Hydroxy compounds through acetyl derivatives	Mathias[304]
	Alkylphenols through acetyl derivatives	Lindeman[305]
	Ethylene oxide adducts of alcohols	Ludwig[306]
	Fumarate, maleate, acrylates	Dietrich[307]
	Cyclandalate isomers	Vlies[308]
	Oil of seeds	Conway[309]
	Surfactants	Flanagan[210]
	Hydroperoxides	Ueda,[311] Ward,[312] Ikeda[313]
Nitrogen functions	Amine–amine salt ratio	Koch[314]
	Betaine, choline	Chastellain[315]
	Phenylglycine derivatives	Warren[316]
	Dinitrotoluene isomers	Mathias[317]
	Quinidine, hydroquinidine	Tho[318]
	Caffeine, phenacetin	Hollis[319]
	Carbamates	Slomp[320]
Unsaturated functions	Olefins	Robinson[321]
	Methylene, methylidyne end groups	Page[322]
	Polyisoprenes	Chen[323]
Miscellaneous functions	Active hydrogen compounds	Kawazoe[324]
	Acid, base after titrimetric neutralization	Degani[325]
	Bromophenylacetonitriles	Daugherty[326]
	Perfluorovinyl derivatives	Jolley[327]

tures, for it requires that all components be soluble in the medium containing the supporting electrolyte.

Polarographic methods have the advantage of being relatively inexpensive. The apparatus can be assembled from components commonly available in the laboratory; the complete instrument also can be acquired from the supply house. Polarography is a sensitive analytical technique. When the differ-

Table 2.15 Examples of Methods Utilizing Mass Spectrometry

Functional group	Compounds	Reference
Oxygen functions	Primary alcohols	Popova[328]
	Acetylsalicylic acid, phenylsalicylate	Tatematsu[329]
	Fatty acids through their methyl esters	Hallgren[330]
	Triacontanoic acid	Koller[331]
	Homovanillic acid	Frei[332]
Nitrogen functions	Acetanilide, phenacetin	Baty[333]
	Amino acids	Bertilsson[334]
	Codeine, papaverine	Tatematsu[335]
	Barbituric acids	Costopan-agiotis[336]
Sulfur functions	Thiols	Knof[337]
Unsaturated functions	Olefinic naphthas	Frisque[338]
	Alkenyl phenols	Occolowitz[339]
Miscellaneous functions	Hydrocarbons	Svob[340]
	Isoparaffins	Ferguson[341]
	Hydroaromatics	Shultz[342]

ential pulse mode is employed, microgram amounts of the organic compound can be determined.

Table 2.16 gives some examples of polarographic determinations, including a few indirect methods. For example, primary alcohols are oxidized to the aldehydes, which are then determined polarographically.

2.7.4 Isotopic-Dilution and Radiochemical Methods

Generally speaking, methods that are based on isotopic dilution and/or measurement of radioactivity are designed for the analysis of research samples. Because these techniques entail more elaborate equipment and usually require a longer time to complete the determination than other available methods, they are seldom selected for routine analysis of organic mixtures. Nevertheless, there are occasions when these techniques can be employed to advantage. Some examples are discussed below.

Eilhauer et al.[370] determined aliphatic polyhydric alcohols by using the partially deuterated compounds; the atomic percentage of ^{2}H was obtained by measuring the refractive index of the water produced after combustion. Griffin[371] determined 4,4′-isopropylidenediphenol in commercial bisphenol

Table 2.16 Examples of Polarographic Methods

Functional group	Compounds	Reference
Oxygen functions	Aldehydes	Tung[345]
	Carbonyl compounds	Prevost[346]
	Aromatic ketones	Hsieh[347]
	Primary alcohols, via $PbCrO_4$ oxidation	Kubis[348]
	Alkyl hydroperoxides	Skoog[349]
	Chloro-1,4-benzophenone and related compounds	Mizunoya[350]
Nitrogen functions	Amino acids, via methyleneimines	Zhantalai[351]
	Pyridine, via suppression of Ni(II) wave	Mark[352]
	Nicotinamides	Kemula[353]
	Nicotinic acid	Cernatescu[354]
	Dinitrobenzoic esters, dinitrophenyl ethers	Furst[355]
	3-Nitro-4-hydroxybenzoic esters	Tammilehto[356]
	Nitroimidazoles	Craine[357]
	Nitrated heterocyclics	Vignoli[358]
	Semicarbazones of aliphatic aldehydes	Coulson[359]
	Semicarbazones of biacetyl, mesityloxide	Sucha[360]
	Semicarbazones of substituted benzaldehydes	Fleet[360a]
	N-Oxides	Ma[361]
Sulfur functions	Cysteine–cystine	Kalousek[362]
	Dithiocarbamic acids	Zahradnik[363]
	Sulfanilamide derivatives	Okazaki[364]
Unsaturated functions	Unsaturated diketones	Ryvolova[365]
	Unsaturated fatty acids	Maruta[366]
	Unsaturated compounds, via bromination	Ryabov[367]
	Styrene, via pseudonitrosite	Sedivec[368]
Miscellaneous functions	Benzene homologues via nitration	Sedivec[369]

by using the ^{14}C-labeled compound; the method is applicable over a wide range of concentrations with an accuracy of about 0.5%. Similarly, Burtle et al.[372] used ^{14}C-labeled β-chloropropionic acid to determine its content in mixtures with propionic and α-chloropropionic acids. Sorensen[373] analyzed chlorophenoxyacetic acids in mixtures using radioactive ^{36}Cl. Abdel-Wahab and El-Kinawi[374] determined edible oils using ^{131}I. Nagase and Baba[375] determined caffeine in the presence of theobromine and phenacetin by using (1-^{14}C)caffeine. Ma and Yang[376] analyzed tannins in plant material by precipitation with uranyl nitrate and measurement of the β-activity.

Margosis et al.[377] determined halogen-containing pharmaceuticals by means of neutron activation; the γ-radiation emitted by ^{38}Cl, ^{82}Br, and ^{128}I was measured. Schlesinger et al.[378] utilized neutron-activation analysis to determine the source of manufacture of meprobamate and diamorphine. Tuckerman et al.[379] determined trace elements, including Cl, Br, As and many metals, in antibiotics, salicylates, steroids, and vitamins by neutron activation followed by γ-ray spectrometry. Another application of the neutron beam is the determination of moisture contents of organic materials. For instance, Wang[380] described a method to determine coffee-bean moisture content by neutron scattering. Peck et al.[381] determined the moisture content of lactose by neutron thermalization, which is based on the moderation of fast neutrons by the hydrogen atoms of water to produce slow or thermal neutrons; the neutron count is a rectilinear function of the moisture content of the sample.

2.7.5 Miscellaneous Physical Methods

In this section we present briefly some other physical methods. Compared with the methods discussed in the previous sections, the following are of limited application. Each method, however, may serve a certain specific purpose.

Adey[382] investigated the *cryoscopic method* for assaying pyridine and reported that the depressions of the freezing point of pyridine caused by additions of 2- and 3-picolines, 2,6-lutidine, and benzene were linear, and that the cryoscopic constants were identical. Cryoscopy has been utilized to determine *p*-xylene (Poppwicz[383]), cineole (Nevskaya[384]), and arenes (Tilicheev[385]). The content of phthalimide admixed with phthalic anhydride (Hrivnak[386]) and the principal constituents of technical phenol (Rappen[387]) can be determined by the melting-point method.

Henein and Csonka-Horvai[388] employed *differential thermal analysis* to determine the proportion of camphoramic acid isomers to within 5%. Flora and Almasy[389] studied the thermal analysis of mixtures of maleic and tetrahydrophthalic anhydrides. Ozawa[390] discussed the scope and limitations of quantitative differential thermal analysis.

The *ebullioscopic method* can be used to determine the ethanol or 2-propanol content in liquids, for which a special apparatus has been described (Arzneibuchkommission[391]). Heitler [392] determined the number of ester groups per mole of a polymeric ester by performing ethanolysis in an ebulliometer and observing the change in the boiling point of the ethanol solvent. Martin et al.[393] determined the benzene content (1 to 10%) in toluene by distillation at a specific rate. The British Standards Institution[394] has published methods for the distillation of petroleum products.

Measurements of additive physical properties (e.g., densities) can be used to analyze mixtures (Janik[395]). For instance, Bittenbender[396] described a method for the determination of fat content that is based on *hydrometric measurement* of the relative density of a solution of the fat extracted from the sample with a given volume of heptane. Agarwal and Mene[397] determined carbon tetrachloride–ethyl acetate–ethanol tertiary mixtures by constructing the density–viscosity composition diagram. Angelidis[398] worked out a linear equation for calculating the percentage of water in aqueous ethanol or the amount of water required to obtain a specific mixture. Densitometry was employed by Skelding and Ashbolt[399] to determine methanol in formalin, and by Pohle and Tierney[400] to determine trimethylene glycol in glycerol. Wisniewski et al.[401] used the butyrometer to determine essential oils in ethanolic galenicals; the apparatus is filled with saturated NaCl solution, rotated at 1000 rpm for 30 min, immersed in hot water, and then cooled so the volume of the essential oil can be read. Krutzch[402] determined the ethanol content of wines, tinctures, and so forth by measuring the difference between the initial and final temperatures for a specified expansion of the liquid.

Solubility is the basis of extraction methods for separating organic mixtures (see Chapter 4). If the extract contains only one component, its content may be simply obtained by *weighing* the residue after evaporation of the solvent, such as in the determination of lipids (Hirtz et al.[403]) and triglycerides (Hallaway and Sandberg[404]). If the solvent extracts more than one component from the mixture, it is possible to utilize different solubilities for the determination of certain combinations, for example, benzoic acid and terephthalic acids (Parausanu et al.[405]), diphenyl in polyphenyls (Silverman and Shideler[406]), naphthalene–disulfonic acids (Czerwinski[407]), Ardil and wool (Glynn[408]). Fischer and Auer[409] determined carvone and cinnamaldehyde in volatile oils by means of the critical solution temperature. Comer and Howell[410] used phase-solubility analysis to study pharmaceutical stability. Suri[411] described phase titration of ternary mixtures. Morton and Tinley[412] proposed a distribution method to analyze binary mixtures, based on the fact that, when a mixture of two substances is dissolved in a suitable solvent and shaken with an immiscible solvent, the weight of mixed solutes present in each phase at equilibrium is a function of the composition of the original mixture.

Refractive index has been used to determine components in mixtures in several ways. Shpitalnyi and Volf[413] determined caprolactam and *N*-methylcaprolactam using a calibration graph. Hill[414] determined chloromethylphenoxyacetic acids by a differential refractometric method. Wood et al.[415] described a similar method to determine saturated impurities in aromatics. Ioffe et al.[416] performed refractometric analysis of three-component solutions by the incomplete-extraction technique, which involves measurement of the refractive index of the liquid before and after the partial extraction of one component with a fixed ratio of sample to extractant. Schlosser et al.[416a] recently proposed a refractometric method to determine ethanol in pharmaceutical mixtures.

Application of *polarimetry* to the analysis of mixtures usually involves the change of the optical activity of one component by suitable means. For instance, Kirsten et al.[417] determined malic acid, tartaric acid, and so forth in the presence of other optically active compounds by measuring the difference in the optical rotation of their solution with and without the addition of MoO_4^{2-}. The molybdate complexes were also used to determine mannitol and sorbitol (Hamon et al.[418]), aldonates (Takiura and Yamamoto[419]), and polyhydric alcohols (Kaigorodova and Klabunovskii[420]). Terentev et al.[420] determined a mixture of sucrose and glucose by measuring the optical rotation before and after treatment with $NaBH_4$ at pH 1 to reduce the glucose to the optically inactive sorbitol; the acidic medium prevents the formation of an optically active complex of H_3BO_3 with sorbitol.

Quantitative analysis by *manual separation* is used in special cases. Thus the British Standards Institution[421] has specified this method to all types of fibers that do not form an intimate mixture; it is more accurate than the previously recommended chemical method.

2.8 COMPARISON OF ANALYTICAL METHODS

In view of the great variety of analytical methods presently available, it is generally true that there are at least two different experimental procedures that can be used to determine an organic compound in a given sample. It is also understandable that, when a new analytical instrument or technique has been developed, the workers like to exploit it to the full extent and to show its application to a wide range of compounds in certain fields. Thus, for the purpose of pharmaceutical analysis, Warren et al.[422] studied the application of infrared and ultraviolet spectrometry; fluorescence was studied by

Köchel,[423] nuclear magnetic resonance spectrometry was studied by Alexander and Koch,[424] electron spin resonance, optical rotatory dispersion, and circular dichroism were studied by Branch,[425] and mass spectrometry was studied by Deavin.[426] Keith and Alford[427] reviewed the application of nuclear magnetic resonance spectroscopy in pesticide analysis. Eisenbrand[428] surveyed the use of fluorometry in the analysis of foodstuffs. The application of physical methods to the analysis of mineral oils was reported by Schultze and Kirchoff,[429] their use in the analysis of edible oils and fats was reported by Wolff,[430] and their application to the analysis of explosives was reported by Sinha and Rao.[431]

When a compound can be determined by more than one method, it is of interest to compare the methods. Many studies of this kind have been under-

Table 2.17 Comparison of Methods, Some examples

Type of compound	Methods compared	Reference
Oxygen functions	Malonaldehyde: 11 spectrophotometric and fluorometric methods	Sawicki[432]
	Carbohydrate components of glycoproteins: hydrolysis and direct colorimetry	Montreuil[433]
	Salicylic acid: two colorimetric methods	Thielemann[434]
	Uronic acids: three colorimetric methods	Federico[435]
	Ascrobic acid: titrimetric, colorimetric, electrometric, biological methods	Hajratwala[436]
	Hydroxy compounds: acylation catalysts	Kingston[437]
	Methanol: sources and ages of reagent for colorimetry	Maurice[438]
	Ethanol in pharmaceutical preparations, cosmetics, beverages: distillation, gas chromatography, thin-layer chromatography	Karawya[439]
	Cholesterol: kinetic and end-point techniques for spectrophotometry	Manasterski[440]
	Sennosides: D.A.B. VII and Eur. P. I methods	Jech[441]
	Prednisolone in tablets: ethanol and chloroform extractions	Wagner[442]
	Lactone groups in cellulose: colorimetric methods	Slavik[443]
	Essential oils in drugs: Swiss P. IV and Eur. P. I methods	Kämpf[444]
Nitrogen functions	Aliphatic amines: colorimetric methods	Zhivopistsev[445]

Table 2.17 (*Contd.*)

Type of compound	Methods compared	Reference
	Aromatic amines: titrimetric methods	Matrka[446]
	Primary aromatic amines: potentiometric and visual titrimetry	Trokowicz[447]
	Quaternary ammonium compounds: argentometry iodometry, alkalimetry	Thoma[448]
	Quinidine: two fluorometric methods	Hartel[449]
	Barbiturates: titrimetric, gravimetric, spectrophotometric methods	Rotondaro[450]
	Phenacetin, caffeine, isopropylphenazone: four methods	Haefelfinger[451]
	Metribuzin: three techniques	Betker[452]
Sulfur functions	Thiobarbituric acids: titrimetric and gravimetric methods	Wojahn[453]
	Sulfanilamide: colorimetric methods	MacDonald[454]
	Sulfoxides: titrimetric and infrared methods	Stefanac[455]
Unsaturated functions	Olefinic unsaturation: addition and substitution methods	Murthy[456]
	Unsaturated fatty acids in edible oils: thiocyanogen, lead salt, and gas chromatographic methods	La Croix[457]
Miscellaneous functions	Acidimetry for barbiturates: aqueous and nonaqueous methods	Ryan[458]
	Hydrocarbons, nonolefinic naphthas: gas-chromatographic and distillation methods	Cirillo[459]
	Ethyl groups containing ^{1}H and ^{2}H: nuclear magnetic resonance and mass spectrometry	Saur[460]
	Benzopyrene in plant tissue: thin-layer and gas chromatography	Siegfried[461]
	Phenols: coupling and acylation methods	Goupil[462]
	Phenolic compounds: colorimetric methods	Bernstein[463]
	Chloroacetic acids: colorimetric, polarographic, and Raman spectroscopic methods	Barcelo[464]
	Chlorofluorobenzenes: infrared and nuclear magnetic resonance spectrometry	Molin[465]
	Chlorocresols and chlorotolyloxyacetic acids: ultraviolet and infrared spectrometry	Sjöberg[466]

taken. Some examples are given in Table 2.17. These comparisons are helpful to workers who are called upon to perform analysis on the same compound or similar compounds. It should be recognized, however, that different methods may be based on different principles and also may measure different properties of the compound in question. For example, Rowson[467] reported on the chemical and chromatographic methods for determining glycosides in digitalis leaves and pointed out the poor correlation between the glycoside content and the biological activity as measured by the *British Pharmacopoeia* method. A review[468] has been published on the factors that affect the biological activity of drugs in pharmaceutical formulations; in many instances, the methods used to assay the drug do not measure the biological availability.

Before an official or standard method (see Section 2.4.1) is established, collaborative studies are usually made so that a specified experimental procedure is tested in different laboratories. In the United States most of the collaborative studies on methods for the determination of organic compounds are supervised by the Association of Official Analytical Chemists. The results of such collaborative studies are open to the public and are published from time to time. For instance, in 1976, these reports included determination of chlorinated pesticides in foods (Finsterwalder[469]), endosulfan, tetrasul, and tetradifon in fruits (Mitchell[470]), metoxuron and its formulations (Sauer[471]), and chlorotoluron and its formulations (Heizler et al.[472]).

It should be noted that official methods are not without faults. For example, Wendt[473] reported that collaborative determinations of alcohol, resins, and vanillin made on genuine and adulterated vanilla extracts show that the AOAC methods of analysis may fail to detect adulteration. On the other hand, de la Court et al.[474] presented the results of a collaborative study of 10 laboratories of the gas-chromatographic analysis of the fatty acid composition of three paint samples and concluded that differences among laboratories were real and not fortuitous. Therefore, after the appropriate method has been selected to perform the analysis of an organic mixture, it is still very important to interpret the experimental data judiciously.

REFERENCES

1. C. A. Bicking, in *Kirk-Othmer: Encyclopedia of Chemical Technology*, Revised ed., Vol. 17, Wiley, New York, 1968, p. 744.
2. Committee E.15, *Tentative Recommended Practice for Sampling Industrial Chemicals, Designation 300-66T*, American Society for Testing and Materials, Philadelphia, 1966.
3. Committee E.15, *Sampling of Solids*, American Society for Testing Materials, Philadelphia, 1967.

4. *Recommended Practice for Choice of a Sample Size to Estimate the Average Quality of a Lot or Process, Designation E122-58*, American Society for Testing and Materials, Philadelphia, 1958.
5. *Sampling and Testing Detergents*, B.S. 3762, British Standards Institution, London, 1964.
6. K. Rothwell and C. A. Grant, *Standard Methods for the Analysis of Tobacco Smoke*, Tobacco Research Council, London, 1974.
7. Collaborative Pesticides Analytical Committee, F.A.O., *Plant Prat. Bull.*, **11**, 36 (1963).
8. Y. I. Stakheev and Y. N. Kuznetsov, *Zavod. Lab.*, **36**, 1 (1970).
9. A. A. Benedetti-Pichler, in W. G. Berl (Ed.), *Physical Methods in Chemical Analysis*, Vol. 3, Academic, New York, 1960, p. 184; A. A. Benedetti-Pichler, *Essentials of Quantitative Analysis*, Ronald, New York, 1956, p. 309.
10. E. O. Rowland, *Mineral. Mag.*, **33**, 524 (1963).
11. E. L. Gooden, *J. Agric. Food Chem.*, **10**, 397 (1962).
12. T. S. Ma and V. Horak, *Microscale Manipulations in Chemistry*, Wiley, New York, 1976, Chap. 7.
13. I. N. Ivleva, A. D. Semenov, and U. G. Datsko, *Gidrokhim. Mater.*, **38**, 144 (1964).
14. L. Scholz, *Z. Anal. Chem.*, **202**, 425 (1964).
15. R. P. Farrow, E. R. Elkins, Jr., and L. M. Beacham, III, *J. Assoc. Off. Agric. Chem.*, **48**, 738 (1965).
16. N. Paillard, *Fruits*, **20**, 189 (1965).
17. A. W. Wells, S. M. Norman, and E. P. Atrops, *J. Gas Chromatogr.* **1**, 19 (1963).
18. M. E. Morgan and E. A. Day, *J. Dairy Sci.*, **48**, 1382 (1965).
19. D. Reymond, F. Mueggler-Chavan, R. Viani, L. Vuata, and R. H. Egli, *J. Gas Chromatogr.*, **4**, 28 (1966).
19a. J. Shipton and F. B. Whitfield, *Chem. Ind. (Lond.)*, **1966**, 2124.
20. J. Novak, J. Gelbicova-Ruzickova, and S. Wicar, *J. Chromatogr.*, **60**, 127 (1971).
20a. G. B. Lawless, J. J. Sciarra, and A. J. Monte-Bovi, *J. Pharm. Sci.*, **54**, 273 (1965).
21. B. Y. H. Liu (Ed.), *Fine Particles: Aerosol Generation, Measurement, Sampling and Analysis*, Academic, New York, 1976.
21a. A. Y. Gore, K. B. Naik, D. O. Kildsig, G. E. Peck, V. F. Smolen, and G. S. Banker, *J. Pharm. Sci.*, **57**, 1850 (1968).
22. R. L. Shriner, R. C. Fuson, and D. Y. Curtin, *The Systematic Identification of Organic Compounds*, 5th ed., Wiley, New York, 1964.
23. N. D. Cheronis and T. S. Ma, in L. Meites (Ed.), *Handbook of Analytical Chemistry*, McGraw-Hill, New York, 1963, Section 2, p. 31.
24. N. D. Cheronis, J. B. Entrikin, and E. M. Hodnett, *Semimicro Qualitative Organic Analysis*, 3rd ed., Wiley, New York, 1965, p. 324.
25. N. D. Cheronis and T. S. Ma, *Organic Functional Group Analysis*, Wiley, New York, 1964. pp. 399, 427.
26. D. E. Jordan, *J. Am. Oil Chem. Soc.*, **41**, 500 (1964).

27. J. A. Magnuson and R. J. Cerri, *Anal. Chem.*, **38**, 1088 (1966).
28. J. Sfiras and A. Demeilliers, *Reh., Engl. Ed.*, **1966** (15), 83.
29. O. Gimesi and G. Rady, *Acta Chim. Hung.*, **55**, 25 (1968).
30. N. Brock, H. Druckery, and H. Hamperl, *Arch. Exp. Pathol. Pharmakol.*, **189**, 709 (1938).
31. J. Eisenbrand and G. Becker, *Dtsch. Lebensm Rundsch.*, **63**, 342 (1967).
32. N. Bellen and Z. Bellen, *Chem. Anal. (Warsaw)*., **9**, 617 (1964).
33. F. Pristera, *Appl. Spectrosc.*, **6**, 29 (1952).
34. J. Kracmar and J. Kracmarova, *Cesk. Farm.*, **15**, 16 (1966).
35. W. Horwitz (Ed.), *Official Methods of Analysis of the Association of Official Analytical Chemists*, 12th ed., Association of Official Analytical Chemists, Washington, 1975.
36. *U.S. Pharmacopoeia* Vol. XIX, U.S. Pharmacopoeial Convention, Inc., Rockville, 1975.
37. V. D. Gupta, *Am. J. Hosp. Pharm.*, **33**, 283 (1976).
38. *ASTM Standards*, American Society for Testing and Materials, Philadelphia.
39. H. W. Talen, *Pure Appl. Chem.*, **10**, 190 (1965).
40. W. Lamprecht, *Farbe Lack*, **69**, 672 (1963).
41. C. J. Nunn, *Paint Manuf.*, **38**, 28 (1968).
42. E. V. Egginton and M. J. Graham, *Analyst*, **89**, 226 (1964).
43. T. S. Ma and J. V. Earley, *Mikrochim. Acta*, **1959**, 129; **1960**, 314; **1965**, 170.
44. T. S. Ma and W. L. Nazimowitz, *Mikrochim. Acta*, **1968**, 405.
45. A. Berka and J. Konopasek, *Mikrochim. Acta*, **1968**, 405.
46. H. Flaschka and J. Garrett, *Chemist-Analyst*, **52**, 101 (1963).
47. T. S. Ma and A. S. Ladas, *Organic Functional Group Analysis by Gas Chromatography*, Academic, London, 1976, p. 88.
48. D. M. W. Anderson and S. S. H. Zaidi, *Talanta*, **10**, 691 (1963).
49. D. M. W. Anderson, J. L. Duncan, M. A. Herbich, and S. S. H. Zaidi, *Analyst*, **88**, 353 (1963).
50. D. M. W. Anderson and S. S. H. Zaidi, *Talanta*, **10**, 1235 (1963).
51. D. M. W. Anderson and J. L. Duncan, *Talanta*, **9**, 661 (1962).
52. L. Dulog and K. H. Burg, *Z. Anal. Chem.*, **203**, 184 (1964).
53. S. Siggia and C. R. Stahl, *Anal. Chem.*, **21**, 550 (1955).
54. Y. V. Rashkes, *Zh. Anal. Khim.*, **20**, 238 (1965).
55. N. D. Cheronis and T. S. Ma, *Organic Functional Group Analysis*, Wiley, New York, 1964, pp. 399, 427.
56. H. Flaschka and R. Weiss, *Mikrochim. Acta*, **1968**, 243.
57. V. A. Dukhota and P. N. Fedoseev, *Izv. Vyssh. Uchebn. Zaved. Khim.*, **10**, 141 (1967).
58. N. N. Sharma, *Z. Anal. Chem.*, **162**, 321 (1958).

59. D. A. Dunnery and G. R. Atwood, *Talanta*, **15**, 855 (1968).
60. J. A. Vinson, J. S. Fritz, and C. A. Kingsbury, *Talanta*, **13**, 1673 (1966).
61. D. W. Rogers, D. Lillian, and I. D. Chawla, *Mikrochim. Acta*, **1968**, 722.
62. D. G. H. Ballard, C. H. Bamford, and F. J. Weymouth, *Analyst*, **81**, 305 (1956).
63. J. Maslowska and K. Cedzynska, *Chem. Anal. (Warsaw)*, **19**, 693 (1974).
64. R. Garcia-Villanova and M. C. Lopez-Martinez, *Ars. Pharm.*, **11**, 321 (1971).
65. G. L. Dernovskaya-Zelentsova, *Tr. Vses. Nauchn. Issled. Inst. Khlebopek Prom-st.* **1955** (6), 188.
66. M. Gajewska and Z. Szrajber, *Chem. Anal. (Warsaw)*, **20**, 99 (1975).
67. F. E. Critchfield and J. B. Johnson, *Anal. Chem.*, **29**, 1174 (1957).
68. F. E. Critchfield and J. B. Johnson, *Anal. Chem.*, **28**, 430 (1956).
69. A. F. Ievinsh and E. Y. Gudrinietse, *Zh. Anal. Khim.*, **11**, 735 (1956).
70. B. Budesinsky and J. Korbl, *Chem. Listy*, **52**, 1513 (1958).
71. F. Pellerin, J. A. Gautier, and D. Demay, *Ann. Pharm. Fr.*, **20**, 97 (1962).
72. G. Tokar and I. Simonyi, *Mag. Kem. Foly.*, **64**, 94, 151 (1958).
73. B. Budesinsky, *Collect. Czech. Chem. Commun.*, **26**, 781 (1961).
74. J. P. Tandon, *Z. Anal. Chem.*, **167**, 184 (1959).
75. V. S. Khailov, B. B. Brandt, and G. N. Shcherbova, *Khim. Nauka Prom.*, **2**, 806 (1957).
76. A. I. Busev, N. V. Shelemina, L. I. Teternikov, and T. A. Danilova, *Anal. Lett.*, **1**, 763 (1968).
77. M. Wronski, *Chem. Anal.* (*Warsaw*), **6**, 869 (1961).
78. A. A. Spryskov and O. I. Kachurin, *Izv. Vyssh. Uchebn. Zaved. Khim.*, **1958** (1), 97.
79. A. K. Saxena, *Microchem. J.*, **14**, 430 (1969).
80. B. Budesinsky, *Chem. Listy*, **51**, 259 (1957).
81. A. J. Krol, L. B. Eddy, D. R. Mackey, and A. E. Weber, U.S. Atomic Energy Comm., Report CCC-1024-TR-239, 1957.
82. E. Ruzicka, *Collect. Czech. Chem. Commun.*, **29**, 2244 (1964).
83. R. P. Marquardt and E. N. Luce, *Anal. Chem.*, **31**, 418 (1959).
84. C. O. Huber and H. E. Stapelfeldt, *Anal. Chem.*, **36**, 315 (1964).
85. A. Berka and J. Zyka, *Chem. Listy*, **50**, 314 (1956).
86. J. Rybacek, J. Dolezal, and J. Zyka, *Cesk. Farm.*, **14**, 59 (1965).
87. R. S. Bottei and N. H. Furman, *Anal. Chem.*, **27**, 1182 (1955).
88. L. Nebbia and V. Bellotti, *Chim. Ind.* (*Milan*), **48**, 366 (1966).
89. C. R. Stahl and S. Siggia, *Anal. Chem.*, **29**, 154 (1957).
90. K. Sigler, A. Berka, and J. Zyka, *Microchem. J.*, **11**, 398 (1966).
91. R. H. Cundiff and P. C. Markunas, *Anal. Chem.*, **35**, 1323 (1963).
92. M. Uhniat and T. Zawada, *Chem. Anal. (Warsaw)*, **9**, 701 (1964).
93. N. Konopik, *Ost. Chem. Ztg.*, **55**, 127 (1954).

94. M. Matrka, Z. Sagner, and F. Vondrak, *Chem. Prum.*, **14**, 198 (1964).
95. J. Kracmar and J. Zyka, *Cesk. Farm.*, **7**, 246 (1958).
96. K. L. Malik, *Chem. Ind. (Lond.)*, **1965**, 724.
97. M. Matrka and Z. Sagner, *Chem. Prum.*, **10**, 638 (1960).
98. Z. Jedlinski and J. Paprotny, *Chem. Anal. (Warsaw)*, **8**, 765 (1963).
99. V. P. Mardykin, E. I. Kvasyuk, and P. N. Gaponik, *Zh. Prikl. Khim. Leningr.*, **42**, 947 (1969).
100. D. L. Maricle, *Anal. Chem.*, **35**, 683 (1963).
101. G. J. Patriarche and J. J. Lingane, *Anal. Chim. Acta*, **49**, 25 (1970).
102. G. Patriarche, Thesis; Univ. Libre de Bruxelles, 1963.
103. T. Takahashi and H. Sakurai, *J. Chem. Soc. Jap., Ind. Chem, Sect.*, **63**, 605 (1960).
104. M. Ishidate and M. Masui, *Pharm. Bull., (Tokyo)*, **2**, 50 (1954).
105. G. Bionda, E. Bruno, and A. Bellomo, *Farmaco (Pavia)., Ed. Prat.*, **18**, 530 (1963).
106. W. T. Lippincott and A. Timnick, *Anal. Chem.*, **28**, 1690 (1956).
107. B. R. Ershov, V. L. Pokrovskaya, and S. P. Dvuglov, *Plast. Massy*, **1961** (10), 58.
108. L. M. Shtifman, V. V. Lastovich, and L. G. Kuryakova, *Zavd. Lab.*, **29**, 546 (1963).
109. V. A. Drozdov, R. R. Tarasyants, E. G. Vlasova, and Z. A. Kubyak, *Izv. Vyssh. Uchebn. Zaved. Khim.*, **6**, 960 (1963).
110. A. P. Kreshkov and V. A. Drozdov, *Dokl. Akad. Nauk SSSR*, **131**, 1345 (1960).
111. Y. L. Liu and C. A. Reynolds, *Anal. Chem.*, **34**, 542 (1962).
112. W. R. Post and C. A. Reynolds, *Anal. Chem.*, **36**, 781 (1964).
113. J. W. Miller and D. D. DeFord, *Anal. Chem.*, **29**, 475 (1957).
114. S. P. Agarwal and M. I. Blake, *Anal. Chem.*, **41**, 1104 (1969).
115. M. Ishidate, H. Nishizawa, H. Sano, and I. Horikoshi, *J. Pharm. Soc. Jap.*, **81**, 1303 (1961).
116. D. J. Curran and J. L. Driscoll, *Anal. Chem.*, **38**, 1746 (1966).
117. J. Meluzin, *Chem. Zvesti*, **8**, 22 (1954).
118. O. Manns and S. Pfeifer, *Mikrochim. Acta*, **1958**, 630.
119. K. A. Bogdanov, *Moslob, Zhir. Prom.*, **1958** (8), 18.
120. P. P. Mazzella and T. S. Ma, *Mikrochim. Acta*, **1968**, 253.
121. N. K. Mankovskaya and Z. I. Germanskaya, *Moslob. Zhir. Prom.*, **1962** (3) 29.
122. A. J. Ultee, Jr., and J. Hartel, *Anal. Chem.*, **27**, 557 (1955).
123. A. J. Durbetaki, *Anal. Chem.*, **29**, 1666 (1957).
124. W. M. Sperry, *J. Lipid Res.*, **4**, 221 (1963).
125. Y. Suzuki, M. Sugahara, and I. Anazawa, *Jap. Analyst*, **23**, 1137 (1974).
126. W. Poethke and H. Koehne, *Pharm. Zentralhalle*, **107**, 729 (1968).
127. L. Nebbia and F. Guerrieri, *Chem. Ind. (Lond.)*, **37** (3), 198 (1955).
128. E. Yanson, A. Ievinsh, and E. Gudrinietse, *Uch. Zap. Latv. Univ.*, **14**; 9 (1957).
129. C. P. A. Kappelmeier, *Fette Seifen*, **57**, 229 (1955).

130. R. Zawadzki, *Farm. Pol.*, **32**, 377 (1976).
131. A. Marine-Font and E. Sancho-Riera, *Circ. Farm.*, **31**, 277 (1973).
132. R. M. Engelbrecht, H. E. Moseley, W. P. Donahoo, and W. R. Rolingson, *Anal. Chem.*, **29**, 579 (1957).
133. H. Wachsmuth, *J. Pharm. Belg.*, **8**, 283 (1953).
134. J. R. Markus, *J. Assoc. Off. Anal. Chem.*, **56**, 162 (1973).
135. A. A. Spryskov and T. I. Potanova, *Izv. Vyssh. Uchebn. Zaved. Khim.*, **2**, 41 (1959).
136. M. M. Schachter and T. S. Ma, *Mikrochim. Acta*, **1966**, 55.
137. S. Araki, S. Susuki, and M. Kitano, *Jap. Analyst*, **18**, 608 (1969).
138. T. S. Ma, C. T. Shang, and E. Manche, *Mikrochim. Acta*, **1964**, 571.
139. S. A. Barker, A. B. Foster, I. R. Siddiqui, and M. Stacey, *Talanta*, **1**, 216 (1958).
140. V. Jaunzems, V. N. Sergeeva, and L. N. Mosheiko, *Zh. Anal. Khim.*, **22**, 1257 (1967).
141. G. S. Deshmukh and A. L. J. Rao, *Z. Anal. Chem.*, **194**, 110 (1963).
142. L. Maros, I. Perl, M. Vajda, and E. Schulek, *Anal. Chim. Acta*, **28**, 179 (1963).
143. H. Medzihradsky-Schweiger, *Acta Chim. Acad. Sci. Hung.*, **37**, 239 (1963).
144. J. J. Bailey, *Anal. Chem.*, **39**, 1485 (1967).
145. C. H. Brieskorn and H. Hofmann, *Arch. Pharm., Berl.*, **297**, 577 (1964).
146. B. G. Belenkii and V. A. Orestova, *Izv. Akad. Nauk SSSR, Ser. Khim.*, **1964**, 182.
147. M. Yamagishi and M. Yokoo, *J. Pharm. Soc. Jap.*, **74**, 1231 (1954).
148. L. Maros, I. Perl, M. Vajda, and E. Schulek, *Magy. Kem. Foly.*, **69**, 123 (1963).
149. P. Kozak, I. Slamova, and M. Jurecek, *Mikrochim. Acta*, **1966**, 1024.
150. E. W. Neumann and H. G. Nadeau, *Anal. Chem.*, **36**, 640 (1964).
151. M. Yamagishi, M. Yokoo, and S. Inoue, *J. Pharm. Soc. Jap.* **77**, 1234 (1957).
152. A. S. Ladas and T. S. Ma, *Mikrochim. Acta*, **1973**, 853.
153. C. Hennart and E. Merlin, *Chim. Anal. (Paris)*, **39**, 429 (1957).
154. M. Yamagishi and M. Yokoo, *J. Pharm. Soc. Jap.*, **74**, 961 (1954).
155. C. Bighi, *J. Chromatogr.*, **14**, 348 (1964).
156. F. E. Martin and R. R. Jay, *Anal. Chem.*, **34**, 1007 (1962).
157. I. Lysyj and R. C. Greenough, *Anal. Chem.*, **35**, 1657 (1963).
158. D. F. Hagen, J. L. Hoyt, and W. D. Leslie, *Anal. Chem.*, **38**, 1691 (1966).
159. G. A. Muck, N. R. Sundararajan, J. Tobias, and R. M. Whitney, *J. Dairy Sci.*, **50**, 1983 (1967).
160. E. P. Crowell, W. A. Powell, and C. J. Varsel, *Anal. Chem.*, **35**, 184 (1963).
161. J. Ahmadzadeh and J. H. Harker, *Microchem. J.*, **19**, 279 (1974).
162. E. Sawicki, T. R. Hauser, and S. McPherson, *Anal. Chem.*, **34**, 1460 (1962).
163. M. Pesez, *Bull. Soc. Chim. Fr.*, **1955**, 190.
164. M. N. Gibbons, *Analyst*, **80**, 268 (1955).
165. R. Fabre, R. Truhaut, and A. Singerman, *Ann. Pharm. Fr.*, **12**, 409 (1954).

166. W. R. Eckert, *Fette, Seifen Anstrichm.*, **70**, 329 (1968).
167. I. Wilczynska and Z. Margasinski, *Chem. Anal. (Warsaw)*, **12**, 991 (1967).
168. F. F. Critchfield and J. B. Johnson, *Anal. Chem.*, **29**, 797 (1957).
169. R. Amos, *Anal. Chim. Acta*, **40**, 401 (1968).
170. S. I. Obtemperanskaya and G. P. Borodina, *Vestn. Mosk. Gos. Univ. Ser. Khim.*, **1967**, 88.
171. E. Pfeil and H. J. Goldbach, *Klin. Wochschr.*, **35**, 191 (1957).
172. L. R. Jones and J. A. Riddick, *Anal. Chem.*, **28**, 254 (1956).
173. S. A. Shchukarev, S. N. Andreev, and I. A. Ostrovskaya, *Zh. Anal. Khim. SSSR*, **9**, 354 (1954).
174. Y. E. Shmulyakovskii, USSR Patent 134, 907 (1961).
175. A. Okuhara and T. Yokotsuka, *J. Agric. Chem. Soc. Jap.*, **36**, 320 (1962).
176. E. Demey-Ponsart, J. Faidherbe, R. Vivario, C. Heusghem, and H. van Cauwenberge, *Ann. Endocrinol.*, **15**, 614 (1954).
177. R. W. H. Edwards and A. E. Kellie, *Biochem. J.*, **56**, 207 (1954).
178. J. C. Speck, Jr. and A. A. Forist, *Anal. Chem.*, **26**, 1942 (1954).
179. M. Langejan, *Pharm. Weekbl.*, **92**, 667 (1957).
180. M. Langejan, *Pharm. Weekbl.*, **92**, 693 (1957).
181. E. Sawicki, and H. Johnson, *Chemist-Analyst*, **55**, 101 (1966).
182. D. J. Morgan, *Mikrochim. Acta*, **1958**, 104.
183. I. M. Korenman and A. A. Belyakov, *Zh. Anal. Khim.*, **9**, 220 (1954).
184. E. G. Clair, *Analyst*, **87**, 499 (1962).
185. V. F. Degtyarev and A. I. Kruzhevnikova, *Tr. Ural. Lesotekh. Inst.*, **1969**, 404.
186. V. D. Gupta and A. J. L. de Lara, *J. Pharm. Sci.*, **64**, 2001 (1975).
187. D. J. Miller, *J. Assoc. Off. Agric. Chem.*, **33**, 708 (1953).
188. D. R. Grassetti and J. F. Murray, Jr., *Anal. Chim. Acta.*, **46**, 139 (1969).
189. G. W. Ashworth and R. E. Keller, *Anal. Chem.*, **39**, 373 (1967).
190. E. Giovannini, *Ann. Chim. Roma*, **43**, 736 (1953).
191. D. R. Long and R. W. Neuzic, *Anal. Chem.*, **27**, 1110 (1955).
192. M. W. Scoggins and H. A. Price, *Anal. Chem.*, **35**, 48 (1963).
193. S. Hünig and J. Utermann, *Chem. Ber.*, **88**, 423 (1955).
194. R. U. Lemieux and E. von Rudloff, *Can. J. Chem.*, **33**, 1710 (1955).
195. E. von Rudloff, *Can. J. Chem.*, **33**, 1714 (1955).
196. G. Ciuhandu and M. Mravec, *Z. Anal. Chem.*, **38**, 104 (1968).
197. A. I. Kartashevskii, *Neft. Kh.-Vo.*, **1954** (5), 73.
198. W. Rzeszutko, *Acta Pol. Pharm.*, **31**, 323 (1974).
199. A. A. Belyakov, *Zh. Anal. Khim.*, **23**, 1729 (1968).
200. E. Mergenthaler, *Z. Lebensm.-Unters.*, **119**, 144 (1963).
201. K. S. Lee, D. W. Lee, and J. Y. Hwang, *Anal. Chem.*, **40**, 2049 (1968).

202. A. M. Guyet-Hermann, P. Balatre, and J. Tiercelin, *Bull. Soc. Pharm. Lille*, **1972**, 25.
203. H. S. Tomlinson and F. Sebba, *Anal. Chim. Acta*, **27**, 596 (1962).
204. P. Horak and J. Zyka, *Cesk. Farm.*, **12**, 286 (1963).
205. A. K. Babko, L. V. Markova, and T. S. Maksimenko, *Zh. Anal. Khim.*, **23**, 1268 (1968).
206. G. Almassy and I. Dezso, *Acta Chim. Acad. Sci. Hung.*, **11**, 7 (1957).
207. S. Kertes, *Anal. Chim. Acta*, **15**, 73 (1956).
208. G. G. Guilbault, *Practical Fluorescence*, Dekker, New York, 1973.
208a. M. Pesez and J. Bartos, *Talanta*, **14**, 1097 (1967); **16**, 331 (1969).
208b. J. R. Hazlett and D. O. Kildsig, *J. Pharm. Sci.*, **59**, 570 (1970).
209. J. Barthel, *Thermometric Titrations*, Wiley, New York, 1975.
210. L. H. Greathouse, H. J. Janssen, and C. H. Haydel, *Anal. Chem.*, **28**, 357 (1956).
211. C. A. Reynolds and R. Ledesma, *Chem. Eng. News*, **44** (46), 35 (1966).
212. L. S. Bark and P. Prachuabpaibul, *Anal. Chim. Acta*, **72**, 196 (1974).
213. R. J. N. Harries, *Talanta*, **15**, 1345 (1968).
214. L. S. Bark and J. K. Grime, *Analyst*, **97**, 911 (1972).
215. G. A. Vaughan and J. J. Swithenbank, *Analyst*, **92**, 364 (1967).
216. R. D. Daftary and B. C. Haldar, *Anal. Chim. Acta*, **25**, 538 (1961).
217. A. B. De Leo and M. J. Stern, *J. Pharm. Sci.*, **53**, 993 (1964).
218. A. B. De Leo and M. J. Stern, *J. Pharm. Sci.*, **55**, 173 (1966).
219. R. D. Parker and T. Vlismas, *Analyst*, **93**, 330 (1968).
220. W. L. Everson, *Anal. Chem.*, **36**, 854 (1964).
221. R. A. Greinke and H. B. Mark, Jr., *Anal. Chem.*, **38**, 340 (1966).
222. E. H. Jensen and W. A. Struck, *Anal. Chem.*, **27**, 271 (1955).
223. D. L. Bissett, T. E. Hanson, and R. L. Anderson, *Microchem. J.*, **19**, 71 (1974).
224. W. J. Blaedel and P. L. Petitjean, *Anal. Chem.*, **30**, 1958 (1958)
225. E. C. Toren, Jr. and M. K. Gnuse, *Anal. Lett.*, **1**, 295 (1968).
226. R. G. Garmon and C. N. Reilley, *Anal. Chem.*, **34**, 600 (1962).
227. A. Meller, *Sven. Papperstidn.*, **57**, 741 (1954).
228. I. V. Berzin, *Dokl. Akad. Nauk SSSR*, **99**, 563 (1954).
229. K. C. Grover and R. C. Mehrotra, *Z. Anal. Chem.*, **160**, 274 (1958).
230. F. Willeboordse and R. L. Meeker, *Anal. Chem.*, **38**, 854 (1966).
231. T. E. Hewitt and H. L. Pardue, *Clin. Chem.*, **19**, 1128 (1973).
232. T. I. Munnelly, *Anal. Chem.*, **40**, 1494 (1968).
233. Z. Bellen and B. Sekowska, *Chem. Anal. (Warsaw)*, **2**, 35 (1957).
234. F. Jancik and B. Budesinsky, *Cesk. Farm.*, **6**, 590 (1957).
235. O. D. Shapilov and Y. L. Kostyvkovstii, *Zh. Anal. Khim.*, **29**, 1643 (1974).
236. J. G. Hanna and S. Siggia, *Anal. Chem.*, **34**, 547 (1962).

237. E. Schwerdtfeger, *Z. Lebensm. Unters. Forsch.*, **156**, 266 (1974).
238. S. Siggia, J. G. Hanna, and N. M. Serencha, *Anal. Chem.*, **36**, 227 (1964).
239. M. Ranny, J. Novak, and J. Prachar, *Prum. Potravin.*, **14**, 211 (1963).
240. M. I. Karayannis and E. V. Kordi, *Analyst*, **100**, 168 (1975).
241. H. Flaschka and M. Hochenegger, *Mikrochim. Acta*, **1957**, 587.
242. L. Legradi, *Mag. Kem. Foly.*, **72**, 336 (1966).
243. J. S. Parsons, W. Seaman, and J. T. Woods, *Anal. Chem.*, **27**, 21 (1955).
244. W. Rodziewicz, I. Kwiatowski, and E. Kwiatkowski, *Chem. Anal. (Warsaw)*, 1067, 1305 (1968).
245. M. Krhal, *Kunstoffe*, **45**, 224 (1955).
246. L. Legradi, *Z. Anal. Chem.*, **237**, 426 (1968).
247. P. R. Carter, *J. Sci. Food Agric.*, **5**, 457 (1954).
248. G. G. Guilbault, *Handbook of Enzymatic Methods of Analysis*, Dekker, New York, 1976.
249. H. U. Bergmeyer and A. Haggen, *Z. Anal. Chem.*, **261**, 333 (1972).
250. E. Sawicki and J. D. Pfaff, *Mikrochim. Acta*, **1966**, 322.
251. P. J. Oles and S. Siggia, *Anal. Chem.*, **46**, 911 (1974).
252. J. C. Sheridan, E. P. K. Lau, and B. Z. Senkowski, *Anal. Chem.*, **41**, 247 (1969).
253. T. Kumamaru, *Anal. Chim. Acta*, **43**, 19 (1968).
254. E. A. Reich, M. A. Carroll, A. Post, and J. E. Zarembo, *Microchem. J.*, **20**, 305 (1975).
255. P. L. de Reeder, *Anal. Chim. Acta*, **9**, 140 (1963).
256. R. J. McGowan, *Anal. Chem.*, **35**, 1664 (1963).
257. N. I. Popova and E. E. Vermel, *Izv. Vost. Fil. Akad. Nauk SSSR*, **1957** (9), 74.
258. J. E. Katon and F. F. Bentley, *Spectrochim. Acta*, **19**, 639 (1963).
259. G. G. Dvoryantseva and Y. N. Sheinker, *Zh. Anal. Khim.*, **17**, 883 (1962).
260. M. F. Carroll and W. J. Price, *Perfum. Essent. Oil Rec.*, **55**, 114 (1964).
261. M. P. Filippov and R. Kohn, *Chem. Zvesti.*, **29**, 88 (1975).
262. D. E. Nicholson, *Anal. Chem.*, **31**, 519 (1959).
263. W. Bek, *Afinudad*, **20**, 323 (1963).
264. W. E. Link and K. M. Buswell, *J. Am. Oil Chem. Soc.*, **39**, 39 (1962).
265. A. Jart, *Acta Polytech. Scand., Ser. Chem.*, **44** (1965).
266. E. Lieber, C. N. R. Rao, T. S. Chao, and C. W. W. Hoffman, *Anal. Chem.*, **29**, 916 (1957).
267. A. Jart, *Acta Polytech. Scand., Ser. Chem.*, **42** (1965).
268. V. V. Zharkov, M. I. Bakhitov, and E. V. Kuzenetsov, *Zh. Anal. Khim.*, **29**, 396 (1974).
269. N. Oi and E. Inaba, *J. Pharm. Soc. Jap.*, **87**, 743 (1967).
270. A. Pozefsky and I. Kukin, *Anal. Chem.*, **27**, 1466 (1955).
271. F. G. Pearson and R. B. Stasiak, *Appl. Spectrosc.*, **12**, 116 (1958).

272. F. H. Suydam, *Anal. Chem.*, **35**, 193 (1963).
273. C. T. Malone, S. K. Freeman, and M. H. Jacobs, *J. Chromatogr.*, **30**, 215 (1967).
274. M. H. Kunz, *Z. Chem. Leipz.*, **5**, 333 (1965).
275. A. S. Wexler, *Spectrochem. Acta*, **21**, 1725 (1965).
276. J. M. Hume and G. I. Jenkins, *Appl. Spectrosc.*, **18**, 161 (1964).
277. H. Luther and H. Oelert, *Angew. Chem.*, **69**, 262 (1957).
278. B. M. Mitzner, E. T. Theimer and S. K. Freeman, *Appl. Spectrosc.*, **19**, 169 (1965).
279. C. C. Bard, T. J. Porro, and H. L. Rees, *Anal. Chem.*, **27**, 12 (1955).
280. R. C. Ferguson, *Anal. Chem.*, **36**, 2204 (1964).
281. A. Ito and S. Hideyo, *Ann. Rep. Takamine Lab.*, **7**, 83 (1955).
282. E. P. Crowell and C. J. Varsel, *Anal. Chem.*, **35**, 189 (1963).
283. A. A. Ponomarenko and B. I. Popov, *Zh. Anal. Khim.*, **19**, 1397 (1964).
284. R. Brandt, J. C. Kouines, and N. D. Cheronis, *J. Chromatogr.*, **12**, 380 (1963).
285. H. C. Hollifield and J. D. Winefordner, *Talanta*, **1** , 860 (1965).
286. G. Schwedt, *Clin. Chim. Acta*, **57**, 291 (1974).
287. M. G. M. Balemans and F. C. G. can de Veerdonk, *Experientia*, **23**, 906 (1967).
288. K. F. Harbaugh, C. M. O'Donnell, and J. D. Winefordner, *Anal. Chem.*, **46**, 1206 (1974).
289. B. Fleet, G. F. Kirkbright, and C. J. Pickford, *Talanta*, **15**, 566 (1968).
290. S. P. McGlynn, B. T. Neely, and C. Neely, *Anal. Chim. Acta*, **28**, 472 (1963).
291. D. Lavalette, B. Muel, M. Hubert-Habart, L. Rene, and R. Latarjet, *J. Chim. Phys.*, **65**, 2144 (1968).
292. R. W. Frei, *C. R. C. Crit. Rev. Anal. Chem.*, , 179 (1971).
293. D. P. Mironov and V. V. Zharkov, *Zaved. Lab.*, **29**, 1441 (1963).
294. D. E. Nicholson, *Anal. Chem.*, **32**, 1634 (1960).
295. L. A. Efremova and K. K. Popkov, *Zaved. Lab.*, **29**, 708 (1963).
296. G. W. Ewing, *Instrumental Methods of Chemical Analysis*, 4th ed., McGraw-Hill, New York, 1975, p. 146.
297. R. Belcher, S. L. Bogdanski, S. A. Ghonaim, and A. Townshend, *Anal. Lett.*, **7**, 133 (1974).
298. R. Belcher, S. L. Bogdanski, A. Calokerivos, A. Townshend, and I. Z. Al-Zamil, Abstracts of the 1976 Federation of Analytical Chemistry and Spectroscopy Societies, Nov. 15–19, 1976, Philadelphia, Abstract No. 168.
299. R. Belcher, R. A. Sheikh, S. L. Bogdanski, and A. Townshend, reported at the 1976 Federation of Analytical Chemistry and Spectroscopy Societies, Nov. 15–19, 1976, Philadelphia.
300. G. J. Papariello, H. Letterman, and R. E. Huettemann, *J. Pharm. Sci.*, **53**, 663 (1964).
301. K. Kuroda, G. Hashizume, and K. Fukuda, *J. Pharm. Soc. Jap.*, **87**, 1175 (1967).
302. Y. G. Mamedaliev, I. G. Izmailzade, S. Mirzoyeva, T. Zeinalova, and K. M. Abdullayeva, *Dokl. Akad. Nauk SSSR*, **102**, 529 (1955).

303. H. M. Fales and T. Luukainen, *Anal. Chem.*, **37**, 955 (1965).
304. A. Mathias, *Anal. Chim. Acta*, **31**, 598 (1964).
305. L. P. Lindeman and S. W. Nicksic, *Anal. Chem.*, **36**, 2414 (1964).
306. F. J. Ludwig, *Anal. Chem.*, **40**, 1620 (1968).
307. M. W. Dietrich and R. E. Keller, *Anal. Chem.*, **36**, 2174 (1964).
308. C. van der Vlies, G. A. Bakker, and R. F. Rekker, *Pharm. Weekbl. Ned.*, **101** (5), 93 (1966).
309. T. F. Conway and F. R. Earle, *J. Am. Oil Chem. Soc.*, **40** (7), 265 (1963).
310. P. W. Flanagan, R. A. Greff, and H. F. Smith, *Anal. Chem.*, **35**, 1283 (1963).
311. H. Ueda, *Anal. Chem.*, **35**, 2213 (1963).
312. G. A. Ward and R. D. Mair, *Anal. Chem.*, **41**, 538 (1969).
313. N. Ikeda and K. Fukuzumi, *J. Am. Oil Chem. Soc.*, **51** (8), 340 (1974).
314. S. A. Koch and T. D. Doyle, *Anal. Chem.*, **39**, 1273 (1967).
315. F. Chastellain and P. Hirsbrunner, *Z. Anal. Chem.*, **278**, 207 (1976).
316. R. J. Warren, J. E. Zarembo, D. B. Staiger, and A. Post, *J. Pharm. Sci.*, **65**, 738 (1976).
317. A. Mathias and D. Taylor, *Anal. Chim. Acta*, **35**, 376 (1966).
318. H. N. Tho and G. Sirois, *J. Pharm. Sci.*, **62**, 1334 (1973).
319. D. P. Hollis, *Anal. Chem.* **35**, 1682 (1963).
320. G. Slomp, R. H. Baker, Jr., and F. A. MacKellar, *Anal. Chem.*, **36**, 375 (1964).
321. J. W. Robinson and D. Truitt, *Spectrosc. Lett.*, **2**, 203 (1969).
322. T. F. Page, Jr., and W. E. Bresler, *Anal. Chem.*, **36**, 1981 (1964).
323. H. Y. Chen, *Anal. Chem.*, **34**, 1793 (1962).
324. Y. Kawazoe and M. Ohnishi, *Chem. Pharm. Bull. Jap.*, **12**, 846 (1964).
325. Y. Degani and A. Patchornik, *Anal. Chem.*, **44**, 2170 (1972).
326. K. E. Daugherty, J. I. Stevens, S. R. Kramer, and R. E. Van Doren, *Appl. Spectrosc.*, **22**, 784 (1968).
327. K. W. Jolley and L. H. Sutcliffe, *Spectrochim. Acta*, **24**, 1293 (1968).
328. T. I. Popova, A. A. Polyakova, and K. I. Zimina, *Khim. Tekhnol. Topl. Masel*, **1965**(2), 48.
329. A. Tatematsu and T. Goto, *J. Pharm. Soc. Jap.*, **85**, 624 (1965).
330. B. Hallgren, E. Stenhagen, and R. Ryhage, *Acta Chem. Scand.*, **11**, 1064 (1957).
331. W. D. Koller, *Z. Anal. Chem.*, **260**, 31 (1972).
332. C. G. Frei, F. A. Wiesel, and G. Sedvall, *Life Sci.*, **14**, 2469 (1974).
333. J. D. Baty, P. R. Robinson, and J. Wharton, *Biomed. Mass Spectrom.*, **3**(2), 60 (1976).
334. L. Bertilsson and E. Costa, *J. Chromatogr.*, **118**, 395 (1976).
335. A. Tatematsu and T. Goto, *J. Pharm. Soc. Jap.*, **85**, 778 (1965).
336. A. Costopanagiotis and H. Budzikiewiez, *Mh. Chem.*, **96**, 1800 (1965).

337. H. Knof, R. Large, and G. Albers, *Erdöll Kohle Erdgas Petrochem. Brennst. Chem.*, **29**, 77 (1976).
338. A. J. Frisque, H. M. Grubb, C. H. Ehrhardt, and R. W. Vander Haar, *Anal. Chem.*, **33**, 389 (1961).
339. J. L. Occolowitz, *Anal. Chem.*, **36**, 2177 (1964).
340. V. Svob, *Nafta, Zagr.*, **19**, 549 (1968).
341. W. C. Ferguson and H. E. Howard, *Anal. Chem.*, **30**, 314 (1958).
342. J. L. Shultz, *Spectrosc. Lett.*, **1**, 345 (1968).
343. E. S. Brodskii, I. M. Lukashenko, Y. A. Volko, Y. M. Goldberg, and V. G. Lededevskaya, *Zh. Anal. Khim.*, **28**, 2262 (1973).
344. P. Zuman, *Organic Polarographic Analysis*, Macmillan, New York, 1964.
345. S. C. Tung and E. K. Wang, *Acta Chim. Sin.*, **29**, 1 (1965).
346. C. Prevost and P. Souchay, *Chim. Anal. (Paris)*, **37**, 3 (1955).
347. S. A. K. Hsieh and T. S. Ma, *Mikrochim. Acta*, **1977** *I*, 325.
348. J. Kubis, *Prac. Lek.*, **11**, 465 (1959).
349. D. A. Skoog and A. B. H. Lauwzesha, *Anal. Chem.*, **28**, 825 (1956).
350. Y. Mizunoya, *Jap. Analyst*, **11**, 87, 220, 393 (1962).
351. B. P. Zhantalai and Y. I. Turyan, *Zh. Anal. Khim.*, **23**, 282 (1968).
352. H. B. Mark, Jr. and C. N. Reilley, *Anal. Chem.*, **35**, 195 (1963).
353. W. Kemula and Y. Chodowski, *Rocz. Chem.*, **29**, 839 (1955).
354. R. Cernatescu, M. Poni, and R. Relea, *Stud. Ceret. Stiin*, **4**, 117 (1953).
355. W. Furst, *Pharm. Zentralhalle*, **107**, 184 (1967).
356. S. Tammilehto and M. Perala, *Pharm. Acta Helv.*, **46**, 351 (1971).
357. E. M. Craine, M. J. Parnell, and L. R. Stone, *J. Agric. Food Chem.*, **22**, 877 (1974).
358. L. Vignoli, B. Cristau, F. Gouezo, and C. Fabre, *Chim. Anal.*, **45**, 499 (1963).
359. D. M. Coulson, *Anal. Chim. Acta*, **19**, 284 (1958).
360. L. Sucha, *Collect. Czech. Chem. Commun.*, **33**, 1375 (1968).
360a. B. Fleet, *Anal. Chim. Acta*, **36**, 304 (1966).
361. T. S. Ma, M. R. Hackman, and M. A. Brooks, *Mikrochim. Acta*, **1975** *II*, 617.
362. M. Kalousek, *Collect. Czech. Chem. Commun.*, **19**, 1111 (1954).
363. R. Zahradnik and L. Jensovsky, *Chem. Listy.*, **48**, 11 (1954).
364. Y. Okazaki, *Jap. Analyst*, **11**, 986, 1142, 1239 (1962).
365. A. Revolova, *Chem. Listy*, **50**, 1918 (1956).
366. S. Maruta and F. Iwama, *J. Chem. Soc. Jap., Pure Chem. Sect.*, **76**, 548 (1955).
367. A. V. Ryabov and G. D. Penova, *Dokl. Akad. Nauk SSSR*, **99**, 547 (1954).
368. V. Sedivec and J. Flek, *Collect. Czech. Chem. Commun.*, **25**, 1293 (1960).
369. V. Sedivec, *Chem. Listy*, **51**, 249 (1957).
370. H. D. Eilhauer, G. Krautschick, and I. Kampfer, *Chem. Tech. (Berl.)*, **20**, 491 (1968).

371. L. H. Griffin, *Anal. Chem.*, **34**, 564 (1962).
372. J. G. Burtle, J. P. Ryan, and M. James, *Anal. Chem.*, **30**, 1640 (1958).
373. P. Sorensen, *Anal. Chem.*, **26**, 1581 (1954).
374. M. F. Abdel-Wahab and S. A. El-Kinawi, *Z. Anal. Chem.*, **186**, 364 (1962).
375. Y. Nagase and S. Baba, *J. Pharm. Soc. Jap.*, **81**, 619 (1961).
376. T. S. Ma and M. H. Yang, *J. Chin. Chem. Soc., Ser. II.*, **10**, 86 (1963).
377. M. Margosis, J. T. Tanner, and J. P. F. Lambert, *J. Pharm. Sci.*, **60**, 1550 (1971).
378. H. L. Schlesinger, M. J. Pro, C. M. Hoffman, and M. Cohan, *J. Assoc. Off. Agric. Chem.*, **48**, 1139 (1965).
379. M. M. Tuckerman, L. C. Bate, and G. W. Leddicotte, *J. Pharm. Sci.*, **53**, 983 (1964).
380. J. K. Wang, *Trans. Am. Soc. Agric. Engr.*, **7**, 42 (1964).
381. G. E. Peck, J. E. Christian, and G. S. Banker, *J. Pharm. Sci.*, **53**, 632 (1964).
382. K. A. Adey, *Analyst*, **88**, 359 (1963).
383. M. Popowicz, *Chem. Anal. (Warsaw)*, **2**, 358 (1957).
384. Y. Nevskaya and T. Sumarokova, *Izv. Akad. Nauk Kaz. SSR, Ser. Khim.*, **1962**(2), 27.
385. M. D. Tilicheev, M. S. Borovaya, and V. S. Buk, *Zh. Anal. Khim.*, **11**, 188 (1956).
386. J. Hrivnak, M. Michalek, and Z. Stota, *Chem. Prum.*, **13**, 18 (1963).
387. L. Rappen, *Brennstchemie*, **39**(5), 65 (1958).
388. R. G. Henein and J. Csonka-Horvai, *Acta Chim. Hung.*, **60**, 37 (1969).
389. T. Flora and A. Almasy, *Acta Chim. Hung.*, **54**, 189 (1967).
390. T. Ozawa, *Bull. Chem. Soc. Jap.*, **39**, 2071 (1966).
391. Arzneibuchkommission der DRR., *Zentbl. Pharm. Pharmakother. Lab. Diagn.*, **113**, 1147 (1974).
392. C. Heitler, *Talanta*, **11**, 1081 (1964).
393. F. Martin, J. Courteix, and S. Vertailier, *Bull. Soc. Chim. Fr.*, **1958**, 494.
394. British Standards Institution, *Determination of Distillation of Petroleum Products*, B.S. 4349 (1968).
395. A. Janik, *Chromatographis*, **6**, 514 (1973).
396. C. D. Bittenbender, *J. Food Sci.*, **35**, 460 (1970).
397. M. M. Agarwal and P. S. Mene, *Indian J. Technol.*, **1**, 274 (1963).
398. O. M. Angelidis, *Chem. Anal. (Warsaw)*, **50**, 118 (1968).
399. A. A. Skelding and R. F. Ashbolt, *Chem. Ind. (Lond.)*, **1959**, 1204.
400. W. D. Pohle and S. E. Tierney, *J. Am. Oil Chem. Soc.*, **31**, 203 (1954).
401. W. Wisniewski, S. Jablonski, W. Bielawska, and A. Janowicz, *Acta Pol. Pharm.*, **20**, 181 (1963).
402. J. Krutzsch, *Dtsch. Apoth. Ztg.*, **98**, 609 (1958).
403. J. Hirtz, R. Berret, and B. Rio, *Chim. Anal. (Paris)*, **51**, 434 (1969).
404. B. E. Hallaway and R. Sandberg, *Clin. Chem.*, **16**, 408 (1970).
405. V. Parausanu, A. Constantin, and L. Dima. *Revta. Chim.*, **19**, 481 (1968).

406. L. Silverman and M. E. Shideler, *Anal. Chim. Acta*, **18**, 540 (1958).
407. W. Czerwinski, *Chem. Anal. (Warsaw)*, **1**, 77 (1956).
408. M. V. Glynn, *J. Text. Inst. Manchester Trans.*, **46**, 228 (1956).
409. R. Fischer and W. Auer, *Pharm. Zentralhalle*, **103**, 243 (1964).
410. J. P. Comer and L. D. Howell, *J. Pharm. Sci.*, **53**, 335 (1964).
411. S. K. Suri, *Talanta*, **19**, 804 (1972); **21**, 604 (1974).
412. C. Morton and E. H. Tinley, *J. Pharm. Pharmacol.*, **8**, 967 (1956).
413. A. S. Shpitalnyi and L. A. Volf, *Zavod. Lab.*, **24**, 1489 (1958).
414. R. Hill, *Analyst*, **81**, 323 (1956).
415. J. C. S. Wood, C. C. Martin, and M. R. Lipkin, *Anal. Chem.*, **30**, 1530 (1958).
416. B. V. Ioffe, M. D. Morachevskaya, and N. I. Kholmovskaya, *Zh. Anal. Khim.*, **30**, 676 (1975).
416a. H. Schlosser, H. Wollmann, and G. Eick, *Zentralbl. Pharm.*, **116**, 145 (1977).
417. W. J. Kirsten, *Anal. Chim. Acta*, **27**, 345 (1962).
418. M. Hamon, C. Morin, and R. Bourdon, *Anal. Chim. Acta*, **46**, 255 (1969); *Ann. Pharm. Fr.*, **27**, 283 (1969).
419. K. Takiura and M. Yamamoto, *J. Pharm. Soc. Jap.*, **85**, 606 (1965).
420. L. N. Kaigorodova and E. I. Klabunovskii, *Zh. Anal. Khim.*, **29**, 1217 (1974).
421. British Standards Institution, B.S. 4407 (1969), Amendment No. 3, 31.5.74.
422. R. J. Warren, I. B. Eisdorfer, W. E. Thompson, and J. E. Zarembo, *J. Pharm. Sci.*, **55**, 144 (1966); **57**, 195 (1968).
423. F. Köchel, *Dtsch. Apoth. Ztg.*, **94**, 1095 (1954).
424. T. G. Alexander and S. A. Koch, *J. Assoc. Off. Agric. Chem.*, **48**, 618 (1965).
425. R. F. Branch, *Pharm. J.*, **197**, 607 (1966).
426. J. Deavin, *Pharm. J.*, **205**, 134 (1970).
427. L. H. Keith and A. L. Alford, *J. Assoc. Off. Anal. Chem.*, **53**, 1018 (1970).
428. J. Eisenbrand, *Dtsch. Lebensm. Rundsch.*, **62**, 327 (1966).
429. G. R. Schultze and K. Kirchhoff, *Erdoll Kohle*, **21**, 148 (1968).
430. J. P. Wolff, *Chim. Anal. (Paris)*, **50**, 18 (1968).
431. S. K. Sinha and K. R. K. Rao, *J. Sci. Ind. Res., India*, **22**, 208 (1963).
432. E. Sawicki, *Anal. Chem.*, **35**, 199 (1963).
433. J. Montreuil and N. Scheppler, *Bull. Soc. Chim. Biol.*, **41**, 13 (1959).
434. H. Thielemann, *Sci. Pharm.*, **39**, 148 (1971).
435. L. Federico and M. Ciucani, *Chim. Ind.*, **36**, 598 (1954).
436. B. R. Hajratwala, *Aust. J. Pharm. Sci., NS* **3**, 33 (1974).
437. B. H. M. Kingston, J. J. Garey, and W. B. Hellwig, *Anal. Chem.*, **41**, 86 (1969).
438. M. J. Maurice and B. Veen, *Z. Anal. Chem.*, **163**, 13 (1958).
439. M. S. Karawya, S. H. Hilal, and M. A. Elsohly, *J. Assoc. Off. Anal. Chem.*, **56**, 1467 (1973).
440. A. Manasterski and B. Zak, *Michrochem. J.*, **19**, 8 (1974).

441. J. Jech, *Dtsch. Apoth. Ztg*, **114**, 1978 (1974).
442. J. G. Wagner, J. K. Dale, C. A. Schlagel, P. D. Meister, and R. E. Booth, *J. Am. Pharm. Assoc.*, **47**, 580 (1958).
443. I. Slavik, M. Pasteka, and M. Kucerova, *Faserforsch. Text. Tech.*, **18**, 4 (1967).
444. R. Kämpf, *Mitt. Geb. Lebensmittunters Hyg.*, **67**, 192 (1976).
445. V. P. Zhivopistsev, E. A. Selezneva, and Z. I. Bragina, *Uch. Zap. Perm. Gos. Univ.*, **178**, 203 (1968).
446. M. Matrka, *Chem. Prague*, **10**, 635 (1958).
447. D. Trokowicz, *Chem. Anal. (Warsaw)*, **8**, 107 (1963).
448. K. Thomas, T. E. Ullmann and P. Loos, *Pharmazie*, **18**, 414 (1963).
449. G. Hartel and A. Harjanne, *Clin. Chim. Acta*, **23**, 289 (1969).
450. F. A. Rotondaro, *J. Assoc. Off. Agric. Chem.*, **38**, 809 (1955).
451. P. Haefelfinger, R. Schmidli, and H. Ritter, *Arch. Pharm.*, **297**, 641 (1964).
452. W. R. Betker, C. F. Smead, and R. T. Evans, *J. Assoc. Off. Anal. Chem.*, **59**, 278 (1976).
453. W. Wojahn and W. Wempe, *Arch. Pharm.*, **228**, 1 (1955).
454. R. P. MacDonald and J. Bloompuu, *Mikrochim. Acta*, **1958**, 147.
455. Z. Stefanac and A. Verbic, *Z. Anal. Chem.*, **3** , 113 (1967).
456. B. G. K. Murthy, M. A. Sivasamban, and J. S. Aggarwal, *Indian J. Chem.*, **3**, 33 (1965).
457. D. E. La Croix, A. R. Prosser, and A. J. Sheppard, *J. Assoc. Off. Anal. Chem.*, **51**, 20 (1968).
458. J. C. Ryan, L. K. Yanowski, and C. W. Pifer, *J. Am. Pharm. Assoc.*, **43**, 656 (1954).
459. V. A. Cirillo, D. J. Skahan, B. Hollis, and H. Morgan, *Anal. Chem.*, **34**, 1353 (1962).
460. W. K. Saur, H. L. Crespi, L. Harkness, G. Gorman, and J. J. Katz, *Anal. Biochem.*, **22**, 424 (1968).
461. R. Siegfried, *J. Chromatogr.*, **118**, 270 (1976).
462. R. Goupil and G. Mangeney, *Chim. Anal. (Paris)*, **41**, 18 (1959).
463. L. Bernstein, B. K. Blenkinship, and M. W. Brenner, *Proc. Am. Soc. Brew. Chem.*, **1968**, 150.
464. J. Barcelo and M. P. Jorge, *Inf. Quim. Anal.*, **8**, 198 (1954).
465. Y. N. Molin, A. K. Petrov, G. I. Kulakova, and G. G. Yakobson, *Zh. Anal. Khim.*, **20**, 396 (1965).
466. B. Sjöberg, *Acta Chem. Scand.*, **16**, 269 (1962).
467. J. M. Rowson, *Pharm. Weekbl. Ned.*, **100**, 1500 (1965).
468. Pharmaceutical Society of Great Britain, Science Committee of the Department of Pharmaceutical Science, *Pharm. J.*, **199**, 620 (1967).
469. C. E. Finsterwalder, *J. Assoc. Off. Anal. Chem.*, **59**, 169 (1976).
470. L. Mitchell, *J. Assoc. Off. Anal. Chem.*, **59**, 209 (1976).

471. H. H. Sauer, *J. Assoc. Off. Anal. Chem.*, **59**, 711 (1976).
472. W. Heizler, J. Meier, K. Nowak, R. Suter, and H. P. Bosshardt, *J. Assoc. Off. Anal. Chem.*, **59**, 716 (1976).
473. A. S. Wendt, *J. Assoc. Off. Agr. Chem.*, **46**, 337 (1963).
474. F. H. de la Court, N. J. P. van Cassel, and J. A. M. van der Valk, *J. Chromatogr.*, **72**, 249 (1972).

CHAPTER 3

Determination of Several Components in a Mixture Without Separation

3.1 INTRODUCTORY REMARKS

In this chapter we discuss the general methods by which several compounds in an organic mixture can be determined without the preliminary steps of separation. In the next chapter we deal with the various techniques that may be used when separation is necessary. Generally speaking, an analytical method that does not involve separation is more rapid. Besides, it eliminates the errors due to losses and contamination that may be incurred during the separation process.

We can classify organic mixtures submitted to quantitative analysis into two categories, namely, "known mixtures" and "unknown mixtures." The known mixtures include samples for quality control or stability test, reaction mixtures resulting from well established organic syntheses, products of an industrial process such as sulfonation of hydrocarbons, and so on. In this category the analyst has sufficient knowledge of the constituents in the sample; the analytical problem is to ascertain the quantitative relationship among the various known constituents. By contrast, unknown mixtures are materials about whose composition the analyst has only limited information. This category includes competitor's products, samples for adulteration test, herbal extracts, biological fluids after drug administration, reaction products of a new organic synthesis, and so on. The analyst knows what compound he is to determine in an unknown mixture, but he does not know what other substances may be present. In such a case, it is prudent to perform the total analysis; this usually involves separation. On the other hand, for the analysis of known mixtures, only the compounds to be determined need be considered, and separation is avoided whenever possible.

As is mentioned in Chapter 2, Section 2.5, a variety of methods are available for the determination of organic compounds. Some methods also permit the determination of two or more compounds in the same sample without mutual interference. Occasionally it is also possible to find more than one method to achieve the determinations. For example, Karavaev and Khelevin[1] described a physical and a chemical method to analyze the sulfonic acid mixtures obtained in the sulfonation of dimethylaniline: The physical method is based on the different ultraviolet absorption spectra of the sodium sulfonates measured at 230, 253.5, and 280 nm, while the chemical method is based on the different reaction rates of the sulfonic acids with bromine, which removes the $—SO_3H$ group to form H_2SO_4.

By taking advantage of the concept of Brönsted and Lowry, a vast number of organic compounds can be determined on the basis of their acidic or basic properties, or both. Since the development of nonaqueous acid–base titrimetry in the 1950s, it has become the most simple, convenient, and economical method for the determination of organic compounds. The easy analysis of explosives as described by Sarson[2] may be cited as an example. Tri- and dinitrotoluenes, pentaerythritol tetranitrate, and hexahydrotrinitrotriazine are titrated as acids in isobutyl methyl ketone; nitroglycine, nitrocellulose, mononitrotoluene, and ammonium nitrate are titrated as acids in dimethylformamide; all can be done without hazards. The reader is referred to references 3 and 4 for the details of the nonaqueous titration technique. It is the method of choice for the analysis of mixtures when two or more constituents show different levels of acidity or basicity and thus can be determined simultaneously without separation (see Sections 3.2 and 3.3).

Another method for simultaneous determination is based on the difference of reaction rates exhibited by two compounds when they are treated with the same reagent (see Section 3.5). Still another method utilizes the different half-wave potentials when two reducible compounds are analyzed in the polarograph (see Section 3.6).

Since every organic compound possesses its own spectroscopic characteristics and molecular fragmentation pattern, it should be possible, theoretically, to single out the desired compound by means of the spectrometer and make quantitative measurements. Although this goal is far from being reached, much progress has been achieved, as shown in Sections 3.7 to 3.10.

3.2 SIMULTANEOUS DETERMINATION OF ACIDIC COMPOUNDS BY DIFFERENTIAL TITRATION

3.2.1 Titration in Nonaqueous Media

In inorganic acidimetry, only a few acids are encountered and the analysis is always carried out in aqueous solutions, thus seldom permitting differential determination. Contrastingly, there are numerous organic acids (by the Brönsted-Lowry definition) with a wide range of pK values distinguishable under suitable conditions. Therefore, when two or more acidic compounds are present in a known mixture, attempts should be made to determine them without separation, usually by nonaqueous titrimetry. The problem is to find the proper titrant and the suitable solvent system.

According to Dahmen,[5] for the determination of the total acidity of an organic sample, a basic solvent should be chosen so that its region of acidity lies above the half-neutralization potentials of the acids. For the simultaneous determination of the individual acids, a less basic or an inert solvent is used so that the half-neutralization potentials lie within as wide a range of acidity as possible. As a rule, the weaker the acid, the stronger should be the base used as the solvent. For instance, Maurmeyer et al.[6] observed that phenol can be titrated in dimethylformamide, whereas sulfanilamide must be titrated in pyridine or a more basic solvent. For reasons that have not been fully elucidated, aliphatic ketones serve well for differential titrations. Bruss and Wyld[7] first reported that isobutyl methyl ketone is an excellent medium for the resolution of mixtures of strong, weak, and very weak acids. Studying the determination of carboxylic acids, phenols, enols, imides, and sulfonamides, Yamamura[8] employed acetone or acetonitrile as the solvent and found that a differentiating titration is possible if the potential difference between the beginning of the inflections of the stronger and weaker acids is greater than >100 mV. According to Frisque and Meloche,[9] binary acid mixtures (e.g., benzoic and *m*-nitrobenzoic acids) that give titration curves with a single sharp inflection can often be analyzed by use of the individual pK values in the calculations.

Since most organic acids submitted to titration are weaker than carbonic acid, the carbon dioxide in the atmosphere interferes seriously. At the same time, the titrant must be a very strong base. Salvesen[10] recommended benzalkonium hydroxide in dimethylformamide, claiming that it is much less sensitive to carbon dioxide than either sodium methoxide or lithium methoxide. Recently Rogozinski and Bosshard[11] proposed potassium *tert*-butoxide, which can titrate imidazole and malonic ester in dimethylformamide medium. Hiller[12] advocated conductimetric titration of weak acids using dimethyl sulfoxide as the solvent and the sodium salt of the methylsulfinyl carbanion as titrant. The extensive applications of nonaqueous titration for simultaneous determination of two or more acidic compounds in a sample can be seen in Table 3.1.

3.2.2 Titration in Aqueous Solutions

It should be noted that titration of organic acids in aqueous medium has not been discarded. Since aqueous titrimetry requires only common equipment and easily available reagents, it can be employed with advantage on certain occasions. For example, Furne and Karimova[96] titrated mixtures of aliphatic sulfonic acids and H_2SO_4 with aq. NaOH to obtain the total acid content and they titrated another aliquot with aq. $BaCl_2$ to determine the H_2SO_4 alone. Khudyakova and Kreshkov[97] demonstrated the differential titration of acids having pK values from 4 to 10 using aq. 0.3 N NaOH. Bykova and Ardashnikova[98] analyzed mixtures of terephthalic and *p*-toluic acids with ethanolic KOH by first titrating in acetone medium and then diluting the solution with water; the first potential break corresponds to neutralization of one carboxyl group of terephthalic acid, while the second corresponds to the simultaneous neutralization of the second carboxyl group and the carboxyl group of toluic acid. Datta and Mukherjee[99] determined fulvic acids from soil humus with aq. NaOH or $Ba(OH)_2$. Rao and Srikantan[100] determined weak acids in vegetable tan liquors and reported sharp maxima.

With a derivative potentiometric arrangement, Beranova and Hudecek[101] determined phenols in the presence of acetic and chloracetic acids using aq. $Ba(OH)_2$ as the titrant. Purdie et al.[102] devised a method to determine the composition of mixtures of diastereoisomers of tartaric acid by pH titration using KOH. Schute[103] reported on the photometric titration curves of mixtures of acids using alkali in aqueous medium. Szabo-Akos and Erdey[104] studied simultaneous determination of acids of various strengths in aqueous solutions and concluded that an acid can be titrated selectively in the presence of a second acid if the protonation constants of the conjugate base of the second acid, of the basic titrant, and of the conjugate base of the first acid decrease in that order.

Chromniak[105] performed simultaneous conductometric determination of phenylenediamine or benzidine dihydrochloride and HCl by titration with NaOH. Khudyakova and Vostokov[106] described chronoconductometric determination of salts of aliphatic, aromatic, and heterocyclic bases using 0.3 N NaOH. Shimomura[107] employed aq. NaOH to titrate maleic hydrazide in the presence of maleic acid and *N*-aminomaleimide. Chromy and Groagova[108] recommended potentiometric titration of naphthyl hydrogen sulfates and naphthyl hydrogen phosphates with 0.1 N NaOH.

Table 3.1 Simultaneous Determination of Acidic Compounds by Nonaqueous Titration

Type of compound	Determination of	Titrant	Reference
C, H, O	Anhydride, carboxylic acid	$(C_4H_9)_4NOH$	Lucchesi[13]
	Aliphatic acids	$(C_2H_5)_4NOH$	Smolova[14]
	Carboxylic, phenolic acids	$KOCH_3$	Meurs[15]
	Adipic, sucinic, oxalic, citric, benzoic acids	$(C_2H_5)_4NOH$	Prokopev[16]
	Acidic additives in mineral oil	$(CH_3)_4NOH$	Kahsnitz[17]
	Benzoic acid derivatives	$NaOCH_3$	Jasinski[18]
	Naphthoic acids	$(C_4H_9)_4NOH$	Kreshkov[19]
	Benzoic, hydroxybenzoic, oxalic acids, H_2SO_4	$(C_2H_5)_4NOH$	Bykova[20]
	Benzoic, salicylic acids	$NaOCH_3$	Blake[21]
	Fatty acids in wool wax	$NaOCH_3$	Radell[22]
	Benzoic, salicylic, camphoric acids	$(CH_3)_4NOH$	Rink[23]
	Toluic, terephthalic acids	KOH	Trussell[24]
	Toluic, terephthalic acids	NaOH	Valcha[25]
	Maleic, phthalic acids	NaOH	Malyshev[26]
	Adipic acid, polyesters	$C_4H_9(C_2H_5)_3$-NOH	Fijolka[27]
	Dicarboxylic acids, monoesters	KOH	Kreshkov[28]
	Mono-, di-, tribasic acids	$(C_4H_9)_4NOH$	Cundiff[29]
	Oxalic, succinic acids, H_2SO_4	$(C_4H_9)_4NOH$	Harlow[30]
	Dibasic acids	$(C_2H_5)_4NOH$	Kreshkov[31]
	Oxalic, malonic, azelaic acids	KOH	Kreshkov[32]
	Saturated dibasic acids	$(C_2H_5)_4NOH$	Kreshkov[33]
	Malonic, adipic, phthalic acids	$KOCH_3$	Kreshkov[34]
	Itaconic, citraconic, maleic, fumaric acids	$(C_2H_5)_4NOH$	Kreshkov[35]
	Benzene polycarboxylic acids	$(C_2H_5)_4NOH$	Vasyutinskii[36]
	Isophthalic, terephthalic, glutaric acids	$(CH_3)_4NOH$	Kreshkov[37]
	Mono-, dicarboxylic acids	$(C_2H_5)_4NOH$	Kreshkov[38]
	Oxalic, succinic acids	Coulometric	Jovanovic[39]
	Hydroperoxides, peracids	$NaOC_2H_4NH_2$	Martin[40]

Table 3.1 (*Contd.*)

Type of compound	Determination of	Titrant	Reference
	Naphthyl esters of carboxylic acids	$NaOCH_3$	Groagova[41]
	Malonic esters	$KOCH_3$	Zufgg[42]
	Mono-, disubstituted diethyl malonates	$KOCH_3$	Inczedy[43]
	Phenols, carboxylic acids	$(C_2H_5)_4NOH$	Harlow[44]
	Phenolic and carboxyl compounds in coal	$NaOCH_3$	Arita[45]
	Phenols, naphthols	$(C_2H_5)_4NOH$	Kreshkov[46]
	Phenols	$(C_4H_9)_4NOH$	Crabb,[47] Hummelstedt[48]
	Sterically hindered phenols, bisphenols	$(C_2H_5)_4NOH$	Kreshkov[49]
	Phenolic esters	$(C_4H_9)_4NOH$	Smith[50]
	Phenolic esters	$KOCH_3$	Glenn[51]
	Cresols, naphthols, pyrogallol, etc.	$NaOCH_3$	Karrman[52]
	Polyhydric phenols	$(C_4H_9)_4NOH$	Allen[53]
	Phenolic, carboxyl groups in same molecule	$KOCH_3$	Gyenes[54]
	Cresol isomers	KOH	Ershov[55]
	Naphthols, hydroxynaphthoic acids	$(C_4H_9)_4NOH$	Bykova[56]
	Hydroxyanthraquinones	$NaOCH_3$	Anastasi[57]
Nitrogen-containing	Amino acids	LiOH	Trnkova[58]
	Amino-, nitro-, hydroxybenzoic acids	$NaOCH_3$	Kreshkov[59]
	Carboxyamino acid anhydrides	$NaOCH_3$	Berger[60]
	Paracetamol, aspirin	$(C_4H_9)_4NOH$	Fogg[61]
	Glutethimide, amino-glutethimide	$NaOCH_3$	Agarwal[62]
	Imide oximes	$(C_4H_9)_4NOH$	Rauret[63]
	Amine reineckates	$NaOCH_3$	Jasinski[64]
	Hydroxamic acids	$KOCH_3$	Jaiswal[65]

Table 3.1 (*Contd.*)

Type of compound	Determination of	Titrant	Reference
	Nitro derivatives of aminobenzoic acids	$NaOCH_3$	Kreshkov[66]
	Nitro alcohols	$(C_2H_5)_4NOH$	Aksenenko[67]
	Nitrobenzoic acids	$NaOCH_3$	Kreshkov[68]
	Dinitrophenols, nitrobenzoic acids	$(C_2H_5)_4NOH$	Kreshkov[69]
	Nitroguanidines	$NaOCH_3$	De Vries[70]
	Pyridinecarboxylic acids	$(C_2H_5)_4NOH$	Kondratov[71]
	Pyridinedicarboxylic acids	$(C_2H_5)_4NOH$	Kreshkov[72]
	Barbituric acids	$NaOCH_3$	Swartz[73]
	Barbituric, acetylsalicylic acids	$NaOCH_3$	Lin[74]
Sulfur-containing	Thiols, halonitrobenzenes	$(C_2H_5)_4NOH$	Obtemperanskaya[75]
	Aliphatic, aromatic thiols	$(C_2H_5)_4NOH$	Obtemperanskaya[76]
	Thiols, phenols, carbazoles	$(C_4H_9)_4NOH$	Buell[77]
	Sulfomaleic, maleic anhydrides	$(C_2H_5)_4NOH$	Stognushko[78]
	Toluenesulfonic acids, H_2SO_4	$KOCH_3$	Filimonova[79]
	Sulfanilic, sulfsalicylic acids	KOH	Kreshkov[80]
	Aliphatic sulfonic acids, H_2SO_4	Diphenyl-guanidine	Popov[81]
	Aromatic sulfonic acids, H_2SO_4	$(C_2H_5)_4NOH$	Gribova[82]
	Phenolsulfonic acids, H_2SO_4	Piperidine	Emelin[83]
	Sulfonamides	$NaOCH_3$	de Reeder[84]
	Sulfonamides, imides, phenols,	$(C_4H_9)_4NOH$	Fritz[85]
	Sulfanilylbenzamide, sulfonamides	$NaOCH_3$	Amirjahed[86]
Unsaturated	Acrylonitrile, methacrylic acid, methyl methacrylate	$(C_2H_5)_4NOH$	Denes[87]
	Poly(acrylic acid)		Mandel[88]
	Methacrylic acids	Diphenyl-guanidine	Zarinskii[89]
Miscel-laneous	Halogenated acetic acids	KOH	Kreshkov[90]
	Perfluorodicarboxylic acids	$(C_2H_5)_4NOH$	Chelnokova[91]
	Trialkyl phosphites, H_3PO_3, H_3PO_4	$(C_4H_9)_4NOH$	Deal[92]

Table 3.1 (*Contd.*)

Type of compound	Determination of	Titrant	Reference
	Alkyl phosphates	KOH	Rublev[93]
	Fluorosilanes	$NaOC_2H_5$	Kreshkov[94]
	Triphenylsilanols	$(C_4H_9)_4NOH$	Schott[95]

3.3 SIMULTANEOUS DETERMINATION OF BASIC COMPOUNDS

At the present time, quantitative analysis of organic compounds depending on their basicity is usually carried out in nonaqueous media. Aqueous titrimetry is employed only in special cases. For instance, Krylova et al.[109] determined the primary and secondary amino groups in polyethylene polyamines by high-frequency titration in aqueous medium using 0.1 *N* H_2SO_4. Cantwell and Pietrzyk[110] studied potentiometric titration of monofunctional bases with HCl in aqueous solution containing a strong cation exchange resin (Li^+ form). Gruzdeva et al.[111] titrated pyridine bases with 0.5 *N* HCl in aqueous acetone medium.

The experimental operation of nonaqueous titration for bases is more convenient than that for acids, because it is not necessary to protect the system from atmospheric interference. For this reason, when a compound can be titrated either as an acid or as a base, the latter approach is generally preferred. Numerous pharmaceutical preparations contain basic substances. Perusal of the current pharmacopeias will reveal the prevalence of nonaqueous titrimetry. In a review on nonaqueous titration in pharmaceutical analysis published in 1972, Subert[112] cited 255 references that had appeared during a short period.

3.3.1 Direct Acid–Base Titration Methods

When the relative strengths of the basic compounds in a mixture are known,[113,114] it is possible to find out whether or not an experimental procedure can be worked out for simultaneous determination by differential titration. McCurdy and Galt[115] undertook a comprehensive study of the titration of weak bases, dissolved in dioxane–formic acid mixed solvent, with 0.1 *N* $HClO_4$ in a similar solvent; it was found that resolution of mixtures of bases differing by 1 $pK_{B\,b}$ unit can be achieved. Buell[116] performed differentiating titrations of basic nitrogen compounds in petroleum, using $HClO_4$ in acetonitrile and acetic anhydride media, to establish the pK_a range of these bases as a means of classification. Higuchi et al.[117,118] investigated the titration of amides and other very weak bases in acetic acid; precise titrations are possible for the more basic amides, while less basic compounds are determined with less precision. It should be recognized that the solvents mentioned above may react with the basic compounds and thus vitiate the analytical results. Therefore it is important to control the environment (see Chapter 1, Section 1.1.2.2) For example, if *N*-formyl esters are readily formed by the base in formic acid medium, the titration must be performed at 0°C.[115]

Some of the various techniques proposed to obtain the titration results are discussed below. Rehm and Higuchi[119] described an extrapolation plot for photometric titration of weak bases that involves the use of relatively weak basic indicators; the extrapolations of the stoichiometric end points from the absorbance–titration plots are from results well past the end points. Tamura[120] performed differential determination of nitrogen bases, their hydrochlorides, amino acids, and sulfonamides by the extinction ratio method; the indicators and the wavelengths for measuring the extinctions are reported. Kreshkov and Vasilev[121] determined mixtures of substituted anilines by titration with $HClO_4$ in acetic anhydride medium, using a single wavelength in the visible region. Jasinski and Andrulewicz[122] described photometric titration of weak bases with picric acid; the method is dependent on the formation of a yellow complex between picric acid and the base. Vajgand and Pastor[123] studied derivative polarographic titration of bases in acetic acid with 0.1 *N* $HClO_4$ using antimony or quinhydrone electrodes; the agreement between results obtained with potentiometric titration was within 0.4%. Waligora et al.[124,125] utilized an interface voltaic cell for the potentiometric titration of organic bases in nonconducting media (e.g., anhydrous acetone, benzene, chloroform) with picric, trichloracetic, or *p*-toluenesulfonic acid. Riolo et al.[126] titrated amines, amino alcohols, and *N*-heterocyclics in acetonitrile medium using Lewis acids (e.g., $AlBr_3$, $GaBr_3$, BBr_3); the end points were located either potentiometrically or conductometrically.

Table 3.2 gives examples of nonaqueous titrations of basic compounds. In a few cases it is recommended to add a known excess of standardized acid to the sample and then back-titrate with standardized base. This procedure is particularly suited for some salts[127] (e.g., tartrates) that are poorly soluble in acetic acid and are not readily determined by direct titration with $HClO_4$. The choice of solvent varies with the investigators. For instance, Kreshkov et al.[182,183] reported that ketones possess the best properties for differential

Table 3.2 Differential Nonaqueous Titration of Organic Compounds as Bases

Type of compound	Determination of	Titrant[a]	Reference
Salts	Alkali carboxylates—add $HClO_4$	[CH_3COONa]	Blake[127]
	Acetate, oleate, benzoate, anthranilate—add $HClO_4$	[$(C_2H_5)_4NOH$]	Kreshkov[128]
	Salts of mono-, dibasic acids—add H_2SO_4	[$(C_2H_5)_4NOH$]	Kreshkov[129]
	Salts of acids—add $HClO_4$	[$(C_2H_5)_4NOH$]	Kreshkov[130]
	Sodium barbiturates, benzoate, salicylate	$HClO_4$	Sell[131]
	Acetates of rare-earth metals	$HClO_4$	Kreshkov[132]
	Sodium, potassium, and lithium salts of aliphatic, aromatic acids	$HClO_4$	Yarovenko[133]
	2-, 3-, 4-Hydroxybenzoates	HCl	Kreshkov[134]
	p-Substituted benzoates	HCl	Kreshkov[135]
Nitrogen bases	Aliphatic, aromitic amines	$HClO_4$	Fritz[136]
	Aliphatic amines	$CH_3C_6H_4SO_3H$ etc.	Kwiatkowski[137]
	Aliphatic, aromatic mono-, diamines	$HClO_4$	Kreshkov[138]
	Aromatic primary, secondary, tertiary amines—add methanolic HCl	[KOH]	Demidova[139]
	Aliphatic amines, aniline	$HClO_4$	Kolthoff[140]
	Tertiary amines	$HClO_4$	Fritz[141]
	Primary, secondary, tertiary amines	$HClO_4$	Kreshkov[142]
	Ethanolamine, amine carbonate	HCl	Chang[143]
	Diamines	$HClO_4$	Kreshkov[144]
	Ethylenediamine-NNN′N′-tetra-[2-(-2-propoxyethoxy)ethanol]-derivatives	$HClO_4$	Cribova[145]
	Alkyl and aryl amines, amino acids, amino alcohols	$HClO_4$	Ballczo[146]
	Amino acids, amino phenols	$HClO_4$	Das[147]
	Anisidines, toluidines, chloro-anilines	$HClO_4$	Rodziewicz[148]
	Aromatic amines	$HClO_4$	Hummelstedt[149]

Table 3.2 (*Contd.*)

Type of compound	Determination of	Titrant[a]	Reference
	Aniline, diethylaniline, nitroaniline	$HClO_4$	Minczewski[150]
	Nitrotoluidines	$HClO_4$	Kreshkov[151]
	Nitroanilines	$HClO_4$	Minczewskii[152]
	Butylamine, methylaniline, dimethylaniline	HCl	Bournique[153]
	Primary, secondary, tertiary amides	$HClO_4$	Wimer[154]
	Caprolactam, aminocaproic- [6-aminohexanoic] acid		Giuffre[155]
	Salts of lactams, amino acids	HCl	Galpern[156]
	Amines, alkaloids	HCl	Evstratova[157]
	Pyridine, quinoline derivatives	$HClO_4$	Kondratov[158]
	Pyridinecarboxylic acids	$HClO_4$	Kondratov[159]
	Isoniazid, sodium aminosalicylate	$HClO_4$	Devani[160]
	1,4-Disubstituted piperazines	$HClO_4$	Ciaccio[161]
	Methylxanthines	$HClO_4$	Kashima[162]
	N-Heterocyclics	$HClO_4$	Kreshkov[163]
	Alkaloids	Naphthalene-2-sulfonic acid	Udovenko[164]
	Theobromine, diprophylline	$HClO_4$	Salvesen[165]
	N-Heterocyclics, *N*-oxides	$HClO_4$	Kondratov[166]
	Phenazine, benzophenazine, *N*-oxides	$HClO_4$	Riolo[167]
	Amines, *N*-oxides	$HClO_4$	Bezinger[168]
	Schiff bases	$HClO_4$	Freeman[169]
	Azoles	$HClO_4$	Veibel[170]
	Caffeine, pentetrazol	$HClO_4$	Beyrich[171]
	Bases in tobacco smoke	$HClO_4$	Warner[172]
	Antibiotics Aureomycin, Terrmycin	$HClO_4$	Sideri[173]
	Picrates of aniline, pyridine, antipyrine	$HClO_4$	Jasinski[174]
	Amidopyrine, nicotinate of guaiacyl salicylate, lignocaine hydrochloride	$HClO_4$	Tortolani[175]

Table 3.2 (*Contd.*)

Type of compound	Determination of	Titrant[a]	Reference
	Urea, diethylamine, nitroaniline	$HClO_4$	Shkodin[176]
	Pyridyl-substituted ureas	$HClO_4$	Kondratov[177]
Sulfur compounds	Sulfoxides, alkyl, aryl, and heterocyclicamines, carboxylic amides	$HClO_4$	Bezinger[178]
	Sulfonamides	$HClO_4$	Tajika[179]
	Aminophenyl sulfides, amino-phenyl sulfones	$HClO_4$	Galpern[180]
Miscellaneous	Organophosphorus compounds	$HClO_4$	Lane[181]

[a]Brackets indicate reagent for back titration.

titration and that ethyl methyl ketone is better than acetone or butyl methyl ketone. By contrast, Kokot et al.[148,174] recommended tetraalkoxysilanes as the best media for titrating mixtures of substituted anilines or picrates of nitrogen bases. Whereas $HClO_4$ is the titrant most frequently employed, Kwiatkowski et al.[137] used 0.1 *M* solution of picric acid, *p*-toluenesulfonic acid, sulfuric acid, or hydrochloric acid in dimethyl sulfoxide.

3.3.2 Indirect Acid–Base Titration Methods

In the analysis of some types of mixtures by nonaqueous titrimetry, certain chemical reactions are carried out before the titration can be performed. Some examples are given in Table 3.3, which lists mixtures of basic compounds. It is seen in the table that salicylaldehyde is used as reagent in many cases. By such treatment the basic strength of one or more compounds in the mixture is altered so that the remaining compound can be differentially titrated. No separation procedure (see Chapter 4) is involved. For instance, Malone[196] described the following method for the determination of mixtures of hydrazine and *N*,*N*-dimethylhydrazine. Both compounds are determined by titrating an aliquot of the sample in acetic acid solution, with 0.1 *N* $HClO_4$ in dioxane, to the methyl violet end point. Salicylaldehyde is added to another aliquot and the basic hydrazone formed with the dimethylhydrazine is titrated in a similar manner; a neutral azine is formed with the hydrazine, which is not titrated. Burns and Lawler[197] modified this method to improve the reproducibility by using $HClO_4$ in acetic acid and determining

Table 3.3 Examples of Indirect Nonaqueous Acid–Base Titrimetry

Mixture containing	Treatment prior to titration	Titrant	Reference
Primary, secondary, tertiary amines	Salicyaldehyde	$HClO_4$	Critchfield[184]
	Phthalic or acetic anhydride	$HClO_4$	Galpern[185]
Primary, secondary fatty amines	Salicylaldehyde	$HClO_4$	Jackson[186]
Amino groups in polyamines	Hg acetate, acetic anhydride, salicylaldehyde	$HClO_4$	Strepikheev[187]
Mono-, di-, triethylamines	Salicylaldehyde	HCl	Ormanets[188]
Cyclohexylamines, aniline	Salicylaldehyde	$HClO_4$	Gribova[189]
Mono-, di-, triethanolamines	Acetylacetone	HCl	Gribova[190]
	Salicylaldehyde	$HClO_4$	Gribova[191]
Fatty amines, quaternary ammonium salts	KI, Hg acetate	$HClO_4$	Blakeley[192]
Diethylenetriamine, hydrazines	Salicylaldehyde	HCl	Malone[193]
Methylhydrazine, hydrazine	Salicylaldehyde	$HClO_4$	Hanna[194]
Mono-, dimethylhydrazines	Phenylisocyanate	$HClO_4$	Malone[195]
Dimethylhydrazine, hydrazine	Salicylaldehyde	$HClO_4$	Malone,[196] Burns[197]
Amidopyrine, amethocaine	Hg acetate	$HClO_4$	Dematrescu[198]
Codeine, quinine, ephedrine	Hg acetate	$HClO_4$	Popper[199]
Sulfates of organic bases	Benzidine	$HClO_4$	Gautier[200]

the end points either potentiometrically or spectrophotometrically with crystal violet as indicator.

For the alkalimetric assay of sulfates of organic bases in nonaqueous solvents, Gautier and Pellerin[200] observed that sulfuric acid is too strongly acidic to permit estimation of these salts by direct titration with 0.1 *N* $HClO_4$ in acetic acid. Conductivity experiments show that the sulfate concentration may be sufficiently reduced by the addition of benzidine equivalent to 95% of the sulfate present, forming the insoluble benzidine sulfate. After the precipitate is allowed to settle, the supernatant liquid can be titrated without difficulty.

For some mixtures, prior treatment makes possible simultaneous determination of several constituents without separation. For example, Strepikheev et al.[187] determined primary, secondary, and tertiary amino groups in polyamines by using the following scheme. The sample is dissolved in acetic acid and 6% mercuric acetate is added to prevent interference from chloride ions. Acetone is added to an aliquot of this solution, and the sum of all three amino groups is determined by titration with $HClO_4$. To a second aliquot are added acetic anhydride (to react with the primary and secondary amino groups) and acetone, and to a third aliquot are added salicylaldehyde (to react with primary amino group) and acetone; these solutions are then similarly titrated. Ormanets et al.[188] described a simple method to determine mono-, di- and triethylamines in a six-component mixture obtained in the catalytic deamination of aliphatic amines, using chloroform as the solvent and HCl in isopropyl alcohol as the titrant. It should be pointed out that gas-chromatographic separation (see Chapter 4) may be more convenient in this case; however, gas chromatography requires much preliminary work and considerably more expensive equipment.

3.4 SIMULTANEOUS DETERMINATION BY CHEMICAL METHODS OTHER THAN DIFFERENTIAL ACID–BASE TITRATION

3.4.1 Methods That Use a Single Sample of the Mixture

Besides acid–base titrimetry, all chemical methods for determining individual compounds[3] can be utilized for the analysis of mixtures. While it is not common to find experimental procedures that can achieve total analysis of a mixture by using a single sample, there are such possibilities. Some examples are cited below.

Using Cu^{2+} for precipitation titration of carboxylic acids and salts, Ashworth and Fehringer[201] found that fatty acids from lauric to stearic can be determined potentiometrically in the presence of an equivalent amount of a lower acid, and picric acid can be titrated in the presence of 2,4-dinitrophenol or 2,4,6-tribromophenol; the Cu^{2+} titrant is effective because of its relatively specific precipitating power. Sass[202] used $AgNO_3$ for the potentiometric titration of the potassium salts of long-chained fatty acid mixtures containing up to seven components. The acid sample is titrated with ethanolic KOH against phenolphthalein; then the potassium salt solution is titrated with 0.1 N $AgNO_3$ in 0.25 ml additions against a calomel electrode.

Barcza[202a] determined carboxylic chlorides and their decomposition products (free organic acid and HCl) as follows. The sample in benzene solution is titrated with standardized solution of amidopyrine to a dimethyl yellow end point, giving the HCl content. The solution is then heated with methanol, and the acid chloride is determined by titrating the liberated HCl in the same way. Subsequently the solution is titrated with methanolic KOH using phenolphthalein indication. The free organic acid present is equivalent to the difference between the last titration and the sum of the others.

Bork et al.[203] described the simultaneous amperometric titration of a ketone and an aldehyde with 0.1 *M* $NH_2OH \cdot HCl$ in isopropyl alcohol; the titration curves showed two end points, the first corresponding to the aldehydes and the second to the ketone.

For the determination of acetaldehyde in ethylene oxide, Reid and Salmon[204] delivered the sample into Na_2SO_3 solution and titrated with 0.1 *N* iodine to the starch end point; then they added solid $NaHCO_3$ and continued the titration. The difference between the two titers is equivalent to the acetaldehyde content.

Maros et al.[205] determined aldonic acids and sugar dicarboxylic acids simultaneously by periodate oxidation. Each aldonic acid yields 1 mole of CO_2 and formaldehyde, while each dicarboxylic acid yields 2 moles of CO_2. Thus the determination of the amounts of CO_2 and formaldehyde formed enables the composition of the mixture to be calculated.

Horacek and Pribil[206] determined EDTA and NTA (nitrilotriacetic acid) by potentiometric titration with $FeCl_3$; the first inflection indicates EDTA and the second indicates NTA. Saxena[207] determined glutamic acid, valine, and alanine by first titrating glutamic acid with Na_2WO_4 to a pink color with catechol violet, then titrating alanine with $HAuCl_4$ to a violet color with Congo red, and finally titrating valine with K_2TeO_3 to a red color. For asparagine mixed with valine, Saxena[208] used xylenol orange as indicator, titrated the former acid with $In_2(SO_4)_3$ to a rose color, and continued to titrate the latter acid with K_2TeO_3 to a purple color.

Jurecek et al.[209] described simultaneous determination of *N*-ethyl, amino, and nitro groups by chromic acid oxidation; the *N*-ethyl group yields CH_3-$COOH$, the amino group yields NH_3, and the nitro group yields HNO_3. Belcher et al.[210] determined *O*-methyl and *N*-methyl groups through HI cleavage first at 150° C and then at 360° C. By coupling the reaction vessel with a gas chromatograph, Ma et al.[211,212] demonstrated the simultaneous determination of several *O*-alkyl and of several *N*-alkyl groups using one sample of the mixture.

Polyanskii[213] used a single sample to determine diketene mixed with crotonaldehyde as follows. Water is added to the mixture whereby diketene is converted into acetoacetic acid, which is then titrated with alkali. The solution is acidified with H_2SO_4 and boiled to destroy the acetoacetic acid, leaving behind crotonaldehyde, which is determined by bromine addition.

Filimonova et al.[214] described a method for simultaneous determination of chloroacetyl chloride and disulfur dichloride that involves hydrolysis of the compounds in aqueous acetone, followed by sequential potentiometric titration of the resulting HCl, $ClCH_2COOH$, and H_2S with methanolic KOH.

Jordan[215] observed that trialkylaluminum and dialkylaluminum hydride can be titrated sequentially with 0.2 *M* pyridine in xylene. The indicator phenazine forms red to brown complexes with trialkylaluminum compounds and green to blue complexes with dialkylaluminum hydrides, so that two end points are obtained. For the same mixtures, Shtifman et al.[216] employed quinoline as reagent to form complexes with different conducting properties. During the course of titration with 0.1 *N* quinoline in petroleum ether, the conductivity of the solution first increases because of the formation of $AlR_3C_9H_7N$, passes through a maximum, and then falls because of the formation of the nonconducting $AlR_2H \cdot C_9H_7N$. As the conducting $AlR_2H \cdot 2C_9H_7N$ forms, the conductivity again increases and passes through a second maximum and then decreases once more because of the low conductivity of the added quinoline solution.

Since enzymatic methods are specific, each compound in a mixture is usually determined separately. However, there are cases where several compounds can be determined simultaneously. For instance, Drawert and Kupfer[217] devised a single procedure to determine glucose, fructose, and sucrose in wine as follows. Glucose and fructose are converted into the corresponding 6-phosphates in the presence of hexokinase and ATP. On addition of glucose-6-phosphate dehydrogenase, the glucose-6-phosphate is broken down in the presence of NADP to release NADPH, which can be determined by measuring the absorbance at 340 or 366 nm. Similar reduction of NADP by fructose-6-phosphate occurs on addition of glucosephosphate isomerase. Sucrose is determined after inversion to glucose and fructose.

3.4.2 Complete Analysis Of a Mixture Using Two or More Samples

Generally speaking, most methods for the determination of all constituents in a mixture entail the use of two or more samples of the mixture. Table 3.4 gives some examples of these methods, arranged by the functional groups present in the respective mixtures. It should be mentioned that the successful analysis of certain combinations (e.g., primary, secondary, and tertiary amines) was achieved a long time ago. Because of their importance and common occurrence, however, new procedures have continued to evolve.

A can be seen in Table 3.4, separate determinations of one or more of the components in a mixture generally utilize simple chemical methods that are employed for the determination of single compounds or monofunction systems. These methods are straightforward in operation. On the other hand, sometimes relatively complicated procedures are involved. For

Table 3.4 Complete Analysis of a Mixture Using More than One Sample Without Separation

Mixture containing	Number of samples needed	Reference
	Oxygen Functions	
Monoalkoxy benzenes (a) dialkoxy benzenes (b)	(1) a→CH_3COOH;(2) a+b by bromination	Mikhailova[218]
4-Methoxyphenol (a), quinol (b)	(1) a+b by $K_2Cr_2O_7$; (2) b by NH_4VO_3	Karpov[219]
Formyl group (a), acetyl group (b)	(1) a→HCOOH; (2) b→CH_3COOH	Kan[220]
Acetic acid (a), acetic anhydride (b), vinyl acetate (c), ethylidene diacetate (d) acetaldehyde (e)	(1) a by NH_4OH; (2) b by aniline; (3) c by Br_2; (4) d by saponification; (5) e by $NaHSO_3$	Capitani[221]
Glutaraldehydic acid (a), trimer (b), lactone of monomer (c)	(1) a+b by NaOH, then c by hydrolysis; (2) a by NH_2OH	Mikhno[222]
CH_3OH (a), HCHO (b), HCOOH (c)	(1) b by OI^-; (2) b+c by $BrO_3^- - Br^-$; (3) a+b+c by $Cr_2O_7^{2-}$	Szekeres[223]
Carboxylic acid (a), anhydride (b)	(1) a+b by NaOH; (2) b by morpholine	Johanson[224]
Maleic acid (a), anhydride (b)	(1) a+b by NaOH; (2) b by $NaOCH_3$	Huhn[225]
Maleic acid or anhydride (a), other dibasic acid (b)	(1)a—insoluble anthracene adduct, b by NaOH; (2) a+b by NaOH	Nebbia[226]
acetyl chloride (a) anhydride (b)	(1) a by Ag acetate; (2) b by aniline	Stürzer[227]
Acyl chloride (a), carboxylic acid (b) mineral acid (c)	(1) a+b+c by NaOH; (2) by HOC_2H_4OH, a→HCl, b→ester; (3) b by aniline	Markevich[228]
Aldehyde (a), acetal (b)	(1) a by aniline; (2) a+b by NaOH	Skvortsova[229]
Carbonyl (a), acetal (b)	(1) a+b→dinitrophenylhydrazones; (2) a reduced by $NaBH_4$, then b→hydrazone	O'Connor[230]
Ketone (a), ketal (b)	(1) a+b by NH_2OH after hydrolysis; (2) a by NH_2OH	Skvortsova[231]
Aldehyde (a), ketone (b)	(1) a+b by NH_2OH; (2) b by aniline→H_2O	Petrova[232]

Table 3.4 (*Contd.*)

Mixture containing	Number of samples needed	Reference
HCHO (a), CH_3CHO (b)	(1) a+b by NH_2OH; (2) b by polarography	Sikorska[233]
Epoxyethane (a), HCHO (b), CH_3CHO (c)	(1) a by KI–HI; (2) b+c by NH_2OH; (3) b by Schiff's reagent	Samokov[234]
HCHO (a), $(CH_2O)_n$ (b)	(1) a by dimedone; (2) a+b by I_2 oxidation	Bellen[235]
Glyoxal (a), CH_3CHO (b)	(1) a by cyclohexylamine; (2) a+b by NH_2OH	Simionvici[236]
Glyoxal (a), HCHO (b) glycolaldehyde (c)	(1) a by nitrobenzoic hydrazide; (2) b by napthol; (3) c by polarography	Brodski[237]
Glyoxal (a), glyoxalic acid (b)	(1) a+b by semicarbazide; (2) a by excess NaOH	Salzer[238]
Fatty acid (a), aldehyde (b)	(1) a by NaOH; (2) b→acid, then by NaOH	Metcalfe[239]
HCHO (a), other aldehyde (b), H_2O_2 (c)	(1) a+b by oxidation; (2) a by Schiff's reagent; c by I_2	Satterfield[240]
D-galactose (a), D-arabinose	(1) a by D-galactose dehydrogenase; (2) b by D-arabinose dehydrogenase	Hu[241]
Ethanol (a), butanol (b)	(1) with *Bacillus pasteurianum*, a→C_4H_7COOH, b→CO_2; (2) with AsO_3^{3-}, a→C_4H_7COOH, b→CH_3COOH	Rosenfeld[242]
Acetone (a), ethanol (b), butanol (c)	(1) a by I_2; (2), (3) a+b+c by $K_2Cr_2O_7$ in 30 and 67% H_2SO_4, respectively	Nakhmanovich[243]
HCOOH (a), CH_3OH (b)	(1) a by Ag(III); (2) a+b by Cu(III)	Jaiswal[244]
Ethanol (a), lactic acid (b)	(1) a by $K_2Cr_2O_7$; (2) b by $KMnO_4$→CH_3CHO	Kovacs[245]
Monoglyceride (a), glycerol (b)	(1) a+b by HIO_4 in CH_3OH; (2) b by aq. HIO_4 in $CHCl_3$–H_2O two-phase system	Kruty[246]

Table 3.4 (*Contd.*)

Mixture containing	Number of samples needed	Reference
	Nitrogen Functions	
Primary (a), secondary (b), tertiary (c) amines	(1) a+b+c by $HClO_4$; (2) c by $HClO_4$ after acylation of a+b	Szepesi[247]
Primary (a), secondary (b), tertiary (c) amines	(1) a+b by KIO_3 via substituted thioureas; (2) c by Cl_3CCOOH	Verma[248]
Aniline (a), diphenylamine	(1) a by $HClO_4$; (2) b by $NaNO_2$	Utkin[249]
Aniline (a), methylaniline (b), dimethylaniline (c)	(1) a with naphthol; (2) c with diazotized nitroaniline; (3) a+b with phenol	Belyakov[250]
Dimethylamine (a), its formate (b), formic acid (c)	(1) a, b by $HClO_4$; (2) b, c by KOH	Turyan[251]
Complexans EDTA (a), DTPA (b), TTHA (c)	(1) By $Th(NO_3)_2$, a+b+c at pH 2.5; (2) b+c at pH 5, then a at pH 3; (3) c by $Zn(NO_3)_2$	Pribil[252]
Surfactants cationics (a), nonionics (b)	(1) By Na tetraphenylborate, a with methyl red indicator; (2) b with Congo red	Uno[253]
Amino (a), carboxyl (b) groups in polyamides	(1) a by HCl; (2) b by $(C_4H_9)_4NOH$	Witek[254]
Amine HCl (a); carbamyl chloride (b)	(1) a+b by $AgNO_3$; (2) b by diisobutylamine	Lesiak[255]
p-nitrotoluenes (a), *m*-nitrotoluenes (b)	After Zn–H_2SO_4 reduction, (1) by $NaNO_2$; (2) by $KBrO_3$–KBr	Portnov[256]
Nitro groups (a), nitroso (b) groups in polymers	Coulometrically by Ti^{3+}, (1) a+b in citrate; (2) b in HCl	Mitev[257]
Cyanoethoxy groups (a) methoxy (b) groups in cellulose ethers	(1) a by Kjeldahl; (2) b by Zeisel	Kozlov[258]
1-(2-hydroxypropyl)-theobromine (a), 7-(2-hydroxyethyl)theophylline (b)	(1) a by $Ce(SO_4)_2$; (2) b by $KBrO_3$–KBr at 50°C	Raber[259]
Penicillin (a), dihydro-streptomycin (b)	(1) a by *Staphyococcus aureus*; b by *Bacillus subtilis*	Lewis[260]
	Sulfur Functions	
Thiols (a), disulfides (b)	(1) a by $AgNO_3$; (2) b→thiol, then by $AgNO_3$	Kolchina[261]

Table 3.4 (*Contd.*)

Mixture containing	Number of samples needed	Reference
Alkane thiol (a), dialkyl sulfide (b)	(1) a by I_2; (2) a+b by $KBrO_3$–KBr	Jaselskis[262]
Alkane thiol (a), alkyl disulfide (b)	(1) a by I_2; (2) b→thiol, then by I_2	Jaselskis[262]
Mercaptoacetic acid (a), sulfides (b), sulfites (c), thiosulfate (d)	(1) a+b by Na 2-hydroxymercuri-benzoate; (2) b, after adding *N*-vinyl cyanide; (3) d by I_2; (4) a+b+c+d by I_2	Wronski[263]
Cysteine (a), mercaptoacetic acid (b), cyanide (c), dithiodiacetic acid (d)	(1) a+b by hydroxymercuri-benzoic acid (*e*); (2) a add HCHO–NaOH, then *e*; (3) a+c by $AgNO_3$; (4) d→sulfide, then by *e*	Wronski[263a]
Cysteine (*a*), cystine (*b*)	By *N*-bromosuccinimide, (1) a directly; (2) a+b after enzymatic reduction of b o a	Thibert[264]
N-Substituted aminoethyl-thiosulfonic acids (a), their sodium salts (b). thiols (1)	(1) a, c by $(C_4H_9)_4NOH$ sequentially; (2) b by $HClO_4$	MacDonald[265]
1-Naphthol-3,6,8-trisulfonic acid (a), chromotropic acid (b)	(1) a+b by 4-nitrophenyldiazon-ium chloride; a+b by I_2	Spiliadis[266]
Phenylthiourea (a), mercaptoacetic acid (b), cysteine (c)	By coulometry, a, b, c under different conditions	Santhanam[267]
Thiourea (a), formamidine disulfide (b)	(1) a by KIO_3; (2) b→a by Zn–H_2SO_4, then by KIO_3	Verma[268]
Thiourea (a), xanthate or dithiocarbamate (b)	By IBr, (1) a in aq. H_2SO_4; (2) b in CH_3CN	Verma[269]
Thiourea (a), Na_2S (b), cyanamide (c)	By excess $AgNO_3$, (1) a+b+c; (2) a+c after bringing to pH 1 and expelling H_2S; (3) a after adding conc. HCl and expelling H_2S	Kramareva[270]
	Unsaturated Functions	
α-Methylstyrene (a), α,α-dimethylbenzyl alcohol (b)	By excess $HgSO_4$, (1) a at 0°C; (2) a+b at 100°C	Polyanskii[271]
But-2-ene-1,4-diol (a), but-2-yne-1,4-diol (b)	(1) a by $KBrO_3$–KBr; (2) a+b by Br_2	Sivakova[272]

Table 3.4 (*Contd.*)

Mixture containing	Number of samples needed	Reference
	Miscellaneous Functions	
Phenol (a), resorcinol (b)	(1) a+b by $KBrO_3$–KBr; (2) b by I_2	Sobczak[273]
Phenol (a), quinol (b), benzoquinone (c), maleic anhydride (d)	(1) a+b by $KBrO_3$–KBr; (2) b by Ce^{4+}; (3) c by KI; (4) d by NaOH	Badarinarayana[274]
Di- (a), trichloroacetaldehyde (b)	Simultaneous equations for (1) a+b by excess KOH; (2) a+b by excess I_2	Malhotra[275]
$CHCl_3$ (a), CCl_4 (b)	By Cr^{2+}, (1) a at 80°C; (2) at 20°C	Kiba[276]
di- (a), trialkyl phosphites (b)	(1) a by ethanolic NaOH; (2) b→a in aci	Bernhart[277]
Mercuriacetic acid (a), phenylmercury acetate (b)	Boiling with CH_3COOH–NaCl, then by sodium mercaptoacetate, (1) a+b after Br_2–Na_2SO_3 treatment; (2) b without treatment	Wronski[278]

instance, Nettesheim[279] described the following procedure for the determination of mixtures of aldehydes, organic peroxides, and hydrogen peroxide in aqueous solution. (1) The total peroxide content is obtained by heating a sample with KI in acid medium and titrating the liberated iodine with $Na_2S_2O_3$. (2) In another sample, the free H_2O_2, together with the H_2O_2 combined with the aldehyde, is determined by either (*a*) measurement in a gas burette of the hydrogen evolved when the mixture (primary and secondary alkyl hydroperoxides being absent) is treated with alkaline formaldehyde, or (*b*) measurement of the yellow color produced when the mixture (primary and secondary hydroperoxides being present) is treated with an acid titanyl sulfate solution. The organic peroxides are calculated by difference. (3) Using a third sample, the aldehydes are determined by passing the material through a column containing AgO and titrating the silver salts of the fatty acids so formed with KSCN.

Horner and Jürgens[280] proposed a method for the determination of organic per compounds in mixtures that is based on the observation that both diphenyl sulfide and triethyl- (or triphenyl-) arsine react specifically with certain groups of per compounds. By combining these reactions with the iodometric and acidimetric procedures, it is possible to determine the composition of mixtures of four groups of compounds, namely, binary, tertiary, and quaternary mixtures containing alkyl hydroperoxides, dialkyl peroxides, per acids, and diacyl peroxides.

It is noted from Table 3.4 that there are a few methods in which different samples are treated with the same reagents but under diverse conditions.[271,276] In these cases, separate determination of the individual components is achieved because of the large difference in the reaction rates under the specific experimental procedures. Further discussion of this subject is given in the next section.

3.5 ANALYSIS OF MIXTURES BASED ON DIFFERENTIAL REACTION RATES

Since chemical methods for the determination of organic compounds are based on the reactions of the functional groups,[3] it is evident that, in the analysis of mixtures, all components that possess the same functional group are expected to undergo similar reactions when treated with a particular reagent. Because of differences in molecular structures, however, two compounds carrying the same functional group may exhibit markedly different reaction rates under the specified experimental conditions. In the case of quantitative analysis of single substances, it is customary to choose an extreme condition that ensures that the reaction proceeds to completion irrespective of structural variations. By contrast, the opposite approach is usually taken in the analysis of mixtures when similar compounds are known to be present in the sample; thus an effort is made to exploit the difference in reaction rates so that the ratio of the two compounds possessing the same functional group can be ascertained.

The analytical methods based on differential reaction rates may be classified into two categories. In the first category, separate determination methods, one component of the binary mixture is reacted with the reagent under conditions that cause the reaction to proceed to completion with respect to that compound, while the other component remains practically unaffected. Measurements are made only at the end of the reaction. In the second category, kinetic methods, both components react with the reagent simultaneously and measurements are taken during the course of the experiment; the proportion of the two components in the original sample is then calculated by extrapolation.

3.5.1 Separate Determination Methods

It is obvious that the separate determination method requires the reaction rates of the components in the sample to be greatly different towards the specific reagent under appropriate experimental conditions. Some examples are given below.

According to Jaiswal and Chandra, a mixture containing formic and acetic acids can be analyzed in one of the two ways based on selective oxidation. In one method,[281] formic acid is selectively oxidized by Ag(III) at 30°C in the presence of OsO_4 as catalyst, while both acids are oxidized by Cu(III) at 100°C; the unconsumed Ag(III) in the first aliquot is titrated iodimetrically, while the unconsumed Cu(III) in the second aliquot is titrated by the $NaAsO_2$–iodine procedure. In another method,[282] the sum of formic and acetic acids is determined in the same manner, while formic acid is oxidized with $Ce(SO_4)_2$ in 8 *N* H_2SO_4 by heating for 15 hr in the presence of $K_2SO_4 \cdot Cr(SO_4)_3$ as catalyst. On the other hand, when formic acid is mixed with oxalic acid,[283] both acids are oxidized by $Ce(SO_4)_2$, while formic acid alone is oxidized by Ag(III).

Polyanskii[284] described a method for determining acetaldehyde and paraldehyde in mixtures as follows. An aliquot is reacted with *M* NH_2-$OH \cdot HCl$ for 30 min at ambient temperature and the HCl evolved is titrated with aq. KOH; this procedure gives the content of acetaldehyde. Another aliquot is heated at 90°C for 1 hr with $NH_2OH \cdot HCl$, giving the sum of acetaldehyde and paraldehyde.

Based on differences in rate of oximation, Fowler[285] determined vanillin in the presence of acetovanillone, and *p*-hydroxybenzaldehyde mixed with 4-hydroxy-3,5-dimethoxyacetophenone. Shahak and Bergmann[286] reported that aldehydes react with 1,2-dimethyl-4,5-di(mercaptomethyl)benzene in HCl to produce cyclic thioacetals in 2 to 5 min, while ketones require 30 min in a solution containing $ZnCl_2$ and HCl. In the determination of aldoses with hypoiodide, Collins[287] observed that mannose reacts more slowly than many other hexoses, and secondary reactions become appreciable as the oxidation approaches completion; hence conditions must be established to analyze mixtures.

Hallam[288] investigated the determination of amides by KOH hydrolysis and diffusion into boric acid solution containing methyl red–bromocresol green indicator.[289] They found that formamide needs 3 hr, acetamide requires 5 hr, and *N*-methylacetamide needs 32 hr; thus the rate differences are sufficient for differential analysis. Lohman and Mulligan[290] described methods for the determination of alkanolamides in mixtures as follows. Each mole of fatty alkanolamide can be saponified with alcoholic KOH to give 2 moles of weak base (1 mole of alkanoate and 1 mole of dialkanolamine), while other compounds present in commercial alkanolamides (e.g., amines, amine salts) yield only 1 mole of weak base per mole of compound. The formation of extra base is utilized to follow the saponification of commercial mono- and di-ethanolamides. It has been found that the reaction rate for lauric diethanol-amide is 70 times as fast as that for the monoethanolamide.

The difference in degree of hydrolysis was used by Stroh and Liehr[291] to determine *N*- and *O*-acetyl groups in sugar hydrazine acetates. On the other hand, Malone and Biggers[292] used the difference in rates of acetylation to determine mixtures of methylhydrazine and 1,1-dimethylhydrazine. Methyl-hydrazine is acetylated immediately, whereas 1,1-dimethylhydrazine reacts very slowly.

Nutiu and Bokenyi[293] determined mercaptoacetic and dithiodiacetic acids in admixture using excess 0.1 *N* iodine; only the former reacts in 5% HCl medium, whereas both acids react in 10% HCl. De Reeder[294] analyzed mixtures of sulfonamides by bromometric methods and by nitrite titration of the free aromatic-NH_2 group. Stroehl and Kurzak[295] determined anionic surfactants containing sulfonate- and sulfate-types by hydrolyzing the sulfonates with aq. NaOH and selectively hydrolyzing the sulfates with 3 *M* HCl.

Depending on the number of positions available for bromination, Spliethoff and Hart[296] analyzed certain alkylated phenol mixtures with the $KBrO_3$–KBr reagent, using a limited excess of the reagent and short reaction times. Zelenetskaya et al.[297] described a method to determine *m*- and *p*-cresols in admixture that is based on the fact that the *m*-isomer reacts with $(SCN)_2$ considerably more rapidly than does the *p*-isomer, whereas both compounds are brominated at about the same rate.

3.5.2 Kinetic Methods

The reader is referred to the publications of Siggia et al.[298] on the techniques and applications of kinetic measurements for chemical analysis and those on differential kinetic analysis of mixtures using the continuous flow method.[299] The synergistic effects of two reactions proceeding simultaneously in the same medium deserve attention.[300] In some mixtures the effect of the presence of the faster reacting component is to decrease the rate of reaction of the slower component; this works to advantage for the graphical method, since it increases the difference between the two slopes of the rate plots. On the other hand, for methods that rely on rate constants for calculation of the components, this effect can cause difficulties. Reviews on kinetic methods have been written by Yatsimirskii,[301] Legradi,[302] and Mottola,[303] the last paper being concerned with the determination of organic ligands by their modification of the rate of metal-catalyzed reactions.

Table 3.5 Examples of Kinetic Methods for Binary Mixtures

Function	Compounds	Reaction	Reference
Oxygen	Carbonyl compounds	Oximation or semi carbazone formation	Papa[304]
	Sugars	Tetrazolium reduction	Mark[305]
	Cortisone, hydrocortisone	Tetrazolium reduction	Guttman[306]
	Hydroxyl compounds	With phenyl isocyanate	Willeboordse[307]
	Primary, secondary butyl alcohols	Acylation	Reilley[308]
	Ethanol, propanol	With alcohol dehydrogenase	Mark[309]
	Glycols	Cleavage with Pb^{4+}	Beuson[310]
Nitrogen	Primary amines	With salicylaldehyde	Shresta[311]
	Secondary amines	With methyl acrylate	Shresta[311]
	Diazonium compounds	Thermal decomposition	Siggia[312]
	Pheno-, pento-barbitones	With 0.1 *M* Na_3PO_4	Chafetz[313]
	L-, D-Propicillins	Decomposition by penicillinase	Zuidweg[314]
Sulfur	Methyl, ethyl methanesulfonates	With butanol	Brook[315]
	Sulfonphthaleins	Oxidation by $H_3IO_6^{2-}$	Ellis[316]
Unsaturated	Ethylenic compounds	Bromination or hydrogenation	Siggia[317]
Miscellaneous	Active H in nitro and nitroso compounds	With $LiAlH_4$	Majer[318]
	Phenol, cresol	With $KBrO_3$–KBr	Burgess[319]
	Phenol, chlorophenol	With *N*-(benzene-sulfonyl)-*p*-benzo-quinoneimine	Guilbault[320]
	Monochloroalkanes	With pyridine–dimethylformamide–H_2O	Jordan[321]
	(1-Chloroethyl)benzenes (2-chloroethyl)benzenes	With NaI in acetone	Zelenetskaya[322]
	(±)-, *meso*-Dibromo-succinic acids	Elimination of HBr	King[323]
	Organophosphorus compounds	With indole–sodium perborate	Guilbault[324]
	Hexaalkylditin compounds	With Ag^+	Zaia[325]

Most procedures that have been worked out for the analysis of organic mixtures by kinetic methods deal with two-component systems. Some examples are given in Table 3.5. Methods for more complex mixtures, however, are also available. For instance, Hawk et al.[326] described a method for the analysis of a mixture containing five peroxy compounds (two peroxycarboxylic acids, two diacyl peroxides, and a hydroperoxide); it is based on the partial reduction of the peroxide functions by use of several sulfides, each having a different reducing power, followed by iodimetric determination of residual unreduced peroxides. The concentration of each component in the mixture is calculated by a matrix of five simultaneous rectilinear equations.

Bond et al.[327] investigated a second-order homocompetitive system in which (*a*) each component has a different stoichiometry with the reagent, (*b*) the total initial molar concentration cannot be measured by a total functional group determination, and (*c*) the conventional second-order kinetic plot cannot be made to cover the early period when both components are reacting. A nonempirical procedure is employed, by which the use of calibration graphs and the need for exact reproduction of bath temperature are avoided. This method has been applied to the analysis of cyclopolymethylenenitramine mixtures.

It should be mentioned that kinetic methods of analysis do not always require chemical changes. For example, Siggia et al.[328] demonstrated the determination of mixtures of sugars and of amino acids by dialysis based on differential kinetics. The rates of diffusion of the components in the mixtures are such that standard first-order rate graphs show a rectilinear portion for each component.

3.6 SIMULTANEOUS DETERMINATION BY POLAROGRAPHY

Generally speaking, a functional group that is electrically reducible has a characteristic half-wave potential, although this is frequently altered as a result of the influence of the other part of the molecule.[329] Nevertheless, when two or more compounds containing different functional groups are present in a given analytical sample, it is worthwhile to explore the feasibility of using the polarographic differences to determine the compounds without restorting to the separation process.

When two compounds possess the same functional group, the presence of another function in one of them, but not in the other, may affect the half-wave potential significantly so that it becomes possible to distinguish these compounds polarographically. Thus binary mixtures containing benzophenone and acetophenone can be determined based on the polarographic reduction of the carbonyl group.[330] On the other hand, Hackman et al.[331] investigated the polarographic behaviors of chlordiazepoxide (Librium, a psychotherapeutic drug) and its metabolites, desmethylchlordiazepoxide and demoxepam, all possessing an azomethine and a *N*-oxide function with different neighboring groups, and found that it is not possible to determine these three compounds simultaneously when they are present in a solution.

Table 3.6 gives a number of reports on the polarographic analysis of mixtures. Simultaneous determination has been achieved in most cases, while experimental procedures still have to be worked out for some systems. This table covers most of the reducible functional groups.

Sometimes a condition can be found under which one of the compounds containing a normally reducible group becomes inactive. For example, for a mixture composed of nitrocyclohexane, 1-methyl-1-nitrocyclopentane, and nitrobenzene, Zaitsev and Zaitseva[371] have observed that in *N* NaOH containing 40% (v/v) methanol, nitrocyclohexane does not give a polarographic wave, but nitrobenzene ($E_{1/2} = -0.78$ V) and 1-methyl-1-nitrocyclopentane ($E_{1/2} = -1.09$ V) can be simultaneously determined.

The simultaneous determination of chloramphenicol and chloramphenicol palmitate in mixtures exemplifies the versatility of polarographic analysis. According to Pflegel and Shoukrallah,[381] the polarograms obtained with these two compounds each exhibit a wave corresponding to the reduction of the nitro function to the amino function, but the half-wave potentials differ. This difference varies with pH, being greatest (about 0.14 V) at pH 5.4 to 6.0. For the determination of chloramphenicol palmitate in concentrations of 0.02 to 0.2 m*M* in the presence of 0.2 m*M* chloramphenicol, the recommended supporting electrolyte consists of equal volumes of ethanol and a pH 6.0 buffer. For the determination of concentrations of 0.02 to 0.2 m*M* chloramphenicol in the presence of 0.2 m*M* chloramphenicol palmitate, the use of pH 4.7 buffer and ethanol (1 : 1) is suitable.

When several compounds possess the same reducible functional group but different halogen atoms (or different number of one halogen), it is easy to carry out their simultaneous analysis by polarography. For instance, Elving and Van Atta[340] have found that iodoacetone, bromoacetone, and chloroacetone each show a single reduction wave over the pH range 1.5 to 10 and that the reductions are irreversible and diffusion controlled. A 0.5 *M* acetate buffer (pH 4.6) is used as the supporting electrolyte. The relative error involved in the simultaneous determination of the three haloacetones is within 2%. Kutanina and Berezina[407] have studied the polarographic characteristics of di-, tri- and tetrachloro derivatives of *p*-toluenefulfonic acids. With a basal electrolyte of 0.05 *M* tetraethylammonium hydroxide in 75% ethanol, the 2,6-dichloro compound shows one derivative peak at -2.3 V, the 2,3,6-trichloro compound shows two peaks at -1.95 and -2.15 V, and the 2,3,5,6-tetrachloro compound shows three peaks at -1.45, -2.0, and

Table 3.6 Determination of Mixtures by Polarography

Function	Mixture containing	Conditions, $E_{1/2}$ (V)	Reference
Oxygen	Alkyl phenyl ketone, benzaldehyde		Boyd[332]
	Acetophenone, benzophenone	−1.54, −1.25	Hsieh[330]
	Formaldehyde, acetaldehyde, furfuraldehyde		Shulman[333]
	Benzil, acetylbenzoyl, diacetyl	−0.53, +0.60, +0.74	Rogers[334]
	Dibenzoylmethane, benzoylacetone, acetylacetone	−1.02, −1.22, −1.52	Rogers[334]
	Aromatic aldehydes		Jehlicka[335]
	Terephthalaldehydic acid, *p*-tolualdehyde, terephthal-aldehyde	−0.84, −1.05, −1.29	Miller,[336] Baranowski[337]
	Glyceraldehyde, 1,3-dihydroxy-acetone methylglyoxal		Fedoronko[338]
	Aliphatic aldehydes	Via Girard T derivatives	Fleet[339]
	Iodo-, bromo-, chloroacetones	−0.14, −0.35, −1.15	Elving[340]
	Chloroacetaldehydes		Elving[341]
	Idonic, gluconic acids		Asahi[342]
	Benzenepolycarboxylic acids		Kryukova[343]
	Pyrenic dianhydride, naph-thlene-1,4,5,8-tetracarboxylic acid dianhydride		Duplyakin[344]
	Phthalic acid esters		Ryvolova[345]
	Phthalic acid esters of diols		Jadrnickova[346]
	Cumene hydroperoxide, acetophenone		Vodzinskii[347]
	2-Isopropylnaphthalene hydroperoxide, 2′-aceto-naphthone	−0.85, −1.70	Pestretsova[348]
	Cumene hydroperoxide, dicumyl peroxide		Gregorowicz[349]
	Cyclohexanone peroxides		Antonovskii[350]
	Alkyl hydroperoxides		Hayano[351]
	Anthraquinones		Ashraf[352]

Table 3.6 (*Contd.*)

Function	Mixture containing	Conditions, $E_{1/2}$ (V)	Reference
	Octahydro-, tetrahydro-2-ethylanthraquinone, 2-ethylanthraquinone	−0.40, −0.55, −0.75	Ditsent[353]
	Anthraquinone dyes		Matrka[354]
	Phenanthrenequinone, diphenaldehyde, diphenaldehydic acid	−0.25, −0.95, −1.19	Rusyanova[355]
	Naphthaquinone, maleic anhydride	+0.15, −0.8	Kurata[356]
Nitrogen	Aromatic amines	Oxidation at Pt anode	Bezuglyi[357]
	Benzidine, phenylenediamine, etc.		Dvorak[358]
	Tyrosine, levdopa	+0.4, +1.3	Kitagawa[359]
	Pyridine derivatives		Volke[360]
	2,2′-, 4,4′-Bipyridyls		Balybin[361]
	Picolinaldehyde, picolinic acid	−0.68, −1.24	Marciszewski[362]
	3-Picoline derivatives		Serazetclinova,[363] Onopko[364]
	Dimetridazole, furazolidine		Slamnik[365]
	Complexans		Hoyle[366]
	Quinoline, isoquinoline derivatives		Burghardt[367]
	Purines, pyrimidines		Smith[368]
	Basic dyes		Thurel[369]
	Nitriles		Bobrova[370]
	Nitrobenzene, 1-methyl-1-nitrocyclopentane	−0.78, −1.09	Zaitsev[371]
	4-Nitrocatchol, 4-nitroanisole, 4-nitrophenol	−0.54, −0.8, −0.94	Burgschat[371a]
	Cyclohexyldinitrophenols		Maruyama[372]
	2-Nitrocyclohexanol, nitrocyclohexane		Zaitsev[373]
	Nitroglycerin, 2,4-dinitrotoluene, phthalic ester		Ligtenberg[374]
	Nitrobenzoic acid derivatives		Fleszar[375]

Table 3.6 (*Contd.*)

Function	Mixture containing	Conditions, $E_{1/2}$ (V)	Reference
	Nitroimidazoles		Dumanovic[376]
	Nitroazoles		Dumanovic[377]
	Nitrocresol, fenitrothion		Gras[378]
	Nitroanisole, chloronitrobenzene, nitrophenol		Honda[379]
	2-Acetamido-3-hydroxy-4′-nitropropiophenone, 2-amino-1-(4-nitrophenyl)propane-1,3-diol		Dumanovic[380]
	Chloramphenicol, chloramphenicol palmitate		Pflegel[381]
	Benzoquinone monoxime, dioxime	−0.28, −0.48	Kotok[382]
	Nitrostyrene, phenylacetaldehyde oxime		Chursina[383]
Sulfur	Cysteine, methionine	In excess of HCHO	Zhantalai[384]
	Thiols, thioethers, thiophens		Holzapfel[385]
	Sulfide, thiols, vitamin B_{12}		Youssefi[386]
	Sulfonic acid, sulfonyl chloride, sulfonate derivatives of naphthquinonediazide		Fodiman[387]
	Aminophenyl sulfones, nitrodiphenyl sulfones		Galpern[388]
	Dialkyldithiocarbamates		Halls[389]
	Benzyl thiocyanate, isothiocyanate		Sugii[390]
	Thiobarbituric acids		Zuman[391]
	Dialkyl dithiophosphates		Makens[392]
Unsaturated	Octachlorocyclopentene, hexachlorocyclopentadiene		Feoktistov[393]
	Hexachlorobutadiene, octachlorocyclopentene, hexachlorocyclopentadiene		Lyalikov[394]
	Crotonaldehyde, acetaldehyde	−1.3, −1.6	Taneeva,[395] Sokolskii[396]

Table 3.6 (*Contd.*)

Function	Mixture containing	Conditions, $E_{1/2}$ (V)	Reference
	Maleic acid, fumaric acid	+1.42, +1.62	Bellen[397]
	Maleic anhydride, naphthaquinone		Barendrecht[398]
	cis-, *trans*-Cinnamic, *cis*, *trans*-acontic, *cis*-, *trans*-crotonic acids		Markman[399]
	cis-, *trans*-Substituted cinnamic acids		Brand[400]
	Acrylamide, unsaturated dicarboxylate	−1.37, −0.6 to −0.9	Myagchenkov[401]
	Acrylates, methacrylates		Bezuglyi[402]
Miscellaneous	CCl_4, $CHCl_3$, CH_2Cl_2		Filimonova[403]
	Ethylmercury chloride, methoxyethylmercury chloride		Shirota[404]
	Butylchlorotin compounds		Issleib[405]
	Organotin compounds		Bork[406]

−2.35 V, while the parent acid and the 2-chloro compound are not reduced on the dropping mercury electrode. From a derivative polarogram, together with a normal polarogram, which shows a combined step for the sum of the reducible compounds, the amounts of the three chloro compounds can be determined.

3.7 SIMULTANEOUS DETERMINATION BY ABSORPTION SPECTROPHOTOMETRY

3.7.1 General Considerations

In this section we discuss the methods that utilize spectral absorption in the ultraviolet, visible, and infrared regions to determine two or more compounds simultaneously. It is noted that the ultraviolet and infrared methods generally measure the compounds as they exist in the sample, while the spectrophotometric procedures in the visible region usually involve chemical reactions to produce the colored species to be measured.

The ideal situation for the constituents of a mixture to be simultaneously determined is that each compound absorbs at a different wavelength without mutual interference. Then the number of wavelengths to be measured is the same as the number of compounds to be determined. However, this is seldom the case. Machek and Lorenz[408] discussed the possibility of analyzing binary mixtures (e.g., progesterone and estradiol benzoate) by measurement of u.v. absorption at two wavelengths and ternary combinations (e.g., vitamins B_1, B_6, and B_{12}) at three wavelengths, and the calculation of the composition of mixtures containing four substances. Kats and Rozkin[409] examined the quantitative criterion for choice of optimum wavelengths in the analysis of multicomponent mixtures. Using benzene, toluene, and chlorobenzene as testing compounds, extinction coefficients are determined for each component at various wavelengths, and factors relating these values to the total absorption of the sample at each wavelength are calculated. For each component, the optimum wavelength for ensuring minimum error is that at which the factor is a maximum. Przybylski[410] proposed a method for increasing the accuracy and precision by measuring the absorbance of the mixture at a number of selected wavelengths, more than the number of components to be determined. Botten[411] applied the principle of double-wavelength spectroscopy to the simultaneous determination of isophthalic and terephthalic acids.

Calculations based on simultaneous equations are commonly used in spectrophotometric analysis of mixtures. For example, Birr and Zieger[412] observed that the absorbances for styrene at 204 and 247 nm and for vinyl chloride at 204 nm all obey Beer's law, and they derived an equation by means of which vinyl chloride could be determined with the accuracy of $\pm 3\%$ even in the presence of a tenfold excess of styrene, Sokolov[413] determined mixtures of *p*- and *m*-diacetylbenzenes by measuring the absorbances at the maxima (256 and 223 nm, respectively, for the two isomers) and calculating from simultaneous equations. Melder[414] described similar methods for the determination of isomeric cresols and naphthols. Kharitonov and Leshchev[415] analyzed mixtures containing chlorobenzene–sulfonic and disulfonic acids using measurements at 12 wavelengths in the range 230 to 285 nm and solving a system of 12 equations by computer by the method of least squares. Vasilev and Pankova[416] applied the linear-programming and least-squares method to analyze acetone solutions containing 4 isomeric benzenehexachlorides, with measurements at 20 frequencies between 447 and 856 cm^{-1}.

A unique method has been described by Jordan[417] for the determination of saturated and α-unsaturated carbonyl compounds in complex systems at a single wavelength. It is based on the fact that the extinction per microgram of carbonyl group of an alkaline solution of the 2,4-dinitrophenylhydrazone changes rectilinearly with the wavelength of maximum absorption as the ratio of saturated to α-unsaturated carbonyl groups in the sample changes. The procedure involves scanning the spectrum from 400 to 460 nm exactly 6 min after the addition of KOH solution to an ethanolic solution of the 2,4-dinitrophenylhydrazone.

Pernarowski et al.[418] have developed a method for determining binary mixtures that is based on absorbance ratios. For two absorbing solutes that obey Beer's law in a transparent solvent, the ratio of the extinctions observed at two wavelengths is a nonlinear function of the composition, but is independent of the dilution. If one wavelength is that of an isosbestic point, the function is rectilinear. This method has been tested with several combinations of pharmaceuticals, for example, caffeine and phenacetin in chloroform at 250 and 265 nm, benzocaine and procaine hydrochloride in water at 285 and 311 nm, phenobarbitone and salicylic acid in chloroform at 5.75 and 7.70 μm, and sulfathiazole and sulfapyridine in KBr discs at 7.00 and 7.55 μm. The coefficient of variation is $<1\%$ in the ultraviolet region and $<3\%$ in the infrared region.

3.7.2 Simultaneous Ultraviolet Spectrophotometry

Table 3.7 gives examples of simultaneous determinations that involve spectrophotometric measurements in the range between 200 and 400 nm. With the exception of a few cases (e.g., fatty alcohols, which are determined after conversion into the corresponding nitrites), the ultraviolet absorptions of the compounds are measured directly.

Since the equipment for ultraviolet spectrophotometry usually can be used for measuring absorbance in the visible region as well, there are procedures for simultaneous determinations involving measurements in both regions. Manchon and Chariot[471] have presented a general method for

Table 3.7 Simultaneous Determinations Using the Ultraviolet Spectrophotometer

Function	Mixture containing	Wavelengths (nm)	Reference
Oxygen	Acetaldehyde, acetone	290–300	Nishino[419]
	Furfuraldehyde, vanillin	277, 347	Christofferson[419a]
	Acetone, diacetone	253, 285	Basinski[420]
	Acetone, diacetone, mesityl oxide	253, 267, 275	Basinski[421]
	Ribose, 2-deoxyribose	261, 277	Garrett[422]
	Mannose, glucose, xylose	280, 310, 340	Scott[422a]
	Glucose, fructose, sucrose	283 (40, 80°C)	Garrett[423]

Table 3.7 (*Contd.*)

Function	Mixture containing	Wavelengths (nm)	Reference
	Benzoic acid, anhydride, or chloride	270–300	Bogatkov[424]
	Benzoic, salicylic acids	270, 310	Anderson[425]
	Benzoic, hydroxybenzoic acids	225, 260, 300	Mantel[426]
	Benzoic, isophthalic, terephthalic acids	210, 230, 242, 254, 256	Kodner[427]
	Isophthalic, terephthalic acids	207, 242, 269	Ciecierska[428]
	Isophthalic, terephthalic acids	230, 240	Filippov[429]
	Phthalic, isophthalic, terephthalic acids	275, 289, 298	Butina[430]
	Phenyl salicylate, salicylic, gentisic acids	303, 314, 330	Pernarowski[431]
	Ester, carboxylic acid	204	Roy[432]
	Fatty alcohols	Via alkyl nitrites	Meluzova[433]
	Furfuryl alcohol, furaldehyde	217, 278	Soltovets[434]
	Di-, trihydroxycholates	250–410	Rautureau[435]
	Hydrocortisone, nitrofurazone	242, 286, 370	Bukowska[436]
Nitrogen	Aniline, phenylhydroxylamine, nitrobenzene	234, 240, 252, 280	Zozulya[437]
	2-, 3-, 4-Aminopyridines	228, 248, 260	Iovchev[438]
	3-, 4-Picolines, 2,6-lutidine	260, 252, 270	Bragilevskaya[439]
	Benzamide, benzonitrile		Astle[440]
	(3-Cyano-2-hydroxy-4-methoxymethyl-6-methyl-pyridine, (3-cyano-2-hydroxy-4-methoxymethyl-6-methyl-4-picoline	230	Blazicek[441]
	2-Deoxy-5-iodouridine, 2-Deoxy-5-iodouracil	290, 310	Kakeme[442]
	Cinchonine, quinine	236, 250	Vincze[443]
	Papaverine HCl, nicotinic acid	270, 309	Milch[444]
	Phenylephrine, amethocaine	241, 311	Salvesen[445]
	Barbiturates, salicylates	239, 253, 259, 296	Bjerre[446]
	Hydantoin, phenobarbitone	255, 270	Huerta[447]
	Theophylline, phenobarbitone	252, 274	Yokoyama[448]

Table 3.7 (*Contd.*)

Function	Mixture containing	Wavelengths (nm)	Reference
	Tetracycline, anhydro-tetracycline	357, 391	Pernarowski[449]
	Melamine, ammeline, ammelide, cyanuric acid	215, 230, 216, 231.5	Boitsov[450]
Sulfur	Thiols, disulfides	With *N*-ethylmaleimide	DeMarco[451]
	o-, *m*-, *p*-Toluenesulfonic acids	Many	Cerfontain[452]
	Arylsulfonic acids	Many	Arends[453]
	Hydroxylbenzene-2-, hydroxylbenzene-4-sulfonic acids, hydroxylbenzene-2,4-disulfonic acids	234, 278, 286	Spryskov[454]
	Bromobenzene-4-, bromobenzene-3-, bromobenzene-2-sulfonic acids, bromobenzene-2,5-disulfonic acid	230, 270, 274, 280	Potapova[455]
	Sulfamethazine, sulfamethyl-thiazole	243, 285	Milch[456]
	Sulfadimidine, sulfathiourea		Milch[457]
	Sulfadimethylpyrimidine, sulfaguanidine		Milch[458]
	Sulfacetamide, *N*-benzoyl-sulfanilamide, sulfathiazole	220, 235, 380	Mody[459]
Unsaturated	*cis*-, *trans*-Stilbene		Ish-Shalom[460]
	Crotonaldehyde, acetaldehyde, vinyl acetate		Capitani[461]
	Ethylenic ketones		Bonnier[462]
Miscellaneous	Benzene, toluene	254, 268	Zöllner[463]
	o-, *m*-, *p*-Ethyltoluenes	271, 272, 273	Díyarov[464]
	Ethylbenzene, *o*-, *m*-, *p*-xylenes	261, 271, 272, 274	Przybylski[465]
	Benzene, cumene	258, 260, 268	Benecze[466]
	Toluene, cumene	240–275	Shapovalov[467]
	Anthracene, carbazole		Swietoslawska[468]
	Polycylic hydrocarbons		Schmidt[469]
	1-, 2-Napthols		Rao[470]

determining organic compounds in biological samples in which several compounds respond to the same color reactions with the reagents used. Measurements of absorbance are made first on the sample solution itself and then on the sample solution after addition of a known concentration of each constituent being determined. Calculations are then made by a system of equations, which, based on the principle of addition, take account of the absolute extinction coefficient of mixtures of the compounds in pure solution and in the sample solution. Thus, for the analysis of a mixture of tyrosine and tryptophan in hydrolyzates of hemoglobin, the former is determined with the Folin-Ciocalteu reagent and the latter from its ultraviolet measurement.

3.7.3 Simultaneous Colorimetry

Table 3.8 gives examples of simultaneous determination that depends on spectrophotometry in the visible region. In some procedures more than one reagent is employed in order to develop different colors for measurement. Hence the absorption is measured in different solutions, but a separation step is not necessary.

Table 3.8 Analysis of Mixtures by Absorptiometry in the Visible Region

Function	Mixture containing	Reagent used	Reference
Oxygen	Saturated, α-unsaturated carbonyls	Dinitrophenylhydrazine	Jordan[472]
	Aldehydes	Guaiacol	Gregorowicz[473]
	Ketonic acids	Dinitrophenylhydrazine	Mendelowitz[474]
	Formaldehyde, dimethylacetal	Phenylenediamine; phenylhydrazine	Legradi[475]
	Acetaldehyde, methyl vinyl ketone	Fusine; phenylene-diamine	Hata[476]
	Furfuraldehyde, methylfurfuraldehyde	Benzidine; NH_2-$NH_2 \cdot HCl$	Rodionova[477]
	Benzaldehyde, benzophenone, cyclohexyl phenyl ketone	Dinitrophenylhydrazine	Zyzynski[478]
	Benzanthrone, dibenzanthronyl	H_2SO_4	Whetsel[478a]
	Hexoses, pentoses	Anisidine	Goodwin[479]
	Lactose, invert sugar	Picric acid	Tateo[480]

Table 3.8 (*Contd.*)

Function	Mixture containing	Reagent used	Reference
	Sucrose, invert sugar	Triphenyltetrazolium chloride	Carruthers[481]
	Benzoic, salicylic, fatty acids	Basic dyes	Korenman[482]
	Citric, aconitic acids	Acetic anhydride +pyridine	Nekshorocheff[483]
	Hydroxy aromatic acids	Nitrophenyldiazonium fluoroborate	Striegler[484]
	Anhydrides, esters	Hydroxylamine	Goddu[485]
	Aliphatic alcohols	Diazotized sulfanilic acid	Legradi[486]
	Pentyl alcohols	$NaNO_2$	Shmulyakovskii[487]
	Primary, secondary, tertiary alcohols	$NaNO_2$	Kapkin,[488] Grechukhina[489]
	Cholesterol, triperpene alcohols	$H_2SO_4 + (CH_3CO)_2O$	Luddy,[490] Duewell[491]
	Anthraquinone, benzanthrone	H_2SO_4	Whetsel[492]
Nitrogen	Primary, secondary amines	Fluorodinitrobenzene	McIntire[493]
	Primary, secondary, tertiary amines	Methyl orange	Silverstein[494]
	Tertiary, quaternary amines	Aconitic anhydride; chloranil	Sass[495]
	Betaine, hexamine	Reineckate	Masse[496]
	Benzidine, *o*-toluidine	NaOCl	Florea[497]
	Aromatic primary, secondary amines	Benzoquinonedichlorimine	Uno[498]
	o-, *p*-Phenylenediamines	$CuCl_2$ + triphenylphosphine	Hashmi[499]
	Amino sugars	4-Dimethylaminobenzaldehyde	Stewart-Tull[500]
	4-Aminosalicylate, 3-aminophenol	HNO_2	Jensen[501]
	Hydrazine, dimethyl hydrazine	Salicylaldehyde	Bailey[502]

Table 3.8 (*Contd.*)

Function	Mixture containing	Reagent used	Reference
	o-, *p*-Dinitrobenzenes	Ascorbic acid	Morita[503]
	2,4-, 2,6-Dinitrotoluenes	$(C_2H_5)_4NOH$	Kaminskii[504]
	Polynitroaromatic compounds	Ethylenediamine	Glover[505]
	Nitrobenzaldehydes	Phenethylamine	Potapov[506]
	Nitrophenols	Iminobispropylamine	Schrier[507]
	Nitrophenols	Via aminophenols	Obtemperanskaya[508]
	Styphnic, picric acids	KOH	Gitis[509]
	Nitrofurazone, nitrofurylacrylamide	NaOCl, phenylhydrazine	Ikeda[510]
	Pyridine, picolines	NaOCl + NaCN + + barbituric acid	Asmuo[511]
	Pyrazoles	Sodium amminoprusside + $NaNO_2$	Peyre[512]
	Indole, tryptophan, anthranilic acid	4-Dimethylaminobenzaldehyde	Kupfer[513]
	Ephedrine, codeine	$CuSO_4$; 4-dimethylaminobenzaldehyde	Bican-Fister[514]
	Uric acid, glucose	Tungstophosphate	Laessig[515]
	Urea, citrulline	Biacetyl monoxime	Moore[516]
Sulfur	Mercapto, disulfide groups	Trisglycine–EDTA	Beveridge[517]
	Sulfamates	Benzoquinone	Benson[517a]
Miscellaneous	*o*-, *m*-Cresols	*o*-Nitrobenzaldehyde	Wehle[518]
	Catechol, quinol, resorcinol	3-Methylbenzothiazolin-2-one hydrazone	Kamata[519]
	Phenol, quinol, catechol	Amidopyrine, etc.	Perelshtein[520]
	Chlorobenzene, benzene	Nitration, then KOH	Mirjolet[521]
	Chloral hydrate, trichloroacetic acid, trichloroethanol	Pyridine + KOH	Friedman[522]
	Di-, trichloroacetic acids	Naphthoresorcinol	Bellen[523]

It should be recognized that the equipment for measuring the intensity of visible colors is considerably less expensive than that for ultraviolet or infrared absorptiometry. In certain cases, a colorimeter fitted with filters may serve the purpose. For example, Wehle[518] observed that *o*-cresol gave a red color with *o*-nitrobenzaldehyde in the presence of $MgCl_2$ and HCl while *m*-cresol gave a green color, and he performed simultaneous determination of the binary mixture using a colorimeter with S53 and S61 filters.

3.7.4 Simultaneous Infrared Spectrophotometry

Infrared spectroscopy is now commonly used for the detection and identification of known organic compounds based on the fact that every compound has its own "fingerprint" molecular spectrum in the i.r. region. For quantitative analysis, however, it is the absorption due to the functional groups (e.g., C=O, O—H, C≡N) that serves the purpose. Studies have been published on the use of i.r. spectrophotometry for the simultaneous determination of several groups of compounds in a sample. For instance, Saier and Hughes[524] described the determination of oxygenated materials as group types (alcohols, acids, aldehydes, ketones, ethers) in a mixture; absorbances are measured at six wavelengths and substituted into a matrix to give the concentrations. Korcek et al.[525] utilized i.r. spectra to calculate acid and ester values of the oxidation products in lubricating oils. Rochkind[526] analyzed gas mixtures (alkanes, alkenes, alkynes, ethers, aldehydes, ketones, aromatic compounds) by condensing the sample mixed with nitrogen by pulsed deposition onto a halide substrate (e.g., CsI) at $-253°C$. Simple, selective spectra with narrow bands, and rectilinear calibration graphs were obtained; the accuracy was within $\pm 5\%$. Egorov and Petrov[527] proposed a method to determine the degree of branching of C_{12} to C_{16} paraffins that is based on the observation that the intensities of the absorption bands at 3.42 and 3.51 μm decrease as the branching in paraffin hydrocarbons increases, while the intensity of the band at 3.38 μm increases with the degree of branching.

Table 3.9 gives examples of the determination of compounds in mixtures by simultaneous i.r. absorptiometry. In a few cases, the absorption bands used for the individual compounds are sufficiently apart. More frequently, however, the bands overlap and simultaneous equations have to be worked out to calculate the compositions.

Table 3.9 Analysis of Mixtures by Infrared Absorptiometry

Function	Mixture containing	Measured at	Reference
Oxygen	Benzil, benzoin, hydrobenzoin	1680, 3460, 3590 cm^{-1}	Falkner[528]
	Ketone, carbinol		Washburn[529]
	Ketosteroids		Rashkes[530]
	Acetic acid, anhydride	7.72, 8.89 μm	Sato[531]
	Substituted 1,3-dioxanes	670–5000 cm^{-1}	Wunderlich[532]
	Mono-, di- tri-substituted oxiranes	11.4, 11.8, 13.0 μm	Bomstein[533]
	Isopropyl-, butyl-2,4-dichloro-phenoxyacetates		Witmer[534]
	Triglycerides, cholesteryl esters		Freeman[535]
	Ethanol, isopropanol, propanol	8.87, 9.87, 11.36 μm	Specht[536]
	Primary, secondary, tertiary alcohols	7090, 7067, 7042 cm^{-1}	Habermehl[537]
Nitrogen	Primary, secondary amines		Lohman[538]
	Primary, secondary, tertiary amine salts	1960, 2770, 2380 cm^{-1}	Thompson[539]
	trans-, *cis*-4-Aminocyclo-hexylacetic acid	640, 725 cm^{-1}	Titova[540]
	Aniline, *N*-ethylaniline		Whetsel[541]
	4-Bromobenzyl, 4,4′-dibromo-dibenzyl amines	1.5293, 2.0223 μm	Mravec[542]
	Glutamic, oxoglutaric acids		Estes[543]
	Acetanilide, phenacetin	695, 827 cm^{-1}	Oi[544]
	Pyridines	10–15 μm	Cook[545]
	2-, 3-, 4-Picolines, 2,6-lutidine, 2-ethylpyridine		Knobloch[546]
	Tetrahydroquinoline, quinoline	9.15, 10.66 μm	Pobiner[547]
	Morphine, codeine	Via acetylation	Genest[548]
	Papaverine, thebaine, narcotine	1160, 1602, 1767 cm^{-1}	Bakre,[549] Merlis[550]
	Papaverine, dihydroxy-codeinone	5.79, 6.25 μm	Salvesen[551]
	Triaryl-*sym*-triazines		Spencer[552]
	Propazine, atrazine, simazine	1020, 1059, 1112 cm^{-1}	Yakutin[553]

Table 3.9 (*Contd.*)

Function	Mixture containing	Measured at	Reference
	Azo derivatives of 2-pyrazolin-5-one	250–4000 cm^{-1}	Yasuda[554]
	o-, *m*-, *p*-Phthalonitriles		Shmulyakovskii[555]
	Di-(2-cyanoethyl)ether, 2-cyanoethanol, vinyl cyanide	8–11 μm	Dupre[556]
	Nitro alkenes, dinitroparaffins, nitro alcohols		Morgan[557]
	4-, 2-, 3-Nitrotoluenes; 2,4-dinitrotoluene	8.49, 8.70, 12.49, 9.39 μm	Pristera[558]
	4-, 2-, 3-Ethylnitrobenzenes	612, 1033, 1250 cm^{-1}	Hamburg[559]
	1,5-, 1,8-Dinitronaphthalenes	11.24, 11.96 μm	Kamada[560]
	4-, 3-, 2-Chloronitrobenzenes	9.14, 12.56, 12.94 μm	Oi,[561] Pozdyshev[562]
	Cyclohexyldinitrophenols		Kamada[563]
	2,4-Diureidotoluene, toluene-2,4-diamine	1360, 855 cm^{-1}	Eröss[564]
Sulfur	1-, 2-Naphthalenesulfonic acids	12.48, 13.53 μm	Ito[565]
	Alk-1-enesulfonates, alkylbenzenesulfonates, sodium alkylsulfates	524, 640, 655 cm^{-1}	Hashimoto[566]
	Ethylbenzylanilinesulfonic acids		Wilson[567]
	4-, 2-Toluenesulfonamides	12.3, 13.1 μm	Gong[568]
Unsaturated	Ethylene, ethylene oxide	10.54, 11.48 μm	Wolsky[569]
	2-Methylbut-2-ene, *tert*-pentyl alcohol	2740, 930 cm^{-2}	Shaburov[570]
	Terminal methylene, *cis*-alkene	1.62, 2.14 μm	Goddu[571]
	Dicyclopentadiene, cyclopentadiene	677, 1344 cm^{-1}	Szewzyk[572]
	Methylenecyclohexane, methylcyclohexenes	700, 889, 1045 cm^{-1}	Pinchas[573]
	Methylstyrene, dimethylbenzyl alcohol, cumene	861, 891, 1054 cm^{-1}	Terentev[574]
	α-Pinene, limonene, cineole	786, 888, 1078 cm^{-1}	Soares[575]

Table 3.9 (*Contd.*)

Function	Mixture containing	Measured at	Reference
	2-, 3-, 4-Vinyltoluenes	Via ethyltoluenes 752, 780, 813 cm^{-1}	Shmulyakovskii[576]
	Vinyl acetate, ethylene–vinyl acetate		French[577]
	Polyisoprenes		Schmalz[578]
	Mono-, dialkylacetylenes	1070–1116 cm^{-1}	Kendall[579]
	Diketen, butyric acid, anhydride		Ushakova[580]
Miscellaneous	Alkane, benzene, toluene		Leimer[581]
	Alkylbenzenes		Rashkes[582]
	Aromatic hydrocarbons		Martin[583]
	Isomeric terphenyls		Parellada[584]
	Biphenyl, phenyl ether		Hidalgo[585]
	p-, *o*-, *m*-Xylenes; ethylbenzene	8.950, 10.175, 11.425, 10.375 μm	Kraczkiewicz[586]
	o-, *m*-Xylenes, ethylbenzene, nitrobenzene		Moseley[587]
	o-, *m*-, *p*-Cymenes, *p*-*tert*-butyltoluene		Mikhailenko[588]
	Disubstituted naphthalenes	11–14 μm	Kamada[589]
	1,5-, 1,8-Disubstituted naphthalenes	1650–2000 cm^{-1}	Constantine[590]
	1-, 2-Methylnapthalenes		Sergienko[591]
	Phenols		Goddu[592]
	p-, *o*-Cyclophenols	12.13, 13.32 μm	Tanaka[593]
	Cresols, mono-, di, tri-*tert*-butylphenols		Pliev[594]
	Naphthol, naphthyl carbonate	3545, 1790 cm^{-1}	Stanescu[595]
	Fluorocarbons		Ayscough[596]
	Chlorobenzenes		Lukasiewicz-Ziarkowska[597]
	Tri-, tetrachlorobenzenes		Stanescu[598]

Table 3.9 (*Contd.*)

Function	Mixture containing	Measured at	Reference
	Hexachlorocyclohexanes	569, 847, 858 cm^{-1}	Lukasiewicz-Ziarkowska[599]
	Penta-, tetra-, trichlorophenols, chloranil	1418, 1446, 1472, 1695 cm^{-1}	Lomakina[600]
	Chloroaliphatic acids		Katon[601]
	Dibutyl hydrogen phosphate, tributyl phosphate		UKAEA[602]
	Triisobutyl and tributyl phosphates		UKAEA[603]
	Ferrocene mono-, dicarboxylic acids		Wolfarth[604]

3.8 SIMULTANEOUS DETERMINATION BY FLUOROMETRY, PHOSPHORIMETRY, OR RAMAN SPECTROMETRY

While the application of fluorescence in organic analysis has been increasing recently, there are only few reports on the use of fluorometry for the simultaneous determination of two or more compounds in mixtures. Hall[605] described procedures for the simultaneous determination of chlortetracycline, tetracycline, and demethylchlortetracycline in blood plasma. Chalmers and Wadds [606] studied the spectrofluorometric analysis of mixtures of the principal opium alkaloids. Kost et al.[607] determined lysergic acid and ergometrine when present together at 240 and 365 nm, respectively, with excitation at 292 nm; similarly, tryptophan and lysergic acid were analyzed at 330 and 440 nm. Several groups of workers investigated the simultaneous fluorometric determination of catecholamines; the combinations include adrenaline and noradrenaline (Persky and Ruston,[608] Vochten et al.[609]), adrenaline, noradrenaline, and dopamine (Riotte et al.[610]), and noradrenaline, dopamine, and serotonin (Ansell and Beisen,[611] Schlumpf et al.[612]).

Phosphorimetry was employed by Acuna et al.[613] for the determination of benzoic and toluic acids. The phosphorescence excitation and emission spectra of the solution of the acids in isopentane–ethyl ether frozen to a glass at 77° K were recorded on a spectrophotometer fitted with a phosphorescence accessory. The excitation and emission wavelengths were set on the spectrophosphorimeter, and the intensities were recorded for standard solutions of benzoic and *o*-, *m*-, and *p*-toluic acids. The calibration graphs were rectilinear in the 0.1 μM to 1 mM range.

Nakamura[614] analyzed mixtures containing 1-butene, *cis*- and *trans*-2-butenes, and isobutene by Raman spectroscopy; the key lines are at 1643, 1661, 1679, and 1658 cm^{-1}, respectively. Jorge and Barcelo[615] studied the Raman spectra of mixtures of mono-, di-, and trichloroacetic acids and found linear relationships between the log of the ratio of intensities and the log of the ratio of concentrations of pairs of acids, these relationships are unaffected by the presence of varying amounts of the third acid.

3.9 SIMULTANEOUS DETERMINATION BY NUCLEAR MAGNETIC RESONANCE SPECTROMETRY

Various mixtures have been analyzed by means of nuclear magnetic resonance spectrometry, as can be seen in Table 3.10. Presently, the proton resonance is generally used. With the development of instrumentation, other nuclei also may be employed. For example, Konishi et al.[625] proposed quantitative analysis of mixtures of alkylphenols by treating methyl- to hexyl phenols with trifluoroacetic anhydride for 5 hr (nonyl and decyl phenols for 24 hr) and measuring the ^{19}F n.m.r. spectra at 56.4 MHz.

Table 3.10 Analysis of Mixtures by Nuclear Magnetic Resonance Spectrometry

Function	Mixture containing	Signals used	Reference
Oxygen	Methanol, ethanol	3.4, 3.7 ppm	Sterk[616]
	Isobutyrates of trimethylpentane-1,3-diol		White[617]
Nitrogen	Diaminotoluenes		Mathias[618]
	Procaine, dimethocaine	1.23 ppm	Neville[619]
	Hydrazine, methyl-, dimethyl-hydrazines		MacDonald[620]
	Dinitrobenzenes		Kreshkov[621]
	2,4-Dinitro; 2,3,4-, 2,3,6-, 2,4,5-trinitrotoluenes		Gehring[622]
Unsaturated	Ethylene–vinyl acetate, ethylene–acrylate		Keller[623]
	Styrene, acrylonitrile		Cernicki[624]
Miscellaneous	Alkylphenols	^{19}F	Konishi[625]
	o-, *m*-, *p*-Cresols		Wainai[626]

While n.m.r. spectrometry entails expensive capital investment, there are occasions when it is more convenient and more accurate than other methods for simultaneous determinations. For instance, Wainai and Suzuki[626] found that mixtures of *o*-, *m*-, and *p*-cresols gave well-separated n.m.r. peaks for the three types of methyl groups when the test solution contained 11 mole % of cresols, 22 mole% of methanol, and 67 mole% of water; the peak area was proportional to the amount of each isomer, with standard deviations of 0.019, 0.024, and 0.015%, respectively.

3.10 SIMULTANEOUS DETERMINATION BY MASS SPECTROMETRY

As in infrared spectrometry, every organic compound exhibits a specific mass spectrum (fragmentation pattern) suitable for identification and quantitation. In contrast to infrared absorptiometry, however, application of mass spectrometry to the simultaneous determination of two or more components in a sample is rather rare. This is due to several reasons: (*a*) high cost of equipment and maintanance, (*b*) restriction to volatile substances, and (*c*) difficulty in feeding a representative sample of the mixture into the system. The situation may be improved as new techniques are developed. Some reports on the use of the mass spectrometer for the analysis of mixtures are discussed below.

Popova et al.[627] described a method to determine the various types of alcohols in mixtures (primary alcohols with an alkyl radical of normal structure having branching at the γ-carbon, primary alcohols with branching at the β-carbon, secondary and tertiary alcohols) based on a catalog of mass spectra; the mean relative error was 10 to 15% when the components were present to the extent of at least 10%. Hageman and van Katwijk[628] reported satisfactory results in quantitative mass-spectrometric analysis of known mixtures including one combination that contained methyl, ethyl, *n*-propyl, and iso-propyl alcohols and water, and another mixture that was composed of nine carboxylic acids. Langer et al.[629] described a procedure to determine alcohols, glycols, amines, phenols, and water in complex mixtures. Zyakun et al.[630] studied the analysis of mixtures of steroids using prednisolone and hydrocortisone as test compounds. The mixture was introduced into a vaporizer for difficultly volatile substances, and the mass spectrum was obtained at a temperature below the melting point of the sample. Integrated peak intensities at m/e 360 and 362 were measured, and the contents of each constituent were calculated by use of an equation. Hirt[631] determined methyl and phenylchlorosilanes in mixtures using the mass numbers 128, 148, 210, and 217; calibrations were based on known mixtures and the coefficient of variation was <0.5 mole % for all the components present.

A recent development in the commercial mass spectrometers involves their coupling with gas chromatographs. Thus, for the analysis of organic mixtures, separation and quantitation are dependent on gas chromatography, while the mass spectra identify the respective compounds. This technique is discussed in Chapter 4.

REFERENCES

1. B. I. Karavaev and R. N. Khelevin, *Izv. Vysch. Uchebn. Zaved. Khim.*, **9**, 257 (1966).
2. R. D. Sarson, *Anal. Chem.*, **30**, 932 (1958).
3. N. D. Cheronis and T. S. Ma, *Organic Functional Group Analysis*, Wiley, New York, 1964, pp. 396, 424.
4. I. Gyenes, *Titration in Non-aqueous Media*, Iliffe, London, 1968.
5. E. A. M. F. Dahmen, *Chim. Anal. (Paris)*, **40**, 378 (1958).
6. R. K. Maurmeyer, M. Margosis, and T. S. Ma, *Mikrochim. Acta*, **1959**, 177.
7. D. B. Bruss and G. E. A. Wyld, *Anal. Chem.*, **29**, 232 (1957).
8. S. S. Yamamura, Ph.D. Thesis, Iowa State College, 1957.
9. A. Frisque and V. W. Meloche, *Anal. Chem.*, **26**, 468 (1954).
10. B. Salvesen, *Medd. Norsk. Farm. Selsk.*, **23**, 177 (1961).
11. S. E. Rogozinski and H. R. Bosshard, *Anal. Chem.*, **45**, 2436 (1973).
12. L. K. Hiller, Jr., *Anal. Chem.*, **42**, 30 (1970).
13. C. A. Lucchesi, L. W. Kao, G. A. Young, and H. M. Chang, *Anal. Chem.*, **46**, 1331 (1974).
14. N. T. Smolova, T. I. Burmistrova, and A. P. Kreshkov, *Zh. Anal. Khim.*, **30**, 1805 (1975).
15. N. van Meurs and E. A. M. F. Dahmen, *Anal. Chim. Acta*, **19**, 64 (1958).
16. B. V. Prokopev, T. V. Kashik, and R. I. Verkhoturova, *Izv. Nauchno-Issled. Inst. Neftuglekhim. Sint. Irkustsk. Univ.*, **8**, 125 (1966).
17. R. Kahsnitz and G. Möhlmann, *Erdoel Khole*, **20**, 861 (1967).
18. T. Jasinski and Z. Kokot, *Chem. Anal. (Warsaw)*, **12**, 809 (1967).
19. A. P. Kreshkov, L. N. Bykova, N. A. Kazaryan, and E. S. Rubtsova, *Izv. Vyssh. Uchebn. Zaved. Khim.*, **9**, 72 (1966).
20. L. N. Bykova and N. A. Kazaryan, *Tr. Komiss. Anal. Khim. Akad Nauk SSSR*, **13**, 309 (1963).
21. M. T. Blake, *J. Am. Pharm. Assoc.*, **46** (5), 287 (1957).
22. J. Radell and E. T. Donahoe, *Anal. Chem.*, **26**, 590 (1954).
23. M. Rink and M. R. Riemhofer, *Pharm. Ztg.*, **107**, 462 (1962).

24. F. C. Trussel and R. E. Lewis, *Anal. Chim. Acta*, **34**, 243 (1966).
25. J. Valcha, *Chem. Zvesti*, **11**, 347 (1957).
26. A. I. Malyshev, *Zavod. Lab.*, **28**, 927 (1962).
27. P. Fijolka and I. Lenz, *Plaste Kautsch.*, **7**, (4), 169 (1960).
28. A. P. Kreshkov, N. A. Kazaryan, and K. N. Shulunova, *Zh. Anal. Khim.*, **23**, 1199 (1968).
29. R. H. Cundiff and P. S. Markunas, *Anal. Chem.*, **28**, 792 (1956).
30. G. A. Harlow and G. E. A. Wyld, *Anal. Chem.*, **30**, 69 (1958).
31. A. P. Kreshkov, L. N. Bykova, and N. T. Smolova, *Zh. Anal. Khim.*, **19**, 156 (1964).
32. A. P. Kreshkov, *Wiss. Z. Tech. Hochsch. Chem., Leuna-Merseburg*, **6**, 255 (1964).
33. A. P. Kreshkov, L. N. Bykova, and N. T. Smolova, *Izv. Vyssh. Uchebn. Zaved. Khim.*, **10**, 504 (1967).
34. A. P. Kreshkov, N. S. Aldarova, N. T. Smolova, and M. V. Slavgorodskaya, *Zh. Anal. Khim.*, **24**, 1100 (1969).
35. A. P. Kreshkov and I. I. Kudreiko, *Tr. Mosk. Khim.-Tekhnol. Inst.*, **1969**, 232.
36. A. I. Vasyutinskii and A. A. Tkach, *Zh. Anal. Khim.*, **24**, 911 (1969).
37. A. P. Kreshkov, N. T. Smolova, N. S. Aldarova, and N. A. Gabidulina, *Zh. Anal. Khim.*, **26**, 2456 (1971).
38. A. P. Kreshkov and L. G. Yarmakovskaya, *Zh. Anal. Chem.*, **29**, 572 (1974).
39. M. S. Jovanovic and B. D. Vucurovic, *Z. Anal. Chem.*, **266**, 117 (1973).
40. A. J. Martin, *Anal. Chem.*, **29**, 79 (1957).
41. A. Groagova and V. Chromy, *Analyst*, **95**, 548 (1970).
42. H. E. Zaugg and F. C. Gargen, *Anal. Chem.*, **30**, 1444 (1958).
43. J. Inczedy and O. Gimesi, *Acta Chim. Acad. Sci. Hung.*, **31**, 347 (1962).
44. G. A. Harlow and D. B. Bruss, *Anal. Chem.*, **30**, 1833 (1958).
45. S. Arita, K. Takeshita, and T. Kato, *J. Chem. Soc. Jap., Ind. Chem. Sect.*, **64**, 192 (1961).
46. A. P. Kreshkov, L. N. Bykova, and Z. G. Blagodatskaya, *Zh. Anal. Khim.*, **23**, 123 (1968).
47. N. T. Crabb and F. E. Critschield, *Talanta*, **10**, 271 (1963).
48. L. I. I. Hummelstedt and D. N. Hume, *Anal. Chem.*, **32**, 1792 (1960).
49. A. P. Kreshkov, Y. A. Gurvich, and G. M. Galpern, *Zh. Anal. Khim.*, **28**, 2440 (1973).
50. B. Smith and A. Haglund, *Acta Chem. Scand.*, **14**, 1349 (1960).
51. R. A. Glenn and J. T. Peake, *Anal. Chem.*, **27**, 205 (1955).
52. K. J. Karrman and G. Johansson, *Mikrochim. Acta*, **1956**, 1573.
53. J. Allen and E. T. Geddes, *Pharm. Pharmacol.*, **9**, 990 (1957).
54. I. Gyenes, *Magy. Kem. Foly.*, **63**, 95 (1957).
55. B. N. Ershov and V. L. Pokrovskaya, *Plast. Massy.*, **1961** (7), 65.

56. L. N. Bykova, N. A. Kazaryan, and E. S. Rubtsova, *Vestn. Tekh. Ekon. Inf. Nauchno-Issled. Inst. Tekh-Ekon. Issled. Gos. Kom. Khim. Prom. Pri. Gosplane SSSR*, **1964** (8), 28.
57. A. Anastasi, V. Gallo, and E. Mecarelli, *Mikrochim. Acta*, **1956**, 252.
58. M. Trnkova and M. Voldan, *Cesk. Farm.*, **12**, 182 (1963).
59. A. P. Kreshkov, Y. Y. Mikhailenko, and L. A. Tumovskii, *Zh. Anal. Khim.*, **19**, 1293 (1964).
60. A. Berger, M. Sela, and E. Katchalski, *Anal. Chem.*, **25**, 1554 (1953).
61. A. G. Fogg, P. J. Sausins, and J. R. Smithson, *Anal. Chim. Acta*, **49**, 342 (1970).
62. S. P. Agarwal and M. I. Blake, *J. Pharm. Sci.*, **54**, 1668 (1965).
63. G. Rauret and J. Garcia-Monjo, *Inf. Quim. Anal.*, **27**, 72 (1973).
64. T. Jasinski and H. Smagowski, *Chem. Anal.*, **8**, 525 (1963).
65. R. P. Jaiswal and S. G. Tandon, *J. Indian Chem. Soc.*, **47**, 755 (1970).
66. A. P. Kreshkov, V. I. Vasilev, and L. A. Tumovskii, *Tr. Mosk. Khim.-Tekhnol. Inst.*, **1965** (48), 39.
67. V. M. Aksenenko and E. G. Aksenenko, *Zavod. Lab.*, **34**, 535 (1968).
68. A. P. Kreshkov and L. A. Tumovski, *Tr. Mosk. Khim.-Tekhnol. Inst.*, **1968** (58), 294.
69. A. P. Kreshkov, L. N. Bykova, M. S. Rusakova, and N. A. Kazaryan, *Zavod. Lab.*, **28**, 11 (1962).
70. J. E. De Vries, S. Schiff, and E. St. Glair-Gantz, *Anal. Chem.*, **27**, 1814 (1955).
71. V. K. Kondratov and E. G. Novikov, *Zh. Anal. Khim.*, **22**, 1881 (1967).
72. A. P. Kreshkov, L. N. Bykova, and Z. G. Blagodatskaya, *Plast. Massy*, **1968** (6), 58.
73. C. J. Swartz and N. E. Foss, *J. Am. Pharm. Assoc.*, **44**, 217 (1955).
74. S. L. Lin and M. I. Blake, *J. Pharm. Sci.*, **55**, 781 (1966).
75. S. I. Obtemperanskaya and T. A. Egorova, *Zh. Anal. Khim.*, **24**, 1439 (1969).
76. S. I. Obtemperanskaya and T. A. Egorova, *Vestn. Mosk. Gos. Univ., Ser. Khim.*, **1969** (5), 115.
77. B. E. Buell, *Anal. Chem.*, **39**, 762 (1967).
78. D. P. Stognushko, G. A. Tember, Y. I. Usatenko, and S. A. Panaeva, *Zh. Anal. Khim.*, **30**, 1442 (1975).
79. M. M. Filimonova and B. F. Filimonov, *Zh. Anal. Khim.*, **24**, 773 (1969).
80. A. P. Kreshkov, L. N. Balyatinskaya, and S. V. Vasileva, *Zh. Anal. Khim.*, **24**, 1732 (1969).
81. A. I. Popov and P. A. Zagorets, *Tr. Mosk. Khim.-Tekhnol. Inst.*, **1968** (57), 48.
82. E. A. Gribova, *Zavod. Lab.*, **27**, 154 (1961).
83. E. A. Emelin, G. N. Svistunova, and Y. A. Tsarfin, *Zavod. Lab.*, **28**, 548 (1962).
84. P. L. de Reeder, *Anal. Chim. Acta*, **10**, 413 (1954).
85. J. S. Fritz and S. S. Yamamura, *Anal. Chem.*, **29**, 1079 (1957).

86. A. K. Amirjahed and M. I. Blake, *J. Pharm. Sci.*, **63**, 696 (1974).
87. F. Denes, N. N. Asandei, and C. I. Simionescu, *Anal. Chem.*, **40**, 629 (1968).
88. M. Mandel, *Eur. Polym. J.*, **6**, 807 (1970).
89. V. A. Zarinskii and I. A. Gurev, *Zh. Anal. Khim.*, **19**, 37 (1964).
90. A. P. Kreshkov and I. I. Kudreiko, *Zh. Anal. Khim.*, **24**, 1300 (1969).
91. M. N. Chelnokova and L. N. Dubrovina, *Zh. Anal. Khim.*, **23**, 1076 (1968).
92. V. Z. Deal and G. E. A. Wyld, *Anal. Chem.*, **27**, 47 (1955).
93. V. V. Rublev and Y. A. Bulatova, *Zh. Anal. Khim.*, **24**, 1106 (1969).
94. A. P. Kreshkov, V. A. Drozdov, and A. D. Romanova, *Zh. Anal. Khim.*, **24**, 1407 (1969).
95. G. Schott and E. Popowski, *Z. Chem. Leipz.*, **8**, 113 (1968).
96. I. V. Furne and L. A. Karimova, *Zavod. Lab.*, **42**, 26 (1976).
97. T. A. Khudyakova and A. P. Kreshkov, *Zh. Anal. Khim.*, **22**, 1153 (1967); *Zh. Prikl. Khim. Lening.*, **40**, 2448 (1967).
98. L. N. Bykova and V. D. Ardashnikova, *Tr. Mosk. Khim.-Tekhnol. Inst.*, **1968** (58), 280.
99. C. Datta and S. K. Mukherjee, *J. Indian Chem. Soc.*, **45**, 555 (1968).
100. P. H. Rao and B. S. Srikantan, *J. Indian Chem. Soc.*, *Ind. Ed.*, **17**, 229 (1954).
101. D. Beranova and S. Hudecek, *Chem. Listy*, **49**, 1723 (1955).
102. N. Purdie, M. B. Tomson, and G. K. Cook, *Anal. Chem.*, **44**, 1525 (1972).
103. J. B. Schute, *Pharm. Weekbl. Ned.*, **105**, 1 (1970).
104. Z. Szabo-Akos and L. Erdey, *Magy. Kem. Lapja*, **26**, 87 (1971).
105. E. Chromniak, *Chem. Anal. (Warsaw)*, **12**, 1039 (1967).
106. T. A. Khudyakova and V. M. Vostokov, *Zh. Prikl. Khim. Lening.*, **41**, 764 (1968).
107. S. Shimomura, *J. Pharm. Soc. Jap.*, **78**, 589 (1958).
108. V. Chromy and A. Groagova, *Analyst*, **95**, 552 (1970).
109. L. P. Krylova, S. T. Baibaeva, and V. A. Zarinskii, *Zh. Anal. Khim.*, **21**, 383 (1966).
110. F. F. Cantwell and D. J. Pietrzyk, *Anal. Chem.*, **46**, 344 (1974).
111. N. A. Gruzdeva, N. B. Zhilina, and T. K. Teikhrib, *Zavod. Lab.*, **34**, 415 (1968).
112. J. Subert, *Farm. Obz.*, **41**, 445 (1972).
113. J. Clark, D. D. Perrin, *Q. Rev.*, **18**, 295 (1964).
114. D. D. Perrin, *Pure Appl. Chem.*, **1964**, 1.
115. W. H. McCurdy and J. Galt, *Anal. Chem.*, **30**, 940 (1958).
116. B. E. Buell, *Anal. Chem.*, **39**, 756 (1967).
117. T. Higuchi, C. H. Barnstein, H. Ghassemi, and W. E. Perez, *Anal. Chem.*, **34**, 400 (1962).
118. K. A. Connors and T. Higuchi, *Anal. Chem.*, **32**, 93 (1960).
119. C. Rehm and T. Higuchi, *Anal. Chem.*, **29**, 367 (1957).
120. B. Tamura, *Jap. Anal.*, **17**, 683 (1968)

121. A. P. Kreshkov and V. I. Vasilev, *Zh. Anal. Khim.*, **17**, 908 (1962).
122. T. Jasinski and E. Andrulewicz, *Chem. Anal. (Warsaw)*, **9**, 667 (1960).
123. V. Vajgand and T. Pastor, *Glas. Hem. Drus., Beograd.*, **28**, 1 (1963).
124. B. Waligora and M. Paluch, *Chem. Anal. (Warsaw)*, **13**, 421 (1968).
125. B. Kamienski, B. Waligora, and M. Paluch, *Bull. Acad. Pol. Sci., Ser. Sci. Chim.*, **16**, 501 (1968).
126. C. B. Riolo, T. F. Soldi, and R. Perego, *Ann. Chim. (Rome)*, **53**, 1574 (1963); **54**, 552 (1964).
127. M. I. Blake, *J. Am. Pharm. Assoc.*, **46**, 163 (1957).
128. A. P. Kreshkov, A. N. Yarovenko, and I. Y. Zelmanova, *Zh. Anal. Khim.*, **17**, 780 (1962).
129. A. P. Kreshkov, A. N. Yarovenko, and I. Y. Zelmanova, *Zavod. Lab.*, **29**, 295 (1963).
130. A. P. Kreshkov, A. N. Yarovenko, and I. Y. Zelmanova, *Dokl. Akad. Nauk SSSR*, **143**, 348 (1962).
131. E. Sell, *Acta Pol. Pharm.*, **25**, 569 (1968).
132. A. P. Kreshkov, A. N. Yarovenko, and L. A. Bondareva, *Izv. Vyssh. Uchebn. Zaved. Khim.*, **12**, 123 (1969).
133. A. N. Yarovenko and K. A. Komarova, *Tr. Mosk. Khim.-Tekhnol. Inst.*, **1967** (2), 90.
134. A. P. Kreshkov, L. P. Senetskaya, and T. A. Malikova, *Zh. Anal. Khim.*, **22**, 1876 (1967).
135. A. P. Kreshkov, L. P. Senetskaya, and T. A. Malikova, *Tr. Mosk. Khim.-Tekhnol. Inst.*, **1967** (54), 122.
136. J. S. Fritz and C. A. Burgett, *Anal. Chem.*, **44**, 1673 (1972).
137. E. Kwiatkowski, H. Plucinska, and I. Nikel, *Chem. Anal. (Warsaw)*, **14**, 1295 (1969).
138. A. P. Kreshkov, N. S. Aldarova, and A. A. Izyneev, *Tr. Buryat. Kompleksn. Nauchno-Issled. Inst.*, **1966** (20), 233.
139. Z. M. Demidova and N. F. Budyak, *Zh. Anal. Khim.*, **24**, 1253 (1969).
140. I. M. Kolthoff, M. K. Chantooni, Jr., and S. Bhowmik, *Anal. Chem.*, **39**, 1627 (1967).
141. J. S. Fritz and M. O. Fulda, *Anal. Chem.*, **25**, 1837 (1953).
142. A. P. Kreshkov, L. N. Bykova, and N. S. Shemet, *Zh. Anal. Khim.*, **16**, 331 (1961).
143. Y. C. Chang, *Anal. Chem.*, **30**, 1095 (1958).
144. A. P. Kreshkov, L. N. Bykova, and I. D. Pevzner, *Zh. Anal. Khim.*, **19**, 890 (1964); *Dokl. Akad. Nauk SSSR*, **150**, 99 (1963).
145. E. A. Gribova, *Zavod. Lab.*, **34**, 283 (1968).
146. H. Ballczo, *Mitt. Chem. Forsch. Inst. Oesterr.*, **7**, 126 (1953).
147. M. N. Das and S. R. Palit, *J. Indian Chem. Soc.*, **31**, 34 (1954).
148. W. Rodziewicz and Z. Kokot, *Chem. Anal. (Warsaw)*, **11**, 961 (1966).

149. L. E. I. Hummelstedt and D. N. Hume, *Anal. Chem.*, **32**, 576 (1960).
150. J. Minczewski and S. Kiciak, *Chem. Anal. (Warsaw)*, **7**, 975 (1962).
151. A. P. Kreshkov and V. I. Vasilev, *Zh. Anal. Khim.*, **19**, 1508 (1964).
152. J. Minczewski and S. Kiciak, *CheU. Anal. (Warsaw)*, **8**, 239 (1963).
153. R. A. Bournique and G. L. Neuser, *Chemist-Analyst*, **53**, 41 (1964).
154. D. C. Wimer, *Talanta*, **13**, 1472 (1966).
155. L. Giuffre and E. Santacesaria, *Chim. Ind. (Milan)*, **51**, 1341 (1969).
156. G. M. Galpern, V. A. Ilina, L. P. Petrova, and F. P. Sidelkovskaya, *Zavod. Lab.*, **34**, 416 (1968).
157. K. I. Evstratova, V. I. Kurov, N. A. Goncharova, A. I. Ivanova, and V. Y. Solomko, *Zh. Anal. Khim.*, **22**, 1160 (1967).
158. V. K. Kondratov, N. D. Rusyanova, N. V. Malysheva, and L. P. Yurkina, *Zh. Anal. Khim.*, **22**, 1585 (1967).
159. V. K. Kondratov, N. D. Rusyanova, and N. V. Malysheva, *Zh. Anal. Khim.*, **21**, 996 (1966).
160. M. B. Devani and C. J. Shishoo, *J. Pharm. Sci.*, **59**, 90 (1970).
161. L. I. Ciaccio, S. R. Missan, W. H. McMullen, and T. C. Grenfell, *Anal. Chem.*, **29**, 1670 (1957).
162. T. Kashima, *J. Pharm. Soc. Jap.*, **74**, 1078 (1954).
163. A. P. Kreshkov and N. S. Aldarova, *Zh. Anal. Khim.*, **19**, 537 (1964).
164. V. V. Udovenko and L. A. Vvedenskaya, *Ukr. Khim. Zh.*, **20**, 684 (1954).
165. B. Salvesen, *Medd. Norsk. Farm. Selsk.*, **20**, 21 (1958).
166. V. K. Kondratov and E. G. Novikov, *Zh. Anal. Khim.*, **22**, 587 (1967).
167. C. B. Riolo and E. Marcon, *Ann. Chim. (Rome)*, **46**, 528 (1956).
168. N. N. Bezinger, G. D. Galpern, N. G. Ivanova, and G. A. Semeshkina, *Zh. Anal. Khim.*, **23**, 1538 (1968).
169. S. K. Freeman, *Anal. Chem.*, **25**, 1750 (1953).
170. S. Veibel and L. B. Kuznetsova, *Anal. Chim. Acta*, **65**, 163 (1973).
171. T. Beyrich and G. Schlaak, *Pharmazie*, **24**, 152 (1969).
172. B. R. Warner and W. W. Haskell, *Anal. Chem.*, **26**, 770 (1954).
173. C. N. Sideri and A. Osol, *J. Am. Pharm. Assoc.*, **42**, 688 (1953).
174. T. Jasinski and Z. Kokot, *Chem. Anal. (Warsaw)*, **13**, 111 (1968).
175. G. Tortolani, *Farmaco, Ed. Prat.*, **31**, 272 (1974).
176. A. M. Shkodin, L. P. Sadovnichaya, and V. P. Panchenko, *Zh. Anal. Khim.*, **17**, 540 (1962).
177. V. K. Kondratov and E. G. Novikov, *Zh. Anal. Khim.*, **22**, 1245 (1967).
178. N. N. Bezinger, G. D. Galpern, and M. A. Abdurakhmanov, *Zh. Anal. Khim.*, **16**, 91 (1961).
179. Y. Tajika, *J. Pharm. Soc. Jap.*, **74**, 1125 (1954).
180. G. M. Galpern, V. A. Ilina, A. V. Ivanov, and S. S. Gitis, *Tr. Vses. Nauchno-Issled. Proekt. Inst. Monomerow*, **1**, 143 (1969).

181. E. S. Lane, *Analyst*, **80**, 675 (1955).

182. A. P. Kreshkov and N. S. Aldarova, *Tr. Komiss. Anal. Khim. Akad. Nauk SSSR*, **13**, 315 (1963).

183. A. P. Kreshkov, L. N. Bykova, and N. S. Shemet, *Tr. Mosk. Khim.-Tekhnol. Inst.*, **1961**, (32), 327.

184. F. C. Critchfield and J. B. Johnson, *Anal. Chem.*, **29**, 957 (1957).

185. G. D. Galpern and N. N. Bezinger, *Zh. Anal. Khim.*, **13**, 603 (1958).

186. J. E. Jackson, *Anal. Chem.*, **25**, 1764 (1953).

187. Y. A. Strepikheev, A. A. Zalikin, and A. L. Chimishkyan, *Zh. Anal. Khim.*, **18**, 1262 (1963).

188. V. Ormanets, V. A. Tronova, and K. V. Topchieva, *Zh. Anal. Khim.*, **17**, 1109 (1962).

189. E. A. Gribova and E. S. Levin, *Zavod. Lab.*, **25**, 38 (1959).

190. E. A. Gribova and Y. P. Bavrina, *Zavod. Lab.*, **40**, 942 (1974).

191. E. A. Gribova and E. Y. Khmelnitskaya, *Zavod. Lab.*, **31**, 417 (1965).

192. S. J. H. Blakeley and V. J. Zatka, *Anal. Chim. Acta*, **74**, 139 (1975).

193. H. E. Malone and R. E. Barron, *Anal. Chem.*, **37**, 548 (1965).

194. J. G. Hanna and E. J. Kuchar, *Anal. Chem.*, **37**, 1116 (1965).

195. H. E. Malone and D. M. W. Anderson, *Anal. Chim. Acta*, **47**, 363 (1969).

196. H. E. Malone, *Anal. Chem.*, **33**, 575 (1961).

197. E. A. Burns and E. A. Lawler, *Anal. Chem.*, **35**, 802 (1963).

198. E. Demetrescu, P. Grintescu, V. Florea, and C. Ivan, *Rev. Chim. Buchar.*, **15**, 113 (1964).

199. E. Popper, E. Galfalvi, and V. Kiss, *Framacia Buchar.*, **20**, 469 (1972).

200. J. A. Gautier and F. Pellerin, *Ann. Pharm. Fr.*, **12**, 505 (1954).

201. M. R. F. Ashworth and R. Fehringer, *Anal. Chim. Acta*, **35**, 111 (1966).

202. C. Sass, *Fette, Seifen Anstrichm.*, **61**, 93 (1959).

202a. L. Barcza, *Talanta*, **10**, 503 (1963).

203. V. A. Bork, L. A. Shvyrkova, and O. Faizulaev, *Zh. Anal. Khim.*, **29**, 1844 (1974).

204. V. W. Reid and D. G. Salmon, *Analyst*, **80**, 602 (1955).

205. L. Maros, I. Perl, and E. Schulek, *Mag. Kem. Foly.*, **67**, 527 (1961).

206. J. Horacek and R. Pribil, *Talanta*, **16**, 1495 (1969).

207. O. C. Saxena, *Microchem. J.*, **14**, 343 (1969).

208. O. D. Saxena, *Microchem. J.*, **14**, 385 (1969).

209. M. Jurecek, J. Kalousova, and P. Kozak, *Mikrochim. Acta*, **1968**, 1313.

210. R. Belcher, M. K. Bhatty, and T. S. West, *J. Chem. Soc.*, **1958**, 2393.

211. T. S. Ma and A. S. Ladas, *Organic Functional Group Analysis by Gas Chromato-Graphy*, Academic, London, 1976, p. 47.

212. M. M. Schachter and T. S. Ma, *Mikrochim. Acta*, **1960**, 55.

213. N. G. Polyanskii, *Zh. Anal. Khim.*, **19**, 121 (1964).

214. M. M. Filimonova, I. A. Kocherovskaya, and A. V. Yazlovitskii, *Zh. Anal. Khim.*, **28**, 2068 (1973).

215. D. E. Jordan, *Anal. Chem.*, **40**, 2150 (1968).

216. L. M. Shtifman, S. V. Syaztsilo and G. G. Larikova, *Tr. Komiss. Anal. Khim., Akad. Nauk SSSR*, **13**, 325 (1963).

217. F. Drawert and G. Kupfer, *Z. Anal. Chem.*, **211**, 89 (1965).

218. N. N. Mikhailova, L. M. Lazorenko, and Z. K. Timokhina, *Zavod. Lab.*, **35**, 27 (1969).

219. O. N. Karpov, I. G. Gakh, and V. T. Lysyak, *Zh. Anal. Khim.*, **22**, 472 (1967).

220. M. Kan, F. Suzuki, and H. Kashiwagi, *Microchem. J.*, **42**, (1964).

221. C. Capitani and P. Imperiale, *Chim. Ind. (Milan)*, **36**, 606 (1954).

222. S. D. Mikhno, I. A. Solunina, V. A. Devyatnin, and V. W. Berezovskii, *Zh. Anal. Khim.*, **22**, 1419 (1967).

223. L. Szekeres and E. Kardos, *Z. Anal. Chem.*, **193**, 271 (1963).

224. J. B. Johanson and G. L. Funk, *Anal. Chem.*, **27**, 1464 (1955).

225. H. Huhn and E. Jenckel, *Z. Anal. Chem.*, **163**, 427 (1958).

225. H. Huhn and E. Jenckel, *Z. Anal. Chem.*, **163**, 427 (1958).

226. L. Nebbia and F. Guerrieri, *Chim. Ind. (Milan)*, **39**, 17 (1957).

227. K. Stürzer, *Z. Anal. Chem.*, **216**, 409 (1966).

228. V. S. Markevich, *Zavod. Lab.*, **34**, 1064 (1968).

229. A. B. Skvortsova, L. N. Petrova, and E. N. Novikova, *Zh. Anal. Khim.*, **17**, 896 (1962).

230. J. G. O'Connor and M. S. Norris, *Anal. Chem.*, **36**, 1391 (1964).

231. A. B. Skvortsova and A. A. Zelenetskaya, T. N. Korchagina, and L. N. Petrova, *Zh. Anal. Khim.*, **22**, 1565 (1967).

232. L. N. Petrova, A. B. Skvortsova, and E. N. Novkova, *Zh. Anal. Khim.*, **18**, 131 (1963).

233. D. Sikorska and K. Hetnarska, *Chem. Anal. (Warsaw)*, **5**, 1063 (1960).

234. S. M. Samoilov, V. N. Andrievakii, and I. L. Kotlyarevskii, *Izv. Akad. Nauk SSSR, Otd. Khim. Nauk*, **1962**, (2), 261.

235. Z. Bellen, *Chem. Anal. (Warsaw)*, **4**, 13 (1951).

236. R. Simionovici, C. Titei, N. Budisteanu, and F. M. Albert, *Z. Anal. Chem.*, **240**, 386 (1968).

237. M. S. Brodskii and A. I. Malyshev, *Zavod. Lab.*, **31**, 672 (1965).

238. F. Salzer, *Z. Anal. Chem.*, **146**, 260 (1955).

239. L. D. Metcalfe and A. A. Schnitz, *Anal. Chem.*, **29**, 1676 (1957).

240. C. N. Satterfield, R. E. Wilson, R. M. Leclair, and R. C. Reid, *Anal. Chem.*, **26**, 1792 (1954).

241. A. S. L. Hu and S. Grant, *Anal. Biochem.*, **25**, 221 (1968).

242. B. Rosenfeld, *Annalen*, **607**, 144 (1957).

243. B. M. Nakhmanovich and N. A. Pryanishnikova, *Zavod. Lab.*, **23**, 165 (1957)]

244. P. K. Jaiswal, *Chim. Anal. (Paris)*, **52**, 987 (1970).
245. E. Kovacs and K. Kokai, *Z. Anal. Chem.*, **204**, 16 (1964).
246. M. Kruty, J. B. Segur, and C. S. Miner, Jr., *J. Am. Oil Chem. Soc.*, **31**, 466 (1954).
247. G. Szepesi and S. Gorog, *Acta Pharm. Hung.*, **41**, 157 (1971).
248. B. C. Verma and S. Kumar, *Analyst*, **99**, 498 (1974).
249. B. N. Utkin, *Zavod. Lab.*, **33**, 1381 (1967).
250. A. A. Belyakov and N. V. Gorbyleva, *Zh. Anal. Khim.*, **12**, 545 (1957).
251. Y. I. Turyan, V. G. Baranova, and V. A. Aliferova, *Zh. Anal. Khim.*, **18**, 121 (1963).
252. R. Pribil and V. Vesely, *Chem.-Anal.*, **56**, 83 (1967).
253. T. Uno and K. Miyajima, *Chem. Pharm. Bull. Jap.*, **11**, 193 (1963).
254. E. Witek, *Polimery*, **19**, 81 (1974).
255. T. Lesiak and L. Szczepkowski, *Chem. Anal. (Warsaw)*, **15**, 165 (1970).
256. M. A. Portnov and B. I. Tomilov, *Zh. Anal. Khim.*, **12**, 402 (1957).
257. S. Mitev and P. K. Agasyan, *Zh. Anal. Khim.*, **29**, 970 (1974).
258. M. P. Kozlov and I. I. Ermilova, *Zh. Anal. Khim.*, **20**, 755 (1965).
259. H. Raber, *Sci. Pharm.*, **37**, 208 (1969).
260. D. C. Lewis and G. Sykes, *U. Pharm. Pharmacol.*, **5**, 933 (1953).
261. N. A. Kolchina and G. M. Kondrateva, *Zh. Anal. Khim.*, **24**, 1884 (1969).
262. B. Jaselskis, *Anal. Chem.*, **31**, 928 (1959).
263. M. Wronski, *Chem. Anal. (Warsaw)*, **7**, 851 (1962).
263a. M. Wronski, *Analyst*, **88**, 562 (1963).
264. R. J. Thibert, M. Sawar, and J. E. Carroll, *Mikrochim. Acta*, **1969**, 615.
265. J. C. MacDonald, *Anal. Chem.*, **37**, 1170 (1965).
266. A. Spiliadis, D. Bretcanu, and E. Badica, *Rev. Chim. Buchar.*, **8**, 296 (1957).
267. K. S. V. Santhanam and V. R. Krishnan, *Z. Anal. Chem.*, **234**, 256 (1968).
268. B. C. Verma and S. Kumar, *Zh. Anal. Khim.*, **29**, 1240 (1974).
269. B. C. Verma and S. Kumar, *Talanta*, **21**, 612 (1974).
270. T. V. Kremareva and V. M. Shulman, *Zh. Anal. Khim.*, **23**, 750 (1968).
271. N. G. Polyanskii, S. M. Markevich, E. D. Safronenko, and M. M. Buzlanova, *Tr. Komiss, Anal. Khim., Akad. Nauk SSSR*, **13**, 93 (1963).
272. R. N. Sivakova and V. I. Lyubomilov, *Zh. Anal. Khim.*, **23**, 311 (1968).
273. N. Sobczak, *Chem. Anal. (Warsaw)*, **8**, 613 (1963).
274. M. C. Badarinarayana, S. H. Ibrahim, and N. R. Kuloor, *Curr. Sci.*, **35**, 169 (1966).
275. O. P. Malhotra and V. D. Anand, *Z. Anal. Chem.*, **160**, 10 (1958).
276. T. Kiba and K. Terada, *J. Chem. Soc. Jap.*, **75**, 196 (1954).
277. D. N. Bernhart and K. H. Rattenbury, *Anal. Chem.*, **28**, 1765 (1956).
278. M. Wronski, *Chem. Anal. (Warsaw)*, **7**, 1011 (1962).
279. G. Nettesheim, *Z. Anal. Chem.*, **191**, 45 (1962).
280. L. Horner and E. Jürgens, *Angew. Chem.*, **70**, 266 (1958).

281. P. K. Jaiswal and S. Chandra, *Chim. Anal. (Paris)*, **51**, 493 (1969).
282. P. K. Jaiswal and S. Chandra, *Microchem. J.*, **14**, 289 (1969).
283. S. Chandra, *Chim. Anal. (Paris)*, **51**, 575 (1969).
284. N. G. Polyanskii, V. A. Fedorov, and V. K. Sapozhanikov, *Zh. Anal. Khim.*, **30**, 1809 (1975).
285. L. Fowler, *Anal. Chem.*, **27**, 1686 (1965).
286. I. Shahar and E. D. Bergmann, *J. Chem. Soc., Org.*, **1966**, 1005.
287. J. R. Collins, *Anal. Chim. Acta*, **9**, 500 (1953).
288. H. E. Hallam, *Analyst*, **80**, 552 (1955).
289. T. S. Ma and G. Zuazaga, *Ind. Eng. Chem., Anal. Ed.*, **14**, 280 (1942).
290. F. H. Lohman and T. F. Mulligan, *Anal. Chem.*, **41**, 243 (1969).
291. H. H. Stroh and H. Liehr, *J. Prakt. Chem.*, **29**, 8 (1965).
292. H. E. Malone and R. A. Biggers, *Anal. Chem.*, **1964** 1037.
293. R. Nutiu and A. Bokenyi, *Revta Chim.*, **20**, 637 (1969).
294. P. L. de Reeder, *Anal. Chim. Acta*, **9**, 314 (1953).
295. G. W. Stroehl and D. Kurzak, *Talanta*, **16**, 135 (1969).
296. W. L. Spliethoff and H. Hart, *Anal. Chem.*, **27**, 1492 (1955).
297. A. A. Zelenetskaya, N. N. Nikitina and G. P. Borodina, *Tr. Vses. Nauchno-Issled. Inst. Sint. Nat. Dushistykh Veshch.*, **1965**, (7), 175.
298. S. Siggia, in I. M. Kolthoff and P. E. Elving (Eds.), *Treatise on Analytical Chemistry*, Part II, Vol. 11, Wiley, New York, 1965, p. 94.
299. J. G. Hanna and S. Siggia, *Anal. Chem.*, **36**, 2022 (1964).
300. S. Siggia and J. G. Hanna, *Anal. Chem.*, **36**, 228 (1964).
301. K. B. Yatsimirskii, V. K. Pavlova, and V. I. Skuratov, *Zavod. Lab.*, **31**, 525 (1965).
302. L. Legradi, *Magy. Kem. Lapja*, **22**, 488 (1967).
303. H. A. Mottola, *Anal. Chim. Acta*, **71**, 443 (1974).
304. L. J. Papa, J. H. Patterson, H. B. Mark, Jr., and C. N. Reilley, *Anal. Chem.*, **35**, 1889 (1963).
305. H. B. Mark, Jr., L. M. Barkes, and D. Pinkel, *Talanta*, **12**, 27 (1965).
306. D. E. Guttman, *J. Pharm. Sci.*, **55**, 919 (1966).
307. F. Willeboordse and F. E. Critchfield, *Anal. Chem.*, **36**, 2270 (1964).
308. C. N. Reilley and L. J. Papa, *Anal. Chem.*, **34**, 801 (1962).
309. H. B. Mark, Jr., *Anal. Chem.*, **36**, 1668 (1964).
310. D. Benson and N. Fletcher, *Talanta*, **13**, 1207 (1966).
311. I. L. Shresta and M. N. Das, *Anal. Chim. Acta*, **50**, 135 (1970).
312. S. Siggia, J. G. Hanna, and N. M. Serenscha, *Anal. Chem.* **35**, 575 (1963).
313. L. Chafetz, J. R. Robinson, and R. J. Petrick, *J. Assoc. Off. Agric. Chem.*, **48**, **48**, 1068 (1965).

314. M. H. Zuidweg, J. de Flines, and H. Weissenburger, *Pharm. Weekbl. Ned.*, **102**, 939 (1967).
315. A. J. W. Brook and K. C. Munday, *Analyst*, **94**, 909 (1969).
316. G. L. Ellis and H. A. Mottola, *Anal. Chem.*, **44**, 2037 (1972).
317. S. Siggia, J. G. Hanna, and N. M. Serencha, *Anal. Chem.*, **35**, 362 (1963).
318. P. Majer and M. Jurecek, *Chem. Zvesti*, **18**, 900 (1964).
319. A. E. Burgess and J. L. Latham, *Analyst*, **91**, 343 (1966).
320. G. G. Guilbault, D. N. Kramer, and E. Hackley, *Anal. Chem.*, **38**, 1897 (1966).
321. D. E. Jordan, *Anal. Chem.*, **40**, 1717 (1968).
322. A. A. Zelenetskaya and N. N. Nikitina, *Tr. Vses. Nauchno-Issled. Inst. Sint. Nat. Dushistukh Veshch.*, **1965** (7), 170.
323. L. C. King and M. Fefer, *Anal. Chem.*, **29**, 1056 (1957).
324. G. G. Guilbault and G. J. Lubrand, *Anal. Chim. Acta*, **43**, 253 (1968).
325. P. Zaia, V. Peruzzo, and G. Lazzogna, *Anal. Chim. Acta*, **51**, 317 (1970).
326. J. P. Hawk, E. L. McDaniel, T. D. Parish, and K. E. Simmons, *Anal. Chem.*, **44**, 1315 (1972).
327. B. N. Bond, H. J. Scullionm and C. P. Conduit, *Anal. Chem.*, **37**, 147 (1965).
328. S. Siggia, J. G. Hanna, and N. M. Serencha, *Anal. Chem.*, **36**, 638 (1964).
329. P. Zuman and C. I. Perrin, *Organic Polarography*, Wiley, New York, 1969.
330. S. A. K. Hsieh and T. S. Ma, *Mikrochim. Acta*, **1977I**, 325.
331. M. R. Hackman, M. A. Brooks, A. J. F. de Silva, and T. S. Ma, *Anal. Chem.*, **46**, 1075 (1974).
332. R. H. Royd and A. R. Amell, *Anal. Chem.*, **28**, 1280 (1956).
333. M. S. Shulman and O. F. Gavrikova, *Tr. Vses. Nauchno Issled. Inst. Spirt. Promsti*, **1955** (5), 176.
334. W. Rogers, Jr. and S. M. Kipnes, *Anal. Chem.*, **27**, 1916 (1955).
335. V. Jehlicka and J. Lakomy, *Chem. Prum.*, **15**, 163 (1965).
336. G. Miller, W. Grabiec-Koska, and N. Paterok, *Chem. Anal. (Warsaw)*, **14**, 1139 (1969).
337. R. Baronowski, Z. Gregorowica, and J. Kulicka, *Mikrochim. Acta*, **1968**, 806.
338. M. Fedoronko, J. Konigstein, and K. Linek, *J. Electroanal. Chem.*, **14**, 357 (1967).
339. B. Fleet and P. N. Kelher, *Analyst*, **94**, 659 (1969).
340. P. J. Elving and R. E. Van Atta, *Anal. Chem.*, **27**, 1908 (1955).
341. P. J. Elving and C. E. Bennett, *Anal. Chem.*, **26**, 1572 (1954).
342. Y. Asahi and F. Kasahara, *Jap. Analyst*, **14**, 619 (1965).
343. G. G. Kryukova, M. S. Rusakova, N. V. Pavelko, and Y. I. Turyan, *Zh. Anal. Khim.*, **25**, 369 (1970).
344. V. K. Duplyakin, F. A. Ivanovskaya, V. A. Serazetdinova, D. K. Semabaev, and B. V. Suvorov, *Izv. Akad. Nauk Kaz. SSSR, Khim.*, **1969**, 82.
345. A. Ryvolova, *Chem. Listy*, **51**, 1201 (1957).

346. L. Jadrinickova and J. Krupicka, *Chem. Prum.*, **18**, 554 (1968).
347. Y. V. Vodzinskii and I. A. Korshunov, *Tr. Khim. Khim.-Tekhnol.*, **1959**, (2), 362.
348. T. A. Pestretsova and G. N. Kirichenko, *Zavod. Lab.*, **36**, 267 (1970).
349. Z. Gregorowicz, J. Cebula, and P. Gdrka, *Z. Anal. Chem.*, **284**, 283 (1977).
350. V. L. Antonovskii and Z. S. Frolova, *Zh. Anal. Khim.*, **19**, 754 (1964).
351. S. Hayano and N. Shinozuka, *Bull. Chem. Soc. Jap.*, **43**, 2039 (1970).
352. M. Ashraf and J. B. Headridge, *Talanta*, **16**, 1439 (1969).
353. V. E. Ditsent, *Zavod. Lab.*, **24**, 951 (1958).
354. M. Matrka and F. Novratil, *Chem. Prum.*, **12**, 498 (1962).
55. N. D. Rusyanova, V. G. Koksharov, and G. F. Belyaeva, *Zh. Anal. Khim.*, **21**, 850 (1966).
356. M. Kurata, *Jap. Analyst*, **4**, 361 (1955).
357. V. D. Bezuglyi and Y. I. Beilis, *Zh. Anal. Khim.*, **20**, 1000 (1965).
358. V. Dvorak, I. Nemec, and J. Zyka, *Microchem. J.*, **12**, 324, 350 (1967).
359. T. Kitagawa and S. Tsushima, *Jap. Analyst*, **20**, 1561 (1971).
360. J. Volke and V. Volkova, *Chem. Listy*, **48**, 1031 (1954).
361. Y. F. Balybin and A. V. Kotova, *Zavod. Lab.*, **33**, 24 (1967).
362. H. Marciszewski, *Chem. Anal. (Warsaw)*, **8**, 775 (1963).
363. V. A. Serazetclinova; A. D. Kagarlitskii, D. K. Sembaev, R. U. Umarova, F. A. Ivanovskaya, and B. V. Suvorov, *Vestn. Akad. Nauk Kaz. SSSR*, **1968**, (2), 68.
364. T. V. Onopko, L. A. Chervoneva, and V. V. Saraeva, *Zh. Anal. Khim.*, **29**, 403 ((1974).
365. M. Slamnik, *Talanta*, **21**, 960 (1974).
366. W. Hoyle, I. P. Sanderson, and T. S. West, *J. Electroanal. Chem.*, **2**, 166 (1961).
367. H. Burghardt, *Dtsch. Apot. Ztg.*, **108**, 1547 (1968).
368. D. L. Smith and P. J. Elving, *Anal. Chem.*, **34**, 930 (1962).
369. J. Thurel and B. Drevon, *Ann. Falsif.*, **54**, 12 (1961).
370. M. I. Bobrova and A. N. Matueeva-Kudasheva, *Zh. Obshch. Khim.*, **28**, 2929 (1958).
371. P. M. Zaitsev and Z. V. Zaitseva, *Zavod. Lab.*, **29**, 656 (1963).
371a. H. Burgschat and K. J. Netter, *J. Pharm. Sci.*, **66**, 60 (1977).
372. M. Maruyama and K. Maruyama, *Jap. Analyst*, **3**, 11 (1954).
373. P. M. Zaitsev and Z. V. Zaitseva, *Zavod Lab.*, **32**, 287 (1966).
374. H. L. Lightenberg, *Chem. Weekbl.*, **58**, 477 (1962).
375. B. Fleszar, *Chem. Anal. (Warsaw)*, **9**, 843 (1964).
376. D. Dumanovic, J. Volke, and V. Vajgand, *J. Pharm. Pharmacol.*, **18**, 507 (1966).
377. D. Dumanovic and J. Ciric, *Talanta*, **20**, 525 (1973).
378. G. Gras, *Trav. Soc. Pharm. Montpellier*, **28**, 243 (1968).
379. K. Honda and S. Kibuchi, *Bull. Chem. Soc. Jap*j, **34**, 529 (1961).

380. D. Dumanovic, J. Volke, and R. Jovanovic, *J. Assoc. Off. Anal. Chem.*, **54**, 884 (1971).
381. P. Pflegel and I. Shoukrallah, *Pharmazie*, **28**, 483 (1973).
382. L. A. Kotok, N. N. Fedorova, and V. D. Bezuglyi, *Zavod. Lab.*, **36**, 1328 (1970).
383. V. M. Chursina, E. F. Litvin, and L. K. Freidlin, *Zh. Anal. Khim.*, **21**, 365 (1966).
384. B. P. Zhantalai, *Biokhimiya*, **29**, 1009 (1964).
385. H. Holzapfel and K. Schöne, *Talanta*, **15**, 391 (1968).
386. M. Youssefi and R. L. Birke, *Anal. Chem.*, **49**, 1380 (1977).
387. Z. I. Fodiman and E. S. Levin, *Zavod. Lab.*, **34**, 1438 (1968).
388. G. M. Galpern, V. A. Ilina, A. V. Ivanov, and S. S. Gitis, *Tr. Vses. Nauchno-Issled. Proekt. Inst. Monomerov*, **1**, 149 (1969).
389. D. J. Halls, A. Townshend, and P. Zuman, *Anal. Chim. Acta*, **41**, 51 (1968).
390. A. Sugii, K. Nagai, M. Matsuo, and K. Kitahara, *Jap. Analyst*, **22**, 102 (1973).
391. P. Zuman, *Chem. Listy*, **48**, 1006 (1954).
392. R. F. Makens, H. H. Vaughan, and R. R. Chelberg, *Anal. Chem.*, **27**, 1062 (1955).
393. L. G. Feoktistov, Y. S. Lyalikov, and A. S. Solonar, *Zh. Anal. Khim.*, **23**, 313 (1968).
394. Y. S. Lyalikov, L. G. Feoktistov, A. S. Solonar, and N. G. Baek, *Zh. Anal. Khim.*, **22**, 1579 (1967).
395. G. V. Taneeva and V. P. Gladyshev, *Zh. Anal. Khim.*, **19**, 138 (1964).
396. D. V. Sokolskii, V. P. Shmonina, and G. V. Taneeva, *Zavod. Lab.*, **30**, 793 (1964).
397. Z. Bellen and B. Sekowska, *Przem. Chem.*, **11**, 523 (1955).
398. E. Barendrecht, *Chem. Weekbl.*, **50**, 785 (1954).
399. A. L. Markman and E. V. Zinkova, *Zh. Obshch. Khim.*, **27**, 1438 (1957).
400. M. J. D. Brandt and B. Fleet, *J. Electroanal. Chem.*, **16**, 341 (1968).
401. V. A. Myagchenkov, V. F. Kurenkov, A. V. Dushechkin, and E. V. Kuznetsov, *Zh. Anal. Khim.*, **22**, 1272 (1967).
402. V. D. Bezuglyi and Y. P. Ponomarev, *Zh. Anal. Khim.*, **23**, 599 (1968).
403. M. M. Filimonova, M. I. Levinski, and Z. D. Gudzenko, *Zavod. Lab.*, **28**, 424 (1962).
404. N. Shirota, M. Kotakemori, and H. Handa, *Ann. Rep. Takamine Lab.*, **9**, 198 (1957).
405. K. Issleib, H. Matschiner, and S. Naumann, *Talanta*, **15**, 370 (1968).
406. V. A. Bork and P. I. Selivokhin, *Plast. Massy*, **1968** (4), 56.
407. L. K. Kutanina and K. G. Berezina, *Zavod. Lab.*, **34**, 1302 (1968).
408. G. Machek and F. Lorenz, *Sci. Pharm.*, **31**, 17 (1963); **34**, 213 (1966).
409. M. D. Kats and M. Y. Rozkin, *Zavod. Lab.*, **38**, 688 (1972).
410. Z. Przybylski, *Chem. Anal. (Warsaw)*, **13**, 453 (1968).
411. D. S. Botten, *UV Spectrum Group Bull.*, **1974** (2), 9.
412. K. H. Birr and G. Zieger, *Z. Anal. Chem.*, **196**, 351 (1963).

413. A. G. Sokolov, *Zh. Anal. Khim.*, **19**, 397 (1964).
414. L. O. Melder, *Tr. Tallin. Politekh. Inst., A*, **1962** (185), 29.
415. V. V. Kharitonov and V. P. Leshchev, *Izv. Vyssh. Uchebn. Aaved. Khim.*, **12**, 1701 (1969).
416. A. F. Vasilev and M. B. Pankova, *Zavod. Lab.*, **38**, 1079 (1972).
417. D. E. Jordan, *J. Am. Oil Chem. Soc.*, **44**, 400 (1967).
418. M. Pernarowski, A. M. Knevel, and J. E. Christian, *J. Pharm. Sci.*, **50**, 943 (1961).
419. Y. Nishino, *Jap. Analyst*, **10**, 83 (1961).
419a. K. Christofferson, *Anal. Chim. Acta*, **31**, 233 (1964).
420. A. Basinski and A. Narebska, *Rocz. Chem.*, **35**, 1131 (1961).
421. A. Basinski and A. Narebska, *Rocz. Chem.*, **35**, 1381 (1961).
422. E. R. Garrett, J. Blanch, and J. K. Seydel, *J. Pharm. Sci.*, **56**, 1560 (1967).
422a. R. W. Scott, *Anal. Chem.*, **48**, 1919 (1976).
423. E. R. Garrett and J. F. Young, *J. Pharm. Sci.*, **58**, 1224 (1969).
424. S. V. Bogatkov, E. Y. Borisova, V. I. Nikolaeva, and E. M. Cherkasova, *Zh. Anal. Khim.*, **23**, 757 (1968).
425. S. Anderson and W. J. Rost, *Drug Stand.*, **28**, 46 (1960).
426. M. Mantel and M. Stiller, *Anal. Chem.*, **48**, 712 (1976).
427. M. S. Kodner, M. P. Filippov, and L. F. Gushchina, *Zh. Vses. Khim. Obshch. im. D.I. Mendeleeva*, **8**, 229 (1963).
428. Z. Ciecierska-Tworek and K. Gorczynska, *Chem. Anal. (Warsaw)*, **14**, 891 (1969).
429. M. P. Filippov, *Zh. Vses. Khim. Obshch. im. D.I. Mendeleeva*, **6**, 706 (1961).
430. I. V. Butina, V. G. Plyusnin, and N. A. Shevchenko, *Zh. Anal. Khim.*, **18**, 1384 (1963).
431. M. Pernarowski, A. M. Knevel, and J. E. Christian, *J. Pharm. Sci.*, **51**, 688 (1962).
432. R. S. Roy, *Anal. Chem.*, **40**, 1724 (1968).
433. G. B. Meluzova, B. P. Kotelnikov, and Z. A. Prokhorova, *Zh. Anal. Khim.*, **17**, 362 (1962).
434. G. N. Soltovets and V. G. Kulnevich, *Zh. Prikl. Khim. Leningr.*, **41**, 435 (1968).
435. M. Rautureau and B. Chevrel, *Pathol. Biol. Semaine Hop.*, **16**, 793 (1968).
436. H. Burkowska and Z. Golucki, *Chem. Anal. (Warsaw)*, **11**, 291 (1966).
437. A. P. Zozulya and N. N. Nikolaeva, *Zavod. Lab.*, **34**, 1061 (1968).
438. A. Iovchev, D. Dzharov, and I. Kanchovska, *C. R. Acad. Bulg. Sci.*, **23**, 957 (1970).
439. M. M. Bragilevskaya, I. E. Kogan, and M. E. Neimark, *Koks Khim.*, **1962** (4), 44.
440. M. J. Astle and J. B. Pierce, *Anal. Chem.*, **32**, 1322 (1960).
441. G. Blazicek, *Chem. Listy*, **60**, 245 (1966).
442. K. Kakemi, T. Arita, and M. Hashi, *J. Pharm. Soc. Jap.*, **84**, 1009 (1964).
443. M. Vincze and Z. Vincze, *Acta Pharm. Hung.*, **43**, 49 (1973).
444. G. Milch, M. Borsai, A. Csontas, and I. Mogacs, *Acta Pharm. Hung.*, **36**, 246 (1966).

445. B. Salvesen, *Medd. Norsk. Farm. Selsk.*, **24**, 1854 (1962).
446. S. Bjerre and C. J. Porter, *Clin. Chem.*, **11**, 137 (1965).
447. J. A. Huerta and M. J. Palacios, *An. Farm. Hosp.*, **9**, 45 (1966).
448. F. Yokoyama and M. Pernarowski, *J. Pharm. Sci.*, **50**, 953 (1961).
449. M. Pernarowski, R. O. Searl, and J. Naylor, *J. Pharm. Sci.*, **58**, 470 (1969).
450. E. N. Boitsov and A. I. Finkelshtein, *Zh. Anal. Khim.*, **17**, 748 (1962).
451. C. De Marco, M. T. Graziani, and R. Mosti, *Anal. Biochem.*, **15**, 40 (1966).
452. H. Cerfontain, H. G. J. Duin, and L. Vollbracht, *Anal. Chem.*, **35**, 1005 (1963).
453. J. M. Arends, H. Cerfontain, I. S. Herschberg, A. J. Prinsen, and A. C. M. Wanders, *Anal. Chem.*, **36**, 1802 (1964).
454. A. A. Spryskov and B. G. Gnedin, *Izv. Vyssh. Uchebn. Zav. Khim.*, **7**, 61 (1964).
455. T. I. Potapova, A. A. Spryskov, and E. P. Kukushkin, *Izv. Vyssh. Uchebn. Zav. Khim.*, **11**, 904 (1968).
456. G. Milch, *Acta Pharm. Hung.*, **32**, 206 (1962).
457. G. Milch, *Acta Pharm. Hung.*, **33**, 257 (1963).
458. G. Milch, M. Borsi, and I. Mogacs, *Acta Pharm. Hung.*, **37**, 6 (1967).
459. S. M. Mody and R. N. Naik, *J. Pharm. Sci.*, **52**, 201 (1963).
460. M. Ish-Shalom, J. D. Fitzpatrick, and M. Orchin, *J. Chem. Educ.*, **34**, 496 (1957).
461. C. Capitani and E. Milane, *Chim. Ind. (Milan)*, **37**, 177 (1955).
462. J. M. Bonnier and G. De Gaudemaris, *Bull. Soc. Chim. Fr.*, **21**, 991 (1954).
463. E. Zöllner and G. Vastagh, *Arch. Pharm.*, **298**, 281 (1965).
464. I. N. Diyarov and M. S. Pevzner, *Zh. Anal. Khim.*, **17**, 102 (1962).
465. Z. Przybylski, *Chem. Anal. (Warsaw)*, **8**, 779 (1963).
466. K. Bencze, *Chem. Zvesti*, **19**, 299 (1965).
467. V. N. Shapovalov, *Nauchn. Tr. Omsk. Med. Inst.*, **1971**, (107), 31.
468. J. Swietoslawska, *Rocz. Chem.*, **30**, 570 (1956).
469. H. Schmidt, *Erdel Kohler*, **21**, 334 (1968).
470. P. S. Rao, N. B. S. N. Rao, and M. P. Reddy, *Indian J. Chem.*, **3**, 408 (1965). (1969).
472. D. E. Jordan, *Anal. Chim. Acta*, **37**, 379 (1967).
473. Z. Gregorowicz and J. Majewska, *Chem. Anal. (Warsaw)*, **15**, 1045 (1970).
474. A. Mendelowitz and J. P. Riley, *Analyst*, **78**, 704 (1953).
475. L. Legradi, E. Pungor, and O. Szabadka, *Acta Chim. Hung.*, **42**, 89 (1964).
476. H. Hata and K. Okada, *Jap. Analyst*, **10**, 165 (1961).
477. Z. M. Rodionova, *Tr. Khim. Khim.-Tekhnol, Gorkii*, **1964**, 2(10), 262.
478. W. Zyzynski, *Acta Pol. Pharm.*, **31**, 53 (1974).
478a. K. B. Whetsel, *Anal. Chem.*, **26**, 1974 (1954).
479. J. F. Goodwin and H. Y. Yee, *Clin. Chem.*, **19**, 597 (1973).

480. F. Tateo, *Ind. Aliment.*, **8**, 71 (1969).
481. A. Carruthers and A. E. Wootton, *Int. Sugar J.*, **57**, 193 (1955).
482. I. M. Korenman, F. R. Sheyanova and S. N. Maslennikova, *Tr. Khim. Khim.-Tekhnol, Gorkii*, **1966**, 1(15), 137.
483. J. Nekshorocheff and J. Wajzer, *Bull. Soc. Chim. Biol.*, **35**, 695 (1953).
484. G. Striegler, A. Gabert, and U. Behrens, *J. Prakt. Chem.*, **17**, 183 (1962).
485. R. F. Goddu, N. F. LeBlanc, and C. M. Wright, *Anal. Chem.*, **27**, 1251 (1955).
486. L. Legradi, *Mikrochim. Acta*, **1970**, 33.
487. Y. E. Shmulyakovskii and A. A. Anisimova, *Zh. Prikl. Spectrosk.*, **2**, 91 (1965).
488. V. D. Kapkin, M. A. Ratomskaya, V. B. Belyanin, and A. N. Bashkirov, *Zh. Anal. Khim.*, **20**, 364 (1965).
489. Z. N. Grechukhina and V. V. Nesmelov, *Zh. Prikl. Khim. Leningr.*, **39**, 2574 (1966).
490. F. E. Luddy, A. Turner, Jr., and J. T. Scanlan, *Anal. Chem.*, **25**, 1497 (1953).
491. H. Duewell, *Anal. Chem.*, **25**, 1548 (1953).
492. K. B. Whetsel, *Anal. Chem.*, **25**, 1334 (1953).
493. F. C. McIntire, L. N. Clements, and M. Sproull, *Anal. Chem.*, **25**, 1757 (1953).
494. R. M. Silverstein, *Anal. Chem.*, **35**, 154 (1963).
495. S. Sass, J. J. Kaufman, A. A. Cardenas, and J. J. Martin, *Anal. Chem.*, **30**, 529 (1958).
496. M. Masse, *Pharm. Acta Helv.*, **33**, 80 (1958).
497. I. Florea, *Rev. Roum. Chem.*, **19**, 157 (1974).
498. T. Uno and M. Yamamoto, *Jap. Analyst*, **17**, 306 (1968).
499. M. H. Hashmi, A. I. Ajmal, A. Rashid, and T. Qureshi, *Mikrochim. Acta*, **1969**, 100.
500. D. E. S. Stewart-Tull, *Biochem. J.* **109**, 13 (1968).
501. R. Jensen and J. Muraine, *Bull. Soc. Pharm. Bordeaux*, **106**, 124 (1967).
502. L. C. Bailey and T. Medwick, *Anal. Chim. Acta*, **35**, 330 (1966).
503. Y. Morita, *J. Chem. Soc. Jap., Pure Chem. Sect.*, **84**, 812 (1963).
504. A. Y. Kaminskii, Z. A. Kozina, and S. S. Gitis, *Zh. Anal. Khim.*, **21**, 1380 (1966).
505. D. J. Glover and E. G. Kayser, *Anal. Chem.*, **40**, 2055 (1968).
506. V. M. Popapov, L. I. Lazutina, and A. P. Terentev, *Zh. Anal. Khim.*, **18**, 1003 (1963).
507. M. Schrier, A. Fono, and T. S. Ma, *Mikrochim. Acta*, **1967**, 218.
508. S. I. Obtemperanskaya and V. K. Zlobin, *Zh. Anal. Khim.*, **29**, 609 (1974).
509. S. S. Gitis, Y. D. Grudtsin, A. Y. Kaminskii, A. V. Ivanov, and S. A. Agapova, *Tr. Vses. Nauchno Proekt. Inst. Monomerov*, **1970**, 2(2), 139.
510. H. Ikeda, *J. Food Hyg. Soc. Jap.*, **3**, 269 (1962).
511. E. Asmus and H. F. Kurandt, *Z. Anal. Chem.*, **149**, 3 (1956).
512. J. P. Peyre and M. Reynier, *Ann. Pharm. Fr.*, **27**, 749 (1969).
513. D. Kupfer and D. E. Atkinson, *Anal. Biochem.*, **8**, 82 (1964).

514. T. Bican-Fister and V. Broz.Kajganovic, *Acta Pharm. Jugosl.*, **15**, 17 (1965).
515. R. H. Laessig and B. J. Basteyns, *Microchem. J.*, **13**, 418 (1968).
516. R. B. Moore and N. H. Kauffman, *Anal. Biochem.*, **33**, 263 (1970).
517. T. Beveridge, S. J. Toma, and S. Nakai, *J. Food Sci.*, **39**, 49 (1974).
517a. G. A. Benson and W. J. Spillane, *Anal. Chem.*, **48**, 2149 (1976).
518. H. Wehle, *Pharmazie*, **20**, 405 (1965).
519. E. Kamata, *Bull. Chem. Soc. Jap.*, **37**, 1674 (1964).
520. E. I. Perelshtein and V. T. Kaplin, *Gidrokhim. Mater.*, **43**, 82 (1967).
521. M. Mirjolet and C. Domange, *Bull. Soc. Pharm. Nancy*, **1959**, (41), 23.
522. P. J. Friedman and J. R. Cooper, *Anal. Chem.*, **30**, 1674 (1958).
523. N. Bellen and M. Jurowska-Wernerowa, *Przem. Chem.*, **11**, 526 (1955).
524. E. L. Saier and R. H. Hughes, *Anal. Chem.*, **30**, 513 (1958).
525. S. Korcek, P. Kalab, J. Baxa, and V. Vesely, *Ropa Uhlie*, **10**, 566 (1968).
526. M. M. Rochkind, *Science*, **160**, 196 (1968).
527. Y. P. Egorov and A. A. Petrov, *Zh. Anal. Khim.*, **11**, 483 (1956).
528. P. R. Falkner, G. Davison, and G. B. Stoker, *Analyst*, **93**, 660 (1968).
529. W. H. Washburn and F. A. Scheske, *Anal. Chem.*, **29**, 346 (1957).
530. Y. V. Rashkes, *Zh. Anal. Khim.*, **20**, 863 (1965).
531. T. Sato, A. Ikegami, and S. Kaneko, *Jap. Analyst*, **10**, 641 (1961).
532. H. Wunderlich, *Z. Anal. Chem.*, **241**, 234 (1968).
533. J. Bomstein, *Anal. Chem.*, **30**, 544 (1958).
534. F. J. Witmer, D. N. Thomas, and J. B. Vernetti, *Anal. Chem.*, **31**, 1280 (1959).
535. N. K. Freeman, *J. Lipid Res.*, **5**, 236 (1964).
536. H. Specht, *Z. Anal. Chem.*, **199**, 201 (1963).
537. G. Habermehl, *Angew. Chem., Int. Ed.*, **3**, 309 (1964).
538. F. H. Lohman and W. E. Norteman, *Anal. Chem.*, **35**, 707 (1963).
539. W. E. Thompson, R. J. Warren, I. B. Eisdorfer, and J. E. Zarembo, *J. Pharm. Soc.*, **54**, 1819 (1965).
540. I. A. Titova and M. G. Belskaya, *Zh. Anal. Khim.*, **1**, 1118 (1966).
541. K. B. Whetsel, W. E. Robertson, and M. W. Krell, *Anal. Chem.*, **32**, 730 (1960).
542. D. Mravec, K. Bencze, and J. Kalamar, *Z. Anal. Chem.*, **246**, 127 (1969).
543. F. L. Estes, A. L. Myers, and S. Briney, *Appl. Spectrosc.*, **24**, 131 (1970).
544. N. Oi, *J. Pharm. Soc. Jap.*, **86**, 850 (1966).
545. G. L. Cook and F. M. Church, *Anal. Chem.*, **28**, 993 (1956).
546. E. Knobloch, *Chem. Listy*, **49**, 268 (1955).
547. H. Pobiner, *Appl. Spectrosc.*, **17**, 79 (1963).
548. K. Genest and C. G. Farmilo, *Anal. Chem.*, **34**, 1464 (1962).
549. V. J. Bakre, Z. Karata, J. C. Barlet, and C. G. Farmilo, *J. Pharm. Pharmacol.*, **11**, 234 (1959).

550. V. M. Merlis, E. I. Kryukova, A. A. Chemerisskaya, E. M. Peresleni, and A. A. Ivanov, *Bull. Narcotics*, **20**, 5 (1968).
551. B. Salvesen, L. Domange, and J. Guy, *Ann. Pharm. Fr.*, **13**, 354 (1955).
552. R. D. Spencer and B. H. Beggs, *Anal. Chem.*, **35**, 1633 (1963).
553. V. I. Yakutin, N. P. Makarova, and S. S. Dubov, *Zavod. Lab.*, **31**, 564 (1965).
555. Y. E. Shmulyakovskii, O. M. Cranskaya, and G. I. Baranova, *Zh. Prikl. Spektrosk.*, **6**, 681 (1967).
556. E. F. Dupre, A. C. Armstrong, E. Klein, and R. T. O'Connor, *Anal. Chem.*, **27**, 1878 (1955).
557. H. Morgan, R. M. Sherwood, and T. S. Washall, *Anal. Chem.*, **38**, 1009 (1966).
558. F. Pristera and M. Halik, *Anal. Chem.*, **27**, 217 (1955).
559. E. Hamburg, *Stud. Cercet. Chim. Cluj, Rom.*, **13**, 255 (1962).
560. H. Kamada and S. Tanaka, *Jap. Analyst*, **4**, 71 (1955).
561. N. Oi and K. Miyazaki, *J. Pharm. Soc. Jap.*, **77**, Lo27 (1957).
562. V. A. Pozdyshev and E. S. Levin, *Zh. Anal. Khim.*, **14**, 128 (1959).
563. H. Kamada and S. Tanaka, *Jap. Analyst*, **2**, 334 (1953).
564. K. Eröss, G. Svehla, L. Erdey, and E. Vazsonyl, *Talanta*, **13**, 767 (1966).
565. A. Ito, *J. Chem. Soc. Jap., Ind. Chem. Sect.*, **60**, 1004 (1957).
566. S. Hashimoto, H. Tokuwaka, and T. Nagai, *Jap. Analyst*, **22**, 559 (1973).
567. C. H. Wilson and M. Dolinsky, *J. Assoc. Off. Agric. Chem.*, **47**, 1153 (1964).
568. A. W. Gong, *Appl. Spectrosc.*, **19**, 196 (1965).
569. O. Wolsky, *Engelhard Ind. Tech. Bull.*, **6**, 107 (1966).
570. M. A. Shaburov, *Tr. Komiss. Anal. Khim., Akad. Nauk SSSR*, **13**, 379 (1963).
571. R. F. Goddu, *Anal. Chem.*, **29**, 1790 (1957).
572. H. Szewczyk, *Chem. Anal. (Warsaw)*, **12**, 709 (1967).
573. S. Pinchas, J. Shabtai, and E. Gil-Av., *Anal. Chem.*, **30**, 1863 (1958).
574. V. A. Terentev, M. A. Shaburov, and A. N. Ivanova, *Neftekhimiya*, **1**, 567 (1961).
575. M. I. Soares and P. G. S. Pereira, *Revta Port. Quim.*, **11**, 26 (1969).
576. Y. E. Shmulyakovskii and O. M. Oranskaya, *Zh. Prikl. Spectrosk.*, **4**, 280 (1966).
577. A. R. French, J. V. Benham, and T. J. Pullukat, *Appl. Spectrosc.*, **28**, 477 (1974).
578. E. O. Schmalz and G. Geiseler, *Z. Anal. Chem.*, **191**, 1 (1962).
579. R. F. Kendall, *Spectrochem. Acta*, **24**, 1839 (1968).
580. A. G. Ushakova, V. V. Zharkov, and V. N. Mironova, *Zavod. Lab.*, **29**, 699 (1963).
581. H. J. Leimer and J. Schmidt, *Chem. Tech. Berl.*, **25**, 99 (1973).
582. Y. V. Rashkes, *Zh. Anal. Khim.*, **17**, 627 (1962).
583. J. M. Martin, Jr., R. W. B. Johnston, and M. J. O'Neal, Jr., *Anal. Chem.*, **26**, 1886 (1954).
584. P. Parellada, J. Bellanato, and A. Hidalgo, *Appl. Spectrosc.*, **18**, 118 (1964).
585. A. Hidalgo, R. Parellada, and J. Bellanato, *An. R. Soc. Esp. Fis. Quim.*, **B61**, 1009 (1965).

586. T. Kraczkiewicz-Biernacka and B. Kontnik, *Chem. Anal. (Warsaw)*, **4**, 97 (1959).
587. H. E. Moseley, *Appl. Spectrosc.*, **18**, 118 (1964).
588. Y. Y. Mikhailenko, N. N. Lebedev, and I. K. Kolchin, *Zh. Anal. Khim.*, **15**, 159 (1960).
589. H. Kamada and S. Tanaka, *Jap. Analyst*, **5**, 98 (1956).
590. P. R. Constantine and R. D. Topsom, *Spectrochim. Acta* **A24**, 1405 (1968).
591. S. R. Sergienko, M. P. Teterina, and Y. A. Bedov, *Tr. Inst. Nefti, Akad. Nauk SSSR*, **10**, 92 (1957).
592. R. F. Goddu, *Anal. Chem.*, **30**, 2009 (1958).
593. S. Tanaka, *Jap. Analyst*, **2**, 228 (1953).
594. T. N. Pliev, *Zh. Anal. Khim.*, **23**, 1703 (1968).
595. G. Stanescu and M. Keul, *Revta Chim.*, **19**, 423 (1968).
596. P. B. Ayscough, *Can. J. Chem.*, **33**, 1566 (1955).
597. Z. Lukasiewicz-Ziarkewska, *Chem. Anal. (Warsaw)*, **11**, 309 (1966).
598. G. Stanescu and O. Radulescu, *Revta Chim.*, **19**, 480 (1968).
599. Z. Lukasiewicz-Ziarkowska, *Chem. Anal. (Warsaw)*, **9**, 527 (1964).
600. M. V. Lomakina, T. M. Shumyacher, A. D. Igoshev, V. P. Khalilov, and A. S. Sobolev, *Zavod. Lab.*, **42**, 430 (1976).
601. J. E. Katon, T. P. Carii, and F. F. Bentley, *Appl. Spectrosc.*, **25**, 229 (1971).
602. United Kingdom Atomic Energy Authority Report PG 402 (W), 1962.
603. United Kingdom Atomic Energy Authority Report PG 438 (W), 1963.
604. E. F. Wolfarth, *Anal. Chem.*, **30**, 185 (1958).
605. D. Hall, *J. Pharm. Pharmacol.*, **28**, 420 (1976).
606. R. A. Chalmers and G. A. Wadds, *Analyst*, **95**, 234 (1970).
607. A. N. Kost, T. V. Koronelli, R. R. Lideman, and R. S. Sagitullin, *Zh. Anal. Khim.*, **20**, 845 (1965).
608. H. Persky and S. Roston, *Science*, **118**, 381 (1953).
609. R. F. C. Vochten, J. Hoste, A. L. Delaunois, and A. F. De Schaepdryver, *Anal. Chim. Acta*, **40**, 443 (1968).
610. M. Riotte, L. Peyrin, M. Vacquier, J. P. Cussac, and D. Naud, *Rev. Eur. Etud. Clin. Biol.*, **15**, 343 (1970).
611. G. B. Ansell and M. F. Beeson, *Anal. Biochem.*, **23**, 196 (1968).
612. M. Schlumpf, W. Lichtensteiger, H. Langemann, P. G. Waser, and F. Hefti, *Biochem. Pharmacol.*, **23**, 2437 (1974).
613. A. U. Acuna, A. Ceballos, and M. J. Molera, *An. Quim.*, **72**, 410 (1976).
614. S. Nakamura, *J. Chem. Soc. Jap., Pure Chem. Sect.*, **83**, 1081 (1962).
615. M. De P. Jorge and J. R. Barcelo, *An. R. Soc. Esp. Fis. Quim.*, **51B**, 125 (1955).
616. B. Sterk, *Z. Lebensm.-Forsch.*, **140**, 154 (1969).
617. H. F. White, C. W. Davisson and V. A. Yarborough, *Anal. Chem.*, **36**, 1659 (1964).
618. A. Mathias, *Anal. Chem.*, **38**, 1931 (1966).

619. G. A. Neville, *Can. Spectrosc.*, **14**, 44 (1969).
620. J. C. MacDonald, *Anal. Chim. Acta*, **44**, 391 (1969).
621. A. P. Kreshkov, A. A. Borisenko, S. I. Petrov, and V. A. Drozdov, *Zh. Anal. Khim.*, **24**, 1278 (1969).
622. D. G. Gehring and G. S. Reddy, *Anal. Chem.*, **40**, 792 (1968).
623. F. Keller and H. Roth, *Plaste Kautsch.*, **15**, 800 (1968).
624. B. K. Cernicki, J. V. Mühl, Z. J. Janovic, and Z. K. Siliepcevic, *Anal. Chem.* **40**, 606 (1968).
625. R. Konishi, Y. Mori, and N. Taniguchi, *Analyst*, **94**, 1002 (1969).
626. T. Wainai and Y. Suzuki, *Jap. Analyst*, **17**, 315 (1968).
627. T. I. Popova, A. A. Polyakova, and K. I. Zimina, *Tr. Komiss. Anal. Khim., Akad. Nauk SSSR*, **13**, 490 (1963).
628. F. Hageman and J. van Katwijk, *Ind. Chim. Belge*, **19**, 391 (1954).
629. S. H. Langer, R. A. Friedel, I. Wender, and A. G. Sharkey, Jr., *Anal. Chem.*, **30**, 1353 (1958).
630. A. M. Zyakun, V. E. Sterkin, G. R. Morozova, and V. M. Adanin, *Izv. Akad. Nauk SSSR, Ser. Khim.*, **1969**, (2), 249.
631. C. A. Hirt, *Anal. Chem.*, **33**, 1786 (1961).

CHAPTER 4

Determination of Several Components Involving Separation

4.1 GENERAL CONSIDERATIONS

In this chapter we discuss the principles and methods employed in the analytical separation of organic mixtures. As mentioned in Chapter 3, when several components in a sample are to be determined, the preferred procedure is one that does not involve separation. Needless to say, this is also true for the determination of only one component in a mixture. Even in the latter case, however, a separation step may be necessary. It should be pointed out that the separation techniques are similar whether only one component or several components are to be isolated from the sample.

The chief concern of analytical separation in quantitative analysis is that the sum of relative amounts of the various components in the separated portions should remain the same as in the original sample. In other words, during the separation process, there should be no loss of part or all of any component that is to be determined subsequently. At the same time, no contaminants that might interfere with the accurate measurement of the desired components should be introduced into the system.

Separation is a common and essential laboratory practice in chemistry.[1-3] It should be recognized, however, that not all separation methods are suitable for quantitative organic analysis[4] and that there is a distinct difference between the objective of analytical separation and that of preparative separation. When a separation experiment is carried out for the purpose of organic preparation (or synthesis), the aim is to obtain the purest possible product at a reasonable yield using the minimum quantity of reagents. In contrast, the first requirement of analytical separation is that the method consistently give the same percent yield, or yields within very narrow limits. This restriction applies to both physical and chemical procedures. For chemical methods, realizing that most organic reactions obey the laws of equilibrium, the analyst usually employs an extremely large excess of the reagent in order to drive the reaction to completion. The cost of the reagent is generally not of much concern, because only minute amounts of the analytical sample is involved. As a matter of fact, it is always desirable to use a minimal quantity of the working material to achieve the separation.[5]

For convenience we divide organic mixtures requiring analytical separation into the following categories, depending on the functional groups[6] present:

1. Mixtures containing acidic functions only.
2. Mixtures containing basic functions only.
3. Mixtures containing one neutral function only.
4. Mixtures containing more than one type of functional group.

Detailed discussion of these four categories is given in Chapters 5 to 7, as well as in Part Two of this book.

4.2 SELECTION OF SEPARATION METHOD

There are numerous methods for separating chemical species. Currently most methods for analytical separation are dependent on the differential migration of the molecules or ions. This dependence is the basis of chromatography, electrophoresis, and liquid–liquid extraction. It should be noted that, in some cases, the chemical species separated are not the original compounds in the organic mixture submitted for analysis. In these cases the compounds have been converted into their derivatives or decomposed and it is the decomposition products that are measured. There are no definite rules in selecting a suitable method to separate and determine a given mixture. The major separation methods for organic analysis are presented in the following sections.

4.3 SEPARATION BY GAS CHROMATOGRAPHY (G.C.)

Since the commercial gas chromatographs appeared a little over two decades ago, g.c. has undergone substantial changes. At present it is probably the most frequently used technique in the analytical separation of organic mixtures.[7,8] While g.c. was originally introduced for the purpose of separating stable volatile organic species, the analysis of numerous nonvolatile compounds has been achieved through the preparation of volatile derivatives, the determination of thermal decomposition (pyrolysis) products under controlled conditions, or the measurement of specific simple compounds produced by reactions with suitable reagents.[9] The reader is referred to the monographs that deal with these subjects in detail.[9a]

The popularity of g.c. is due to its simple and rapid operation, as well as its versatility. It has several desirable features: (*a*) Whereas the quantity of material to be separated is normally in the microgram region, separation by g.c. can be carried out at the milligram level (e.g., analysis of alkoxyl compounds via their alkyl iodides[10,11]) to below the nanogram level. For example, Eaborn et al.[12] performed steroid analysis at the nanogram level via the (halogenmethyl)dimethylsilyl ethers. Benerzra[12a] separated aromatic ketones in the subnanogram range by plasma chromatography. Walle and Ehrsson[13] determined picogram amounts of amino and hydroxy compounds through their heptafluoro butyryl derivatives; rectilinear calibration graphs were obtained in the range of 0.01 to 2 ng and the coefficient of variation was about 4%. (*b*) Gas chromatography permits easy separation of many fractions from complex mixtures. Thus Di Corcia[14] reported separation of mixtures containing up to 20 phenolic compounds in 45 min. Kostanyan et al.[15] determined the contents of 20 compounds obtained in the synthesis of butyric acid. Scora[16] separated 21 components of the oil distilled from the leaves of *Monarda punctata*. Thies and Schuster[17] investigated 47 essential oil constituents. In the resolution of C_4 to C_{12} petroleum mixtures by capillary g.c., Merchant[18] presented a chromatogram of a 91 component mixture. (*c*) Gas chromatography can separate closely related species, including isotopic molecules such as deuterium analogues.[19,20] Therefore gas chromatographic separation is a convenient method to identify adulterants in natural mixtures. For example, Morgantini[21] used this technique to detect the adulteration of butter fat by dolphin oil. (*d*) The conditions for separating a given mixture can be easily varied by changing the column, temperature programming, and so on to obtain the optimal results. Thus, for the analysis of cyclic diolefins, Schmitt and Jonassen[22] evaluated 14 different columns and found that a 20% loading of $AgNO_3$–Carbowax 20M (7 :13) or $AgNO_3$–Carbowax 1540 (2 :3) on 40 to 60 mesh firebrick gave the best separation. (*e*) Because the quantity of sample used is small, explosive compounds such as organic peroxides can be separated without hazards.[23,24]

4.3.1 Separation

Efforts have been made to predict the separation of certain mixtures and the optimal conditions for such separations. For instance, Sawyer and Brookman[25] devised a retention index for a series of aliphatic and aromatic hydrocarbons that is based on the relationship between the logarithm of the retention volume and the free energy of adsorption; this index permits prediction of the retention volume at any temperature and the ideal conditions for the separation of complex mixtures. Schomburg[26] reported the retention indices for the methyl esters of 63 α-branched aliphatic and alicyclic acids on capillary columns of squalene, polyoxypropylene glycol, Ucon oil, and polyoxypropylene glycol sebacate; the relationship between retention index and structure, and the influence of the polarity of the stationary phase on the separation of isomeric esters were studied. Haken[27] investigated the correlation of retention data and structural parameters of carbonyl compounds and demonstrated that the relationship can be represented nomographically by a single plot.

It should be recognized that the separation of a mixture is dependent on many variables. For example, Duvall and Tully[28] recommended separation of phenol and five of its *tert*-butyl derivatives on a column containing Silicone oil 550–Carbowax 4000 (3 :2) on Chromosorb W at 220°C; three ratios of Silicone oil to Carbowax gave maximum separation of the components, but the order of emergence varied in each case. The presence of water frequently has a significant effect on gas chromatographic separation. In the separation of lower alcohols from aqueous solutions, Pepelyaev et al.[28] found that their retention times increased with increasing H_2O content, while the opposite effect was observed in the case of hydrocarbons. Takens[29] studied the relationship between water content of alumina and retention time of hydrocarbons C_1 to C_4 and reported that the retention of ethane depended critically on the H_2O content of the alumina, which was maintained at 2.8%. In the analysis of volatile fatty acids, Ackman and Burgher[30] reported that excessive amounts of water may cause double-peak formation.

For the purpose of quantitative analysis, there should be complete separation of the components to be determined. It should be noted that this cannot always be attained. For instance, Gulko[31] used g.c. to analyze the dehydrogenation products of ethyltoluenes and reported that ethylbenzene, *m*- and *p*-xylenes, and *m*- and *p*-ethyltoluenes were not separated. In the analysis of mixtures containing HCl, acetyl chloride, acetic and chloracetic acids, Sato et al.[32] found that the peaks for HCl and acetyl chloride were not separated. In the analysis of the compounds obtained in the production of butanol, Sobolev[33] found that butyraldehyde and ethyl methyl ketone, and ethyl acetate and acetone were not completely separated. Butaeva and Inshakov[34] investigated the determination of monosaccharides as their methyl esters and found that the separation was not always complete.

4.3.2 Quantitation

The experimental techniques for quantitative analysis by g.c. have been reviewed by Novak.[35] Tranchant[36] has discussed the factors involved in such analysis, namely, measurement of peak areas, response of detectors, use of internal standards, and precision. Comparing various methods for calculating results, Deans[37] recommended recording of ratios of compounds to one another; measurement of peak height was found to give more reproducible results than measurement of peak area.

Sparks[38] described an apparatus and technique that simplify the introduction of replicate samples of a mixture of such complexity that the repetitive sampling technique must be used; concentration changes between injections are minimized. Langlais et al.[38a] modified the injection system to improve reproducibility. Karlsen and Rasmussen[39] investigated quantitative analysis by means of solid-sampling g.c. When the separation of warfarin and octadecane with injection (*a*) by the solid-sampling technique and (*b*) as solution in $CHCl_3$ or CS_2 were compared, the solid-sampling mode combined with electronic integration was found to be superior. For the analysis of polychlorinated biphenyls, Zobel[40] advocated a computer approach that enables the g.c. peaks obtained from a mixture to be resolved and determined in terms of the individual commercial products, which are themselves complex mixtures of compounds.

When retention times are used for quantitative g.c., Seher and Kühnast[41] reported that rectilinear relationships exist between the logarithm of the relative retention times of homologous series (e.g., the methyl esters of fatty acids) and the operating temperature; thus a column operated under constant conditions that has been calibrated for a homologous series can be used for all substances that can be analyzed under the given conditions without decomposition. Because diffusion of the sample originates from a band source and not from a point, Brandt and Lands[42] have found that the method of calculating relative peak area from retention times and peak heights can be improved by using a correction factor derived from the approximate sample bandwidth at zero time; this gives values comparable to those obtained by triangulation.

Gassiot-Matas and Gondal-Bosch[42a] have provided equations relating the relative molar responses to the molar volumes at the boiling points of alkanes, olefins, aromatic hydrocarbons, ketones, alcohols, and esters using the thermal-conductivity detector. Similarly, Ettre and Kabot[43] have shown that, with the flame ionization detector, the relative molar responses of homologous series are linearly proportional to the carbon number, and the relative peak-area values approximate closely the concentration by weight of the individual components.

While internal standards using compounds similar to those to be determined are frequently employed in quantitative g.c., complications sometimes are encountered. For example, in the analysis of (trimethyl)silyl derivatives of sugars, Halpern et al.[44] reported that difficulties have been caused by different rates of etherification of the internal standard and of the sample in the determination of glucose; therefore they recommended inert internal standards (e.g., terphenyl, pyrene).

It should be recognized that the gas chromatogram does not identify the separated components unequivocally. Hence other methods must be applied to confirm the identity of the compounds determined. When the quantity of the sample analyzed is at the milligram (or 0.1 millimole) level,[6] the g.c. effluents may be collected and determined by titrimetry (e.g., for alkoxyl compounds[10]); thermal-conductivity detection is used in this range and also for g.c. columns with high surface coverage.[45] When the amount separated by g.c. is at the microgram level, mass spectrometry is recommended for the identification of the compounds.[46] Murray et al.[47] described a technique to trap the fully resolved components in precooled, packed steel tubes, which were then transferred to the spectrometer. Tal'rose and Grishin[48] proposed the mass-spectrometric determination of the molecular weights of components leaving a g.c. column by placing an effusion chamber between a quick-release valve at the exit of the column and the detector (mass spectrometer); when the component of interest reaches the detector, the valve is shut and the decay of the signal is recorded versus time, the molecular weight being subsequently calculated from the decay curve. Bruner et al.[49] carried out g.c. using a dual detection system comprising a flame ionization detector and a total-ion monitor; the former is for recording of chromatograms and the latter is for mass-spectrometric identification of the peaks. The stream-splitting technique is used by some workers to measure only a portion of the effluent; Bruderrick et al.[50] have described a device to prevent the distortion of the sample composition caused by the difference in mass of its components in such systems. Combination of g.c. and mass spectrometry has been employed to study very complex mixtures, such as more than 150 compounds in an oil derived from strawberries[51] and aromatic hydrocarbons in tobacco smoke.[52] For the analysis of corticosteroids, Baillie et al.[53] observed that the derivatives (e.g., bismethoxime-trimethylsilyl ethers) can by monitored by single characteristic fragment ions. Skinner et al.[54] determined 21 barbiturates in biological samples and reported standard deviations of $<\pm 5\%$ in all compounds in the range 2 to 100 μg per ml of whole blood, serum, or plasma.

4.4 SEPARATION BY LIQUID CHROMATOGRAPHY (l.c.)

This section includes the various techniques of separation that are based on the adsorption or partition of organic compounds on solid supports contained in a tube (column) followed by differential removal of these compounds by suitable liquid-solvent systems. For example. Golkiewicz and Wawrzynowicz[55] separated seven alkaloids of *Fumarie officinalis* by two-stage elution from columns of silica gel with mixed solvents containing increasing concentrations of the active component; the alkaloids of low adsorption affinity (stylopine, sinactine, aurotensine) were separated by elution with propanol–cyclohexane, and then strongly adsorbed alkaloids (e.g., protopine, crytocavine) were separated with CCl_4–methanol, the active component in each case being the alcohol. Pal and Nath[56] separated hyaluronate, chondroitin sulfate, and heparin by adsorption on $Ca_3(PO_4)_2$, $BaSO_4$, or Al_2O_3, followed by sequential elution. Miller[57] reported on the separation of sulfapyrimidines by partition chromatography using a column prepared from a suspension of Celite in 0.1 *N* $KHCO_3$; sulfadimidine, sulfamethyldiazine, and sulfadiazine were eluted in turn from the column with ethyl ether–butanol (9 :1) saturated with 0.1 *N* $KHCO_3$, $CHCl_3$–butanol (5 :1) saturated with 0.1 *N* $KHCO_3$, and ethyl acetate–butanol (3 :2) saturated with H_2O.

Recently high-pressure l.c. has gained popularity. A review was written by Perry[58] covering the theoretical aspects, principal features of the commercial instruments, and detectors. Stewart et al.[59] described the construction of stainless-steel columns that can be operated at pressures up to 6000 psi. For many analytical separations, however, simple apparatus operating at 1000–2000 psi is suitable.[60]. Prepacked silica gel columns of 10 mm inside diameter operating at 50 psi also have been shown to be useful in separating certain mixtures; for example, caffeine, salicylamide, acetaminophen, and aspirin are conveniently resolved and determined by their respective peak heights.[61]

It should be noted that the characteristics of l.c. are easily affected by the operational conditions. Gonnet and Rocca[62] studied the influence of H_2O on the separation of pharmaceutical compounds and found that H_2O appears to suppress certain interaction between the adsorbent and the polar solvents. In the separation of prostaglandins, Valenzuela and Antonini[63] observed that different batches of silicic acid with the same specifications and from the same manufacturer could lead to different patterns.

The gradient-elution technique[5,64] is frequently employed for the separation of complex mixtures. Alternatively, the l.c. column can be eluted successively with solvents of increasing polarity; thus Kulnevich et al.[65] separated the products of autoxidation of 2-furaldehyde by putting the sample into a column of Celite and eluting with heptane, butanol, ethanol, and water in that order. Another approach is to pass the mixture in solution through several columns connected in series.[5] For instance, Brown[66] determined ethylene oxide and ethylene chlorohydrin by packing the first column with Florisil, which adsorbed ethylene chlorohydrin, the second column with Celite and aq. HCl, which converted the ethylene oxide into ethylene chlorohydrin, and the third column again with Florisil; the first and third columns were then eluted with ethyl ether.

Gel permeation and Sephadex were originally developed for the separation of macromolecules and particles.[67] Subsequently, these methods have been applied to the separation of many natural products, such as glycosides,[68] flavonoids,[69] steroids,[70-72] alkaloids,[73] indole compounds,[74] and amino acids and sugars.[75] Separations in Sephadex columns also have been reported for diastereoisomers of carboxylic acids,[76] keto–enol tautomers,[77] benzoic acid from fatty acids,[78] pyridine derivatives,[79] azo dyes,[80] and petroleum oils.[81,82] According to Chang,[83] there is a rectilinear relationship between the elution volumes of hydrocarbons (and alcohols and acids derived therefrom) and the logarithm of their molecular weights.

In one type of l.c. procedure, the column is filled with the dry adsorbent without any l)quid phase (known as "dry-column chromatography"). For example, Bonar[84] separated glycerides of cacao butter on a dry-packed column of paraffin supported on cellulose powder, using acetone–methanol as eluent. Kubeczka[85] separated essential oils using the following method. The essential oil in pentane is placed on a 10 cm column of silica gel and covered with a 1 cm layer of silica gel; the column is percolated with pentane to give fraction 1 and then with benzene to give fraction 2; finally the dry column is extruded and cut into three sections, to be suspended in ethyl ether–methanol to give fractions 3, 4, and 5. It is apparent that such separations cannot provide precise quantitative data. This is also true with centrifugal chromatography; for example, carotenoid mixtures were separated by Pfander et al.[86] after centrifuging the silica gel column at 3000 rpm, extruding the packing, drying, slicing into sections, and eluting with CH_2Cl_2.

4.5 SEPARATION BY ION-EXCHANGE, ION-PAIR, OR COMPLEXATION CHROMATOGRAPHY

Chemical reactions are involved when organic mixtures are separated by ion exchange, ion-pair formation, or complex formation. Generally speaking, ion-exchange methods are applicable to acidic and basic compounds. For instance, Fritz et al. studied the conditions for separation of phenols, carboxylic acids,[87] and aromatic sulfonic acids[88] with anion exchangers.

Lazdins et al.[89] determined 23 pyridine derivatives (acids, aldehydes, aldoximes, etc. obtained by catalytic synthesis from alkylpyridines) using columns packed with cation exchanger KU-2 × 8 (H^+ or $NH_4{}^+$ form) or with anion exchanger AU-17 × 8 (Cl^- or OH^- form). Watanabe and Suzuki[90] separated salicylic acid, cinchophen, and amidopyrine on a column of Amberlite CG-50; the first two compounds were successively eluted with 0.2 *M* acetate buffer of pH 4.6 containing 10% ethanol, and amidopyrine was eluted with 0.1 *M* HCl; quantitative measurements were made at 295, 255, and 250 nm, respectively, and the recovery was >95%. Nerlo et al.[91] analyzed the vitamins of baker's yeast using a column of SD cationite; riboflavine was eluted with acetone–H_2O (1 :1), thiamine was eluted with hot 0.1 *M* HCl in 25% aq. KCl, and then a combination of pyridoxine and nicotinic acid was eluted with conc. NH_4OH. Kaiser[92] separated mono-, di-, hydroxy-, and oxocarboxylic acids by means of high-pressure liquid chromatography on an anion-exchange resin.

It should be noted that certain neutral compounds are amenable to ion-exchange separation techniques. For example, Martinsson and Samuelson[93] separated sugars on columns packed with ion-exchange resin T5C ($SO_4{}^{2-}$ form) or Dowex 50W- X8 (Li^+ from) and operated at 75°C; monosaccharides were eluted with ethanol in the order of increasing number of hydroxyl groups. Walborg and Lantz[94] separated mono-, di-, and trisaccharides by ion-exchange chromatography using Dowex 1- X4 (OH^- form) with boric acid–glycerol buffers as eluents; the eluted saccharides were determined with the aniline–acetic acid–H_3PO_4 reagent, the absorbance being measured at 365 nm. According to Christofferson,[95] mixtures of aldehydes (formaldehyde, acetaldehyde, furfuraldehyde, 5-hydroxymethyl-2-furaldehyde, and vanillin) formed during the sulfite cooking of wood can be separated completely on columns of Dowex 1- X8 ($HSO_3{}^-$ form) with $HSO_3{}^-$ solution of increasing concentration as eluent. Jandera and Churacek[96] have reviewed the methods for ion-exchange chromatography of aldehydes, ketones, ethers, alcohols, polyhydric alcohols, and saccharides. Cassidy and Streuli[97] studied the separation of amides by cation-exchange resin from nonaqueous solvents; acetamides were easily separated from acetanilides because the former were not eluted with methanol–acetonitrile (1 :9); *N*-methyl- and *N*-propylacetanilides were not separated.

Ion-pair chromatography is dependent on the formation of organic salts of large molecular weights. Using the mixture mepyramine maleate–codeine sulfate, Doyle and Levine[98] showed that either component may be selectively eluted, or both may be successively eluted with different solvents. Santi et al.[99] described the separation of ergot and tropane alkaloids by means of high-speed ion-pair partition chromatography using picrate as the counterion; the picrate ion pairs of the alkaloids were injected as a solution in $CHCl_3$ and were eluted with $CHCl_3$ saturated with 0.06 *M* picric acid. Su et al.[100] investigated the separation of 14 sulfa drugs on silica columns using tetrabutylammonium as the counterion and butanol–heptane as the mobile phase; the columns were stable for long periods and could be stripped and reused with little loss in efficiency. Bosly and Bonnard[101] separated drofenine, amidopyrine, and codeine phosphate on a column of Celite impregnated with H_2SO_4. Persson[102] has reviewed the separation of ammonium compounds by ion-pair partition. It may be mentioned that the chromatographic resolution of racemic compounds on columns containing optically active polymers also uses the ion-pair principle, for example, the separation of DL-mandelic acid[103] and DL-tartrate[104] into their optical isomers.

Complexation chromatography was employed by Klimisch and Fox[105] to separate polycyclic aromatic hydrocarbons from their nitrogen-containing heterocyclic analogues; the column was packed with Kieselgel pretreated with $AgNO_3$ solution and elution was effected with 1% acetonitrile solution in hexane, the heterocyclic compounds being retained on the column. Inczedy[106] described the separation of acids as complexes with bivalent metals on anion-exchange resins.

4.6 SEPARATION BY THIN-LAYER OR PAPER CHROMATOGRAPHY

Voluminous literature has been published on thin-layer and paper chromatography. A bibliography of these publications has been compiled. During a 5-year period from 1961 to 1965, nearly 10,000 literature references are cited.[107] Because of inherent difficulties, however, quantitative analysis of organic mixtures by means of paper or thin-layer chromatography is not commonly employed. Nevertheless, these techniques are useful under certain circumstances. A brief discussion is given below. It should be mentioned that the ring-oven technique (e.g., the determination of antioxidants by Sibalic et al.[108]) and thread chromatography as described by Szabo et al.[109] utilize principles similar to those of paper chromatography. Recently Zlatkis and Kaiser[109a] edited a monograph on high-performance thin-layer chromatography, which is defined as the combined action of several factors, including (*a*) an optimized coating material with a separation power superior to the best high-performance liquid chromatographic separation material, (*b*) a new method of feeding the mobile phase, (*c*) a novel procedure for layer conditioning, (*d*) a considerably improved dosage method, and (*e*) a competent data acquisition and processing system.

4.6.1 Applying the Sample and Developing the Chromatogram

Bark et al.[110] described a method for increasing the reproducibility of chromatographic parameters in thin-layer chromatography (t.l.c.) by applying multiple spots simultaneously in <1 min and using a modified saturation chamber; R_F values were reproducible to within ± 0.01 unit. Ouelette and Balcius[111] studied the factors affecting behavior in t.l.c. mathematically and concluded that prediction of R_F values was not possible. According to Honegger,[112] a gradient of activity of the adsorbent (alumina) along a thin-layer plate is easily produced by partical deactivation of the alumina with water. Niederwieser and Brenner[113] proposed partial separation of the components of the mobile phase on the thin layer of the adsorbent to give two solvent fronts to aid the chromatographic separation of the sample mixture.

Bacon[114] constructed an apparatus for quantitative application as streaks in t.l.c. or paper chromatography. Berthold[115] advocated confining the sample, during development, to narrow bands formed by removal of parallel thin lines of the adsorbent; this ensures that the breadth of the spots is constant, and thus the amount of material in the spot can be calculated, after calibration, from the length of the spot. Sources of error in quantitative t.l.c. and paper chromatography were studied by Fairbairn and Relph[116]; errors of up to 20% can occur during application of a solution by means of a conventional syringe, mainly as a result of "creep back" outside the needle and "capillation" inside the needle. A device that delivers samples by rapid ejection overcomes these disadvantages.

The environment may seriously affect t.l.c. results. For example, McGregor and Khan[117] reported that the presence in the laboratory atmosphere of pyridine, HCl, or H_2O vapor led to poor separation and low values in the analysis of urinary 4-hydroxy-3-methoxymandelic acid. In the determination of fatty acids by paper chromatography as copper soaps, Kaufmann and Ahmed[118] found that (*a*) the quality of the paper, (*b*) the purity of the reagents, (*c*) the quality of the water used for removal of excess copper acetate from the chromatogram, and (*d*) losses due to autooxidation of polyenic acids are of importance. Idler et al.[119] investigated the destruction of microquantities of steroids on silica gel by repeated t.l.c.; hydrocortisone was the most subject to destruction.

4.6.2 Methods of Quantitation

For compounds that are colored or can produce colored products when sprayed with suitable reagents, the simplest method of quantitation is by visual comparison against standards. Understandably, this method is the least accurate; thus Scherz[120] reported that sugars or amino acids could be estimated to within ±20%. Baixas et al.[121] measured the spot areas planimetrically in determining sugars by paper chromatography, showing a rectilinear relationship between the logarithm of the weight of sugar versus area of spot; del Campo[122] used a similar technique to analyze explosives by t.l.c. For compounds that fluoresce, Brühl and Schmid[123] described a method for automated planimetric measurement of the fluorescent zones on t.l.c. plates under u.v. light at 366 nm.

Spectrophotometry is the preferred technique of quantitation and can be performed directly on the chromatogram. Lefor and Lewis[124] have reviewed the instrumentation for quantitative evaluation of t.l.c. Ma and Roper[125] described a simple attachment to the Beckman DU spectrophotometer for measuring the absorbance of a paper chromatogram after both sides of the paper are sprayed. Klaus[126] studied the effect of the shape of the spot on the photometric values obtained and pointed out that failure to consider deformation of spots can lead to inaccurate results. To eliminate error due to inhomogeneity of the paper, Hörer and Popescu[127] recommended using two wavelengths; other workers[128,129] advocated reflection instead of transmission measurements. Eich et al.[130] reported on the direct quantitative evaluation of t.l.c. by remission and fluorescence measurements; the mean deviation of results calculated from three measurements of a spot was <2.7%. Libby[131] described quantitative t.l.c. with x-ray emission spectrometry.

The spots on a chromatogram can be removed or eluted with solvents and the solutions obtained are then measured spectrophotometrically; albeit tedious, this technique generally gives accurate results. For example, Frodyma et al.[132] analyzed water-soluble dyes by t.l.c. using reflectance; direct measurements gave reproducibility of ±5%, but this figure was much improved when the spots were scraped from the plate. Lieu et al.[133] described a simple device for removing spots. To reduce extraction errors, de Deyne and Vetters[134] concentrated the spots in small areas, after removal of unwanted adsorbent, or transferred the material to pieces of filter paper by rapid development at right angles to the original direction of development. Krueger et al.[135] used Kieselgel or cellulose sheets for t.l.c., cut out the zones, and extracted with an appropriate solvent in the sheet-elution apparatus. The cutout technique is frequently used in quantitative paper chromatography. For instance, Kay et al.[136] determined amino acids by spraying with ninhydrin; the resulting purple bands were cut out and eluted in closed tubes by shaking with ethanol. Schriefers et al.[137] eluted the paper using a 10 ml syringe with the needle bent to form a hook; the paper was hung on the hook and the eluate was collected in a beaker.

Choulis[138] performed quantitative t.l.c. using an interference refractometer. Siakotos and Rouser[139] photographed the chromatogram after it was charred, and the developed photograph was measured by densitometry. Dittrich[140] exposed the chromatogram to iodine vapor; the iodinated spots were then eluted with KI solution and the iodine was titrated with 0.001 *N* $Na_2S_2O_3$. Pazdera et al.[141] described quantitative procedures for paper chromatography using colorimetry, nonaqueous titrimetry, polarography, and so forth that are suitable for quality control. Szakasits et al.[142] performed quantitative t.l.c. using a flame ionization detector; the chromatogram was scanned at 300 to 450°C by the detector, and the signals, which are proportional to the amount of each separated component, were fed into an electrometer, recorder, and digital integrator. Sarsunova et al.[143] analyzed strychnine and brucine by t.l.c. and precipitation of the separated alkaloids with an acid solution containing ^{131}I, the radioactivity of the precipitates being measured. Neuhoff and Weise[144] determined 0.1 picomole amounts of 4-aminobutyric acid and serotonin by reacting the mixture with ^{14}C-labeled 5-dimethylaminonaphthalene-1-sulfonyl chloride and separating on t.l.c. plates coated with polyamide powder; quantitation was by autoradiography or by scintillation counting.

4.7 SEPARATION BY ELECTROPHORESIS

Compounds that can produce species carrying an electrical charge may be separated by differential migration in an electrical field. For quantitative analysis, electrophoresis is generally performed on paper strips; then the separated components are eluted and determined. While electrophoresis usually deals with macromolecules, many simple compounds containing oxygen, nitrogen, or sulfur functions[6] have been analyzed by this technique.

Sturm and Scheja[145] demonstrated the separation and determination of mixtures containing 15 to 20 phenolic acids by paper electrophoresis at 30 to 60 V per cm in an electrolyte of pyridine–acetic acid–H_2O (1 : 10 : 89) at pH 3.6. Sawicki et al.[146] described the separation and fluorometric analysis of polynuclear phenols. For the analysis of monohydric alcohols and hydroxy acids, Frahn[147] converted them into potassium xanthates by treatment with powdered KOH and CS_2 in dimethylsulfoxide; the xanthate solution was subjected to electrophoresis on filter paper impregnated with a solution of pH 6 to 13 at 4°C at 21 V per cm, using 4-nitrobenzenesulfonate as a standard.

Paper electrophoresis of carbohydrates has been studied extensively. Nagata[148] separated 15 carbohydrates on a horizontal apparatus with 1% $Na_2B_4O_7$ solution; all the migrating species traveled to the anode, and the ratio

$$\frac{\text{distance of the carbohydrate from sucrose}}{\text{distance of D-gluconic acid from sucrose}}$$

was found to be independent of current, voltage, time of electrophoresis, position and volume of drops, and nature and length of the paper strip. On the other hand, Piras and Cabib[149] reported that the use of cetyltrimethylammonium borate buffer instead of tetramethylammonium borate could reverse the order of mobilities of sugars; thus some separations may be effected that are not otherwise possible (e.g., separation of glucose and fructose). Gerlaxhe and Casimir[150] separated glucose and sorbitol using borate buffer at pH 7 or 7.5. Weigel[151] presented the absolute mobilities of reference compounds and relative mobilities of a large number of carbohydrates and derivatives. Other naturally occuring compounds that have been separated by paper electrophoresis include flavins[152] and courmarins.[153]

Most nitrogen compounds that have been analyzed by electrophoretic methods contain the amino or *N*-heterocyclic function. Corbett and Fooks[154] separated phenylenediamines and aminophenols on paper strips or Gelman Sepraphore III gel with *M* acetic acid as electrolyte and applied a potential of 10 to 50 V per cm. Mayer and Westphal[155] separated hexosamines as their anionic molybdate complexes at pH 5. For the separation of basic amino-acids, Jirgl[156] used borate buffer at pH 12; ninhydrin served as the visualization reagent. For the determination of diaminocarboxylic acids formed during the preparation of EDTA, Doran[157] converted them into the Cu(II) complexes, which were separated by low-voltage electrophoresis; the bands were revealed with sodium diethyldithiocarbamate, extracted with ethanol, and measured spectrophotometrically. The analyses of alkaloids,[158] thiamine,[159] and creatine and creatinine[160] by paper electrophoresis have been reported. Tauber[161] described the paper electrophoresis of α-keto acid dinitrophenylhydrazones and their colorimetric determination.

Skelly[162] separated mixtures of biphenyl-4-sulfonic and biphenyl-4,4′-disulfonic acids by continuous electrophoresis in 2% acetic acid and determined the two compounds by u.v. spectrophotometry; a relative error of ±0.7% for the major component was obtained at the 25 mg level. Derlikowski et al.[163] employed ionophoresis to separate 50 pharmaceutical mixtures containing from two to four compounds. Thin-layer electrophoresis has been utilized to separate gibberellins[164] and ergot alkaloids[165]

According to Yoneda and Miura,[166] complete resolution of the racemic trisethylenediamine cobalt(III) complex into its optical antipodes could be achieved by means of electrophoresis. The racemic chlorides and the diastereoisomeric tartrates were subjected to paper electrophoresis in a supporting electrolyte of Na(+)tartrate (0.18 *M*) and $AlCl_3$ (0.12 *M*) for 2.5 hr with a potential gradient of 250 V per 35 cm at 35°C. Migration distances of the trisethylenediamine cobalt(III) cations were 59 and 12 mm for the *d*- and *l*-forms, respectively. It is of interest to note that no resolution occurred when $AlCl_3$ was absent.

4.8 SEPARATION BY EXTRACTION

Solvent extraction for the purpose of quantitative analysis of organic mixtures has become less important since the development of chromatography. However, solvent extraction is still useful on certain occasions and studies have continued to be undertaken. Thus Amin et al.[167] recently described apparatus for the automated quantitative extraction of active ingredients from pharmaceutical preparations. It consists of a chamber enclosing an interchangeable filter funnel in which the sample is placed; the addition of the extracting solvent and the stirring, heating, cooling, and filtration are controlled by an adjustable electronic programmer.

Given below are some examples involving simple extraction procedures in the analysis of mixtures containing two or more compounds. Butina et al.[168] determined isomeric phthalic acids by aqueous extraction, depending on the fractional dissolution of the isomers at a known temperature, followed by titration of each extract with NaOH. Khodzhaev and Ibragimov[169] analyzed a mixture of benzoic, phthalic, trimesic, and trimellitic acids using freshly distilled $CHCl_3$ or acetone for extraction. Van Handel[170] separated sugars, lipids, and glycogen in biological samples by extracting the lipids into $CHCl_3$–CH_3OH (1 :1) and then sugars into 66% ethanol saturated with Na_2SO_4; the glycogen remained adsorbed on the Na_2SO_4. According to Pan,[171] penicillin and penicilloic acid are separated from each other by extraction with isobutyl methyl ketone at two pH values. Bontemps[172] described a method to determine binary mixtures of drugs that have similar color reactions but different partition coefficients between $CHCl_3$ and an aqueous buffer. Wollish et al.[173] determined aspirin, phenacetin and caffeine by extraction and nonaqueous titration. Similarly, Krepinsky and Stiborova[174] analyzvd amidopyrine, phenazone, and caffeine in mixtures by making use of the different solubilities of these compounds in benzene and $CHCl_3$. Welch and Welch[175] reported a solvent-extraction method for the simultaneous determination of noradrenaline, dopamine, 5-hydroxytryptamine, and 5-hydroxyindol-3-ylacetic acid in one mouse brain. Examples of solvent extraction for analyzing mixtures obtained in industrial processes include products of aniline sulfonation,[176] alkanesulfonates,[177] chlorotoluenes,[178] and hydrocarbons in admixture with oxidation products.[179]

Differential analysis[180] by countercurrent distribution is useful for the separation of complex mixtures. For example, Aho et al.[181] fractionated tall-oil fatty acids by countercurrent distribution followed by gas chromatography of the methyl esters of the acids obtained. Saha and Basak[182] separated phenols between cyclohexane and aqueous 10% NaCl. Countercurrent separation is also recommended for the analysis of mixtures of steroids.[183-185] Schönenberger et al.[186] determined amine mixtures and demonstrated that complete separation was not necessary since the composition of the mixture could be obtained by graphical analysis of the overlapping distribution curves. Zinner[187] separated isoniazid and *p*-aminosalicylic acid on a 50 mg sample in 25 steps using isobutanol–water. Hickey and Phillips[188] separated aureomycin from terramycin quantitatively by using a *n*-butanol–aq. 0.01 *N* HCl system. Mold et al.[189] carried out selective separation of polycylic aromatic hydrocarbons with a solvent system containing 1,3,7,9-tetramethyluric acid as complexing agent.

Solvent extraction based on ion-pair formation is a convenient way to separate binary mixtures. Thus Matsui and French[190] separated pharmaceutical amines of different polarity (e.g., diphenohydramine and ephedrine) by extracting (e.g., with benzene containing $<1\%$ isoamyl alcohol) from a buffered aqueous phase an ion pair derived from the amine and an indicator dye; the less polar amine was determined with bromocresol purple at 410 nm, while the two amines together were determined with bromothymol blue. Pellerin et al.[191] determined a number of alkaloids using ethanolic Eriochrome black T at pH 4 to 6 and extracting with $CHCl_3$. Tsubouchi[192] reacted quaternary ammonium compounds with tetrabromophenolphthalein ethyl ester in ethanol at pH 10, then diluted with H_2O, extracted with $C_2H_4Cl_2$, and measured the absorbance of the organic phase at 615 nm against a reagent blank. Hartung and Jewell[193] separated indole, carbazole, and phenazine from petroleum products by forming the perchlorates in 72% $HClO_4$, followed by extraction of the perchlorates into benzene. Borg[194] demonstrated ion-pair extraction in the low-concentration range, with applications to the analysis of biological materials.

When extraction is performed for quantitative analysis, it should be recognized that the solvent and environment may affect the sample. For example, in the separation of farnesylacetone from phytol by conversion of the former into its Girard-T derivative in 15% ethanol, followed by extraction of the unreacted phytol into hexane, Osman and Barson[195] found an appreciable amount of farnesylacetone in the hexane layer as a result of the hydrolysis of the hydrazone; the separation is improved if the reaction is carried out in dimethylsulfoxide medium and the phytol is extracted into light petroleum. Comparing the four extraction procedures for root-translocated dieldrin in plants, Caro[196] reported that the best values were obtained by blending hexane–isopropyl alcohol (2 : 1) and then performing Soxhlet extraction with $CHCl_3$–CH_3OH (1 : 1). Dexheimer and Fuchs[197] investigated the separation of the intimate mixture of two polymers by extraction with a liquid that is a solvent for only one of them; it was demonstrated that the extent of extraction depended on the particle size of the mixed polymer, the molecular weights having little influence.

4.9 SEPARATION BY DISTILLATION

Fractional distillation of a liquid mixture, even when fractionating columns of high efficiency are employed, is generally carried out for the purpose of organic preparation and not for quantitative analysis. On the other hand, simple distillation may be performed on a mixture of liquids or a solution of solid materials as a preliminary step for the subsequent analysis of the desired components. For example, in a procedure for the determination of preservatives (e.g., benzoic, salicylic, sorbic, dehydroacetic acids, *p*-hydroxybenzoates) in food products, Prohl et al.[198] mixed the sample with $MgSO_4$ and citric or tartaric acid, plus (for solid samples) H_2O and then steam distilled; the distillate was acidified and extracted with ethyl ether, and the dried extract was concentrated and then separated and determined by thin-layer chromatography. For the analysis of the volatile components of papaya fruit, Katague and Kirch[199] homogenized the fruit with H_2O, steam distilled the mixture, saturated the distillate with NaCl, extracted with ethyl ether, dried the extract with Na_2SO_4, concentrated under reduced pressure, and injected the residue into the gas chromatograph. To determine the composition of the turpentine oil obtained from larch (*Larix europaea*), Matawowski and Stopinska[200] separated the oil into six fractions by distillation; subsequently, each fraction and the oil itself were analyzed by gas chromatography.

Recently Zhukhovitskii et al.[201] described a method for the separation of mixtures based on distillation under chromatographic conditions (chromadistillation). The sample is carried by the carrier gas into a column packed with inert material (e.g., steel beads), the column temperature decreasing from the inlet to the outlet. It was shown that, at the column outlet, adjacent zones of the pure components of the resolved mixture are recorded. This method permits both qualitative and quantitative analysis of mixtures (e.g., *n*-alkanes).

An apparatus for quantitative fractionation of phenols by steam distillation was proposed by Dumazert and Ghiglione.[202] The device consists of a motor-driven syringe, which feeds water into a microboiler at a constant rate. The microboiler is a stainless-steel U tube, electrically heated and containing a roll of fine-mesh metal cloth to promote instantaneous volatilization. Steam is conveyed through a silicone-rubber tube into glass columns, which are packed with firebrick (315 to 630 μm) and coated with glycerol, cholesterol, or silicone oil and are contained in a thermostatically controlled oven. The effluent is condensed in small fractions with a microcondenser, and phenols are determined colorimetrically.

Certain combinations of organic compounds can be easily separated by means of distillation. The following examples may be cited. Warner and Raptis[203] determined formic acid in the presence of acetic acid by azeotropic distillation of formic acid with $CHCl_3$. For the determination of formaldehyde and methanol, Wenger and Kutschke[204] separated methanol from the formaldehyde by distillation in the presence of chromotropic acid, and, after oxidation, determined it as formaldehyde colorimetrically. Rosenberger and Shoemaker[205] determined the composition of mixtures containing ethylene glycol, glycerol, and water by azeotropic distillation using three entrainers; benzene entrains only water, while tetrachloroethylene forms azeotropes with both water and ethylene glycol, and limonene carries over all three components. Obukhova,[206] separated and determined higher normal fatty acids by distillation of their methyl esters over specific ranges. Fair and Friedrich[207] analyzed alkylphenol mixtures by fractional distillation at 50 torr from a column containing 25 theoretical plates; Murthy et al.[208] employed a similar technique to analyze mixtures containing chlorinated ethanes and ethylenes. Soucek and Frankova[209] separated trichloroethylene from trichloroacetic acid in solution by distillation under reduced pressure and determined it in aqueous solution, while trichloroacetic acid was determined in the residue. With the current emphasis on gas chromatography, these distillation methods have been ignored. They may be useful, however, on occasions when only a few samples are to be analyzed, since these methods do not require preliminary work, such as finding the proper chromatographic column and conditions, and the preparation of calibration graphs.

4.10 OTHER SEPARATION TECHNIQUES

In spite of the tedium of precipitation and filtration, this technique is employed in some procedures for the analysis of organic mixtures. Thus Uksila et al.[210] separated the fatty acids in linseed oil by crystallization. For the separation of toluene-2,4-diamines and toluene-2,6-diamines, Pakhomov and Mitrofanova[211] recently proposed the following method involving cocrystallization with naphthalene. The sample (10 to 40 mg) containing these isomers was dissolved in decane (250 ml); naphthalene (1530 mg) was added to 10 ml of this solution, and the suspension was heated at 40 to 50°C while the solid phase dissolved. The supersaturated solution of naphthalene was stirred at 20°C for 40 min and was then set aside for 20 min; the precipitate formed was separated. Toluene-2,4-diamine in the precipitate was then determined spectrophotometrically by a diazotization and coupling method. Toluene-2,6-diamine remained in the liquid phase.

Quantitative separation of nitrogen bases using ammonium reineckate was reported by Lee[212]; bases with pK_b values sufficiently far apart can be separated by pH adjustment of the reineckate solution. Biehl and Li[213] described two precipitation procedures to determine *o*- and *m*-aminobenzoic acids in mixtures. In one procedure, both isomers are converted into tribromo derivatives with 0.1 *N* KBr–$KBrO_3$, the *o*-isomer being converted into 2,4,6-tribromoaniline and the *m*-isomer into 3-amino-2,4,6-tribromobenzoic acid; the mixture is then made alkaline and the undissolved 2,4,6-tribromoaniline is collected and weighed. In another procedure, the sample solution is treated with $Zn(NO_3)_2$ solution at pH 5.5, whereupon the zinc-*o*-aminobenzoate is precipitated; then the precipitate is dissolved in 4 *N* HCl and analyzed by titration with standardized KBr–$KBrO_3$ solution.

Matsuda et al.[214] separated sugars by sublimation of the sugar derivatives (e.g., acetates of pentoses and hexoses) on a horizontal apparatus; the temperatures of the zones at which the sublimate condensed were found to be closely related to the molecular weights. Free sugars showed no distinct zones, since hemiacetal groups are very labile and readily undergo decomposition.

Clampitt[215] demonstrated that the composition of a linear polyethylene–high-pressure polyethylene mixture could be determined by differential thermal analysis. Three endothermic peaks were shown on the thermogram, at 115, 124, and 134°C. The area under the peak at 134°C was proportional to the content of linear polyethylene in the mixture.

Polyakova et al.[216] described a mass-spectrometric method for determining the group hydrocarbon composition of the products obtained by thermal cracking of paraffin, polymerization of propene, and so on. To avoid the memory effect, the mass spectra were recorded at an analyzer-chamber with a feed-system temperature of 100°C. For each type of sample, the authors give a system of equations from which the content of paraffin, monoolefinic, naphthenic, diene, and cycloolefinic hydrocarbons, and alkyl- and alkenylbenzenes can be determined. The relative error is 10%. The sample required is 0.1 to 3 ml. An analysis takes 1.5 to 2 hr.

REFERENCES

1. B. L. Karger, C. R. Snyder, and C. Horvath, *An Introduction of Separation Science*, Wiley, New York, 1973.

2. E. W. Berg, *Physical and Chemical Methods of Separation*, McGraw Hill, New York, 1963.
3. *Proc. Anal. Chem. Conf. 3rd, Budapest*, **1970**, Vol. 1, *Separation Methods*.
4. L. D. Metcalfe, *Anal. Chem.*, **33**, 1559 (1961).
5. T. S. Ma and V. Horak, *Microscale Manipulations in Chemistry*, Wiley, New York, 1976.
6. N. D. Chernois and T. S. Ma, *Organic Functional Group Analysis*, Wiley, New York, 1964.
7. R. L. Grob (Ed.), *Modern Practice of Gas Chromatography*, Wiley, New York, (1977).
8. Symposium on the Basis of Chromatographic Separation Techniques, With Special Consideration of Gas Chromatography, *Ber. Bunsenges. Phys. Chem.*, **69**, 758 (1965).
9. T. S. Ma and A. S. Ladas, *Organic Functional Group Analysis by Gas Chromatography*, Academic, London, 1976.

9a. K. Blau and G. King (Eds.), *Handbook of Derivatives for Chromatography*, Heydew, London, 1977.

10. M. M. Schachter and T. S. Ma, *Mikrochim. Acta*, **1966**, 55.
11. N. Nikuchi and T. Miki, *Jap. Analyst*, **17**, 1102 (1968).
12. C. Eaborn, C. A. Holder, D. R. M. Walton, and B. S. Thomas, *J. Chem. Soc.*, **1969**, 2502.

12a. S. A. Benezra, *J. Chromatogr. Sci.*, **14**, 122 (1976).

13. T. Walle and H. Ehrsson, *Acta Pharm. Suec.*, **7**, 389 (1970).
14. A. Di Corcia, *J. Chromatogr.*, **80**, 69 (1973).
15. G. G. Kostanyan, L. O. Ustyan, and A. A. Movsisyan, *Armyan. Khim. Zh.*, **23**, 134 (1970).
16. R. W. Scora, *J. Chromatogr.*, **19**, 601 (1965).
17. H. Thies and G. Schuster, *Dtsch. Apoth. Ztg.*, **108**, 1608 (1968).
18. P. Merchant, Jr., *Anal. Chem.*, **40**, 2153 (1968).
19. G. C. Goretti, A. Liberti, and G. Nota, *J. Chromatogr.*, **34**, 96 (1968).
20. M. Possanzini, A. Pela, A. Liberti, and G. P. Cartoni, *J. Chromatogr.*, **38**, 492 (1968).
21. M. Morgantini, *Riv. Stal. Sostanze Grasse*, **40**, 49 (1963).
22. D. L. Schmitt and H. B. Jonassen, *Anal. Chim. Acta*, **49**, 580 (1970).
23. M. Kotani, K. Uetake, and N. Sakikawa, *Jap. Analyst*, **25**, 863 (1976).
24. C. F. Cullis and E. Fersht, *Combust. Flame*, **7**, 185 (1963).
25. D. T. Sawyer and D. J. Brookman, *Anal. Chem.*, **40**, 1847 (1968).
26. G. Schomburg, *J. Chromatogr.*, **14**, 157 (1964).
27. J. K. Haken, *J. Gas Chromatogr.*, **4**, 85 (1966).
28. A. H. Duvall and W. F. Tully, *J. Chromatogr.*, **11**, 38 (1963).

28a. Y. V. Pepelyaev, L. N. Stepanov, and A. P. Tereshchenko, *Zavod. Lab.*, **35**, 551 (1969).

29. W. Takens, Gas, *The Hague*, **83**, 248 (1963).
30. R. G. Ackman and R. D. Burgher, *Anal. Chem.*, **35**, 647 (1963).
31. G. I. Gulko, *Azerb. Neft. Khoz.*, **1968**, (4), 44.
32. T. Sato, A. Ikegami, and Z. Fujino, *Jap. Analyst*, **10**, 854 (1961).
33. A. S. Sobolev, *Tr. Bashlirsk. Nauchno-Issled. Inst. Pererabotke Nefti*, **1963** (6), 171.
34. I. L. Butaeva and M. D. Inshakov, *Tr. Vses. Nauchno-Issled. Inst. Tsellyul.-Bum. Prom.*, **1970** (56), 101.
35. J. Novak, *Chem. Listy*, **62**, 1281 (1968).
36. J. Tranchant, *Riv. Ital. Sostanze Grasse*, **40**, 633 (1963).
37. D. R. Deans, *Chromatographia*, **1968** (5), 187.
38. H. E. Sparks, *J. Gas Chromatogr.*, **6**, 410 (1968).
38a. R. Langlais, R. Schlenkermann, and M. Weinberg, *Chromatographia*, **9**, 601 (1976).
39. J. Karlsen and K. E. Rasmussen, *Medd. Norsk. Farm. Selsk.*, **37**, 51 (1975).
40. M. G. R. Zobel, *J. Assoc. Off. Anal. Chem.*, **57**, 791 (1974).
41. A. Seher and R. Kühnast, *Fette, Seifen Anstrichm.*, **67**, 754 (1965).
42. A. E. Brandt and W. E. M. Lands, *Lipids*, **3**, 178 (1968).
42a. M. Gassiot-Matas and L. Condal-Bosch, *Afinidad*, **20**, 397 (1963).
43. L. S. Ettre and F. J. Kobot, *J. Chromatogr.*, **11**, 114 (1963).
44. Y. Halpern, Y. Houmner, and S. Patai, *Analyst*, **92**, 714 (1967).
45. L. D. Belyakova, A. V. Kiselev, and G. A. Soloyan, *Chromatographia*, **3**, 254 (1970).
46. J. Weber and T. S. Ma, *Mikrochim. Acta*, **1975-II**, 401, **1976-I**, 227.
47. K. E. Murray, J. Shipton, A. V. Robertson, and M. P. Smyth, *Chem. Ind. (Lond.)*, **1971**, 401.
48. V. L. Tal'rose and V. D. Grishin, *Dokl. Akad. Nauk SSSR*, **182**, 1361 (1968).
49. F. Bruner, A. Di Corcia, G. Gorett, and S. Zelli, *J. Chromatogr.*, **76**, 1 (1973).
50. H. Bruderreck, W. Schenider, and I. Halasz, *J. Gas Chromatogr.*, **5**, 217 (1967).
51. W. H. McFadden, R. Teranishi, J. Corse, D. R. Black, and T. R. Mon, *J. Chromatogr.*, **18**, 10 (1965).
52. G. Neurah, J. Gewe, and H. Wichern, *Beitr. Tabakforsch.*, **4**, 247 (1968).
53. T. A. Baillie, C. J. W. Brooks, and B. S. Middleditch, *Anal. Chem.*, **44**, 30 (1972).
54. R. Skinner, E. G. Gallaher, and D. B. Predmore, *Anal. Chem.*, **45**, 574 (1973).
55. W. Golkiewicz and T. Wawrzynowicz, *Chromatographia*, **3**, 357 (1970).
56. M. K. Pal and J. Nath, *Anal. Biochem.*, **57**, 395 (1974).
57. H. M. Miller, *J. Assoc. Off. Anal. Chem.*, **53**, 1100 (1970).
58. S. G. Perry, *Chem. Br.*, **7**, 366 (1971).
59. H. N. M. Stewart, R. Amos, and S. G. Perry, *J. Chromatogr.*, **38**, 309 (1968).
60. R. Stillman and T. S. Ma, *Mikrochim. Acta*, **1973**, 491, **1974**, 641.
61. R. Stillman and T. S. Ma, *Mikrochim. Acta*, **1976I**, 545.

62. C. Gonnet and J. Rocca, *J. Chromatogr.*, **120**, 419 (1976).
63. G. Valenzuela and R. Antonini, *Prostaglandins*, **11**, 769 (1976).
64. M. Knedel and A. Fateh-Moghadam, *Glas-Instr.-Tech.*, **9**, 675 (1965).
65. V. G. Kulnevich, G. F. Muzychenko, and I. A. Vishnyakova, *Zh. Anal. Khim.*, **23**, 1396 (1968).
66. D. J. Brown, *J. Assoc. Anal. Chem.*, **53**, 263 (1970).
67. A. Tiselius, J. Porath, and P. A. Albertsson, *Science*, **141**, 13 (1963).
68. J. H. Zwaving, *J. Chromatogr.*, **35**, 562 (1968).
69. K. M. Johnston, D. J. Stern, and A. C. Waiss, *J. Chromatogr.*, **33**, 539 (1960).
70. T. Seki, *J. Chromatogr.*, **29**, 246 (1967).
71. H. van Baelen, W. Heyns, and P. de Moor, *J. Chromatogr.*, **30**, 226 (1967).
72. F. E. Newsome and W. D. Kitts, *Steroid*, **26**, 215 (1975).
73. C. O. Björling and B. W. Johnson, *Acta Chem. Scand.*, **17**, 2638 (1963).
74. J. A. Anderson, *J. Chromatogr.*, **33**, 536 (1968).
75. L. D. Zelenick, *J. Chromatogr.*, **14**, 139 (1964).
76. G. I. Glover, P. S. Mariano, and S. Cheowtirakul, *Sep. Sci.*, **11**, 147 (1976).
77. R. Haavaldsen and T. Norseth, *Anal. Biochem.*, **15**, 536 (1966).
78. R. L. Bridges, L. R. Fina, and S. L. Tinkler, *J. Chromatogr.*, **39**, 519 (1969).
79. A. J. W. Brooks and R. K. Robertson, *Chem. Ind. (Lond.)*, **1967**, 2110.
80. A. L. Baetz and H. Diehl, *J. Chromatogr.*, **34**, 534 (1968).
81. H. H. Oelert, *Erdoel Kohle*, **22**, 19 (1969).
82. R. C. Talarico, E. W. Albaugh, and R. E. Snyder, *Anal. Chem.*, **40**, 2192 (1968).
83. T. L. Chang, *Anal. Chim. Acta*, **39**, 519 (1967).
84. A. R. Bonar, *Chem. Ind. (Lond.)*, **1965**, 221.
85. K. H. Kubeczka, *Chromatographia*, **6**, 106 (1973).
86. H. Pfander, F. Haller, F. J. Leuenberger, and H. Thommen, *Chromatographia*, **9**, 630 (1976).
87. J. S. Fritz and A. Tateda, *Anal. Chem.*, **40**, 2115 (1986).
88. J. S. Fritz and R. K. Gillette, *Anal. Chem.*, **40**, 1777 (1968).
89. I. Lazdins, I. V. Smorodina, R. Bandere, G. Glemite, and A. Avots, *Zh. Anal. Khim.*, **28**, 396 (1973).
90. H. Watanabe and T. Suzuki, *Jap. Analyst*, **17**, 1264 (1968).
91. H. Nerlo, S. Pawlak, and W. Czarnecki, *Acta Pol. Pharm.*, **26**, 173 (1969).
92. U. J. Kaiser, *Chromatographia*, **6**, 387 (1973).
93. E. Martinsson and O. Samuelson, *J. Chromatogr.*, **50**, 429 (1970).
94. E. F. Walborg, Jr., and R. S. Lantz, *Anal. Biochem.*, **22**, 123 (1968).
95. K. Christofferson, *Anal. Chim. Acta*, **33**, 303 (1965).
96. P. Jandera and J. Churacek, *Chromatogr., Rev.*, **18**, 55 (1974).
97. J. E. Cassidy and C. A. Streuli, *Anal. Chim. Acta*, **31**, 86 (1964).

98. T. D. Doyle and J. Levine, *Anal. Chem.*, **39**, 1282 (1967).
99. W. Santi, J. M. Huen, and R. W. Frei, *J. Chromatogr.*, **115**, 423 (1975).
100. S. Su, A. V. Hartkopf, and B. L. Karger, *J. Chromatogr.*, **119**, 523 (1976).
101. J. Bosly and J. Bonnard, *J. Pharm. Belg.*, **28**, 62 (1973).
102. B. Persson, *Acta Pharm. Suec.*, **8**, 217 (1971).
103. G. Manecke and W. Lamer, *Naturwissenschaften*, **54**, 647 (1967).
104. D. A. Gonci and W. C. Purdy, *Sep. Sci.*, **4**, 243 (1969).
105. H. J. Klimisch and K. Fox, *J. Chromatogr.*, **120**, 482 (1976).
106. J. Inczedy, *Magy. Kem. Lapja*, **23**, 621 (1968).
107. Bibliography of Paper and Thin-layer Chromatography, 1961 to 1965, and Survey of Applications, *J. Chromatogr., Suppl. Vol. (*1968).
108. S. M. Sibalic, V. M. Adamovic, and N. Miletic, *Mikrochim. Acta*, **1967**, 1028.
109. M. Szabo, L. Meszaros, G. Horvath, and F. Sirokman, *Acta Univ. Szeged. Phys. Chem.*, **20**, 487 (1974).
109a. A. Zlatkis and R. E. Kaiser, "High Performance Thin-Layer Chromatography," Department of Chemistry, University of Houston, 1971.
110. L. S. Bark, R. J. T. Graham, and D. McCormack, *Talanta*, **12**, 122 (1965).
111. R. P. Ouelette and J. F. Balcius, *J. Chromatogr.*, **29**, 247 (1967).
112. C. G. Honegger, *Helv. Chim. Acta*, **47**, 2384 (1964).
113. A. Niederwieser and M. Brenner, *Experientia*, **21**, 50 (1965).
114. M. F. Bacon, *J. Chromatogr.*, **16**, 552 (1964).
115. I. Berthold, *Z. Anal. Chem.*, **240**, 320 (1968).
116. J. W. Fairbairn and S. J. Relph, *J. Chromatogr.*, **33**, 494 (1968).
117. R. F. McGregor and M. Khan, *Clin. Chim. Acta*, **14**, 844 (1966).
118. H. P. Kaufmann and A. K. S. Ahmed, *Fette, Seifen, Anstrichm.*, **66**, 757 (1964).
119. D. R. Idler, N. R. Kimball, and B. Truscott, *Steroids*, **8**, 865 (1966).
120. H. Scherz, *Mikrochim. Acta*, **1967**, 490.
121. J. E. Baixas, L. Codern, M. Montagut, and J. E. Straub, *Afinidad*, **20**, 407 (1963).
122. P. del Campo, *Infcion. Quim. Anal. Pura Apl. Ind.*, **20**, 108 (1966).
123. W. Brühl, Jr. and E. Schmid, *Arzneim.-Forsch.*, **20**, 485 (1970).
124. M. S. Lefar and A. D. Lewis, *Anal. Chem.*, **42**, 79A (1970).
125. T. S. Ma and R. Roper, *Mikrochim. Acta*, **1968**, 169.
126. R. Klaus, *J. Chromatogr.*, **16**, 311 (1964).
127. O. L. Hörer and M. Popescu, *Anal. Chim. Acta*, **35**, 6 (1966).
128. W. Braun and G. Kortüm, *Zeiss Inf.*, **1968** (67), 27.
129. G. Ackermann and G. Assmus, *Z. Anal. Chem.*, **200**, 418 (1964).
130. E. Eich, H. Geissler, E. Mutscher, and W. Schunack, *Arzneim.-Forsch.*, **19**, 1895 (1969).
131. R. A. Libby, *Anal. Chem.*, **40**, 1507 (1968).

132. M. M. Frodyma, R. W. Frei, and D. J. Williams, *J. Chromatogr.*, **13**, 61 (1964).
133. V. T. Lieu, R. W. Frei, M. M. Frodyma, and I. T. Fukui, *Anal. Chim. Acta*, **33**, 639 (1965).
134. V. J. R. de Deyne and A. F. Vetters, *J. Chromatogr.*, **31**, 261 (1967).
135. H. Krueger, J. Kurzidim, and R. Mueller, *Chromatographia*, **9**, 211 (1976).
136. R. E. Kay, D. C. Harris, and C. Entenman, *Arch. Biochem. Biophys.*, **63**, 14 (1956).
137. H. Schriefers, H. Thomas, and F. Pohl, *Clin. Chim. Acta*, **8**, 744 (1963).
138. N. H. Choulis, *J. Chromatogr.*, **30**, 618 (1967).
139. A. N. Siakotos and G. Rouser, *Anal. Biochem.*, **14**, 162 (1966).
140. S. Dittrich, *J. Chromatogr.*, **12**, 47 (1963).
141. H. J. Pazdera, W. H. McMullen, L. L. Ciaccio, S. R. Missan, and T. C. Grenfell, *Anal. Chem.*, **29**, 1649 (1957).
142. J. J. Szakasits, P. V. Peurifoy, and L. A. Woods, *Anal. Chem.*, **42**, 351 (1970).
143. M. Sarsunova, J. Tölgyessy, and M. Hradil, *Pharmazie*, **19**, 336 (1964).
144. V. Neuhoff and M. Weise, *Arzneim.-Forsch.*, **20**, 368 (1970).
145. A. Sturm, Jr. and H. W. Scheja, *J. Chromatogr.*, **16**, 194 (1964).
146. E. Sawicki, M. Guyer, R. Schumacher, W. C. Elbert, and C. R. Engel, *Mikrochim. Acta*, **1968**, 1025.
147. J. L. Frahn, *J. Chromatogr.*, **37**, 279 (1968).
148. A. Nagata, *J. Agric. Chem. Soc. Jap.*, **35**, 8 (1961).
149. R. Piras and E. Cabib, *J. Chromatogr.*, **8**, 63 (1962).
150. S. Gerlaxhe and J. Casimir, *Bull. Inst. Agron. Gembloux*, **25**, 265 (1957).
151. H. Weigel, *Adv. Carbohydr. Chem.*, **18**, 61 (1963).
152. P. Cerletti and N. Siliprandi, *Biochem. J.*, **61**, 324 (1955).
153. J. Dutta, H. C. Chakrabortty, and B. K. Barman, *Sci. Cult.*, **28**, 84 (1962).
154. J. F. Corbett and A. G. Fooks, *J. Soc. Cosmet. Chem.*, **18**, 693 (1967).
155. H. Mayer and O. Westphal, *J. Chromatogr.*, **33**, 514 (1968).
156. V. Jirgl, *Anal. Biochem.*, **8**, 519 (1964).
157. M. A. Doran, *Anal. Chem.*, **33**, 1752 (1961).
158. S. N. Tewari, *Pharmazie*, **23**, 58 (1968).
159. D. Siliprandi and N. Siliprandi, *Biochem. Biophys. Acta*, **14**, 52 (1954).
160. J. Fischl, S. Segal, and Y. Yulzari, *Clin. Chim. Acta*, **10**, 73 (1964).
161. H. Tauber, *Anal. Chem.*, **27**, 287 (1955).
162. N. E. Skelly, *Anal. Chem.*, **37**, 1526 (1965).
163. J. Derlikowski, A. B. Narbutt-Mering, E. Perkowski, W. Weglowska, and J. Potajado-Gulinska, *Acta Pol. Pharm.*, **21**, 9 (1964).
164. G. Schneider, G. Sembdner, and K. Schreiber, *J. Chromatogr.*, **19**, 358 (1965).
165. S. Agurell, *Acta Pharm. Suec.*, **2**, 357 (1965).
166. H. Yoneda and T. Miura, *Bull. Chem. Soc. Jap.*, **43**, 574 (1970).

167. M. Amin, Z. Korbakis, and D. Petrick, *Z. Anal. Chem.*, **279**, 283 (1976).
168. I. V. Butina, V. G. Plyusnin, and N. A. Shevchenko, *Izv. Sib. Otd. Akad. Nauk SSSR*, **1962** (6), 68.
169. G. Khodzhaev and A. P. Ibragimov, *Dokl. Akad. Nauk, Uzb. SSR*, **1955** (8), 17.
170. E. van Handel, *Anal. Biochem.*, **11**, 266 (1965).
171. S. C. Pan, *Anal. Chem.*, **26**, 1438 (1954).
172. R. Bontemps, *Pharm. Weekbl.*, **93**, 61 (1958).
173. E. F. Wollish, R. J. Colarusso, C. W. Pifer, and M. Schmall, *Anal. Chem.*, **26**, 1753 (1954).
174. J. Krepinsky and J. Stiborova, *Cesk. Farm.*, **15**, 25 (1966).
175. A. S. Welch and B. L. Welch, *Anal. Biochem.*, **30**, 161 (1969).
176. B. I. Karavaev and V. S. Vorotilova, *Izv. Vyssh. Uchebn. Zaved. Khim.*, **12**, 781 (1969).
177. W. Kupfer, J. Jainz, and H. Kelker, *Tenside*, **6**, 15 (1969).
178. C. Hanson, A. N. Patel, and D. K. Chang-Kakoti, *J. Appl. Chem.*, **18**, 89 (1967).
179. Y. S. Tsybin and R. N. Volkov, *Zavod. Lab.*, **35**, 1032 (1969).
180. L. C. Craig, *Science*, **144**, 3662 (1964).
181. Y. Aho, O. Harva, and S. Nikkala, *Tekn. Kem. Aikak.*, **19**, 190 (1962).
182. N. C. Saha and N. G. Basak, *J. Sci. Ind. Res., India*, **20**, 11 (1961).
183. C. E. Morreal, T. L. Dao, and P. Lonergan, *Anal. Biochem.*, **34**, 352 (1970).
184. M. Vitali, P. del Grande, and G. Pancrazio, *Ann. Chim. Roma*, **48**, 622 (1958).
185. H. Carstensen, *Acta Chem. Scand.*, **9**, 1026 (1955).
186. H. Shönenberger, H. Thies, and K. Borah, *Z. Anal. Chem.*, **251**, 294 (1970).
187. G. Zinner, *Arch. Pharm.*, **288**, 129 (1955).
188. R. J. Hickey and W. F. Phillips, *Anal. Chem.*, **26**, 1640 (1954).
189. J. D. Mold, T. B. Walker, and L. G. Veasey, *Anal. Chem.*, **35**, 207 (1963).
190. F. Matsui and W. N. French, *J. Pharm. Sci.*, **60**, 287 (1971).
191. F. Pellerin, J. A. Gautier, O. Barat, and D. Demay, *Chim. Anal. (Paris)*, **45**, 395 (1963).
192. M. Tsubouchi, *J. Pharm. Sci.*, **60**, 943 (1971).
193. G. K. Hartung and D. M. Jewell, *Anal. Chim. Acta*, **26**, 514 (1962).
194. K. O. Borg, *Acta Pharm. Suec.*, **8**, 1 (1971).
195. S. F. Osman and J. L. Barson, *Anal. Chem.*, **39**, 530 (1967).
196. J. H. Caro, *J. Assoc. Off. Anal. Chem.*, **54**, 1113 (1971).
197. H. Dexheimer and O. Fuchs, *Makromol. Chem.*, **96**, 172 (1966).
198. L. Prahl, R. Engst, and E. Jarmatz, *Nahrung*, **12**, 845 (1968).
199. D. B. Katague and E. R. Kirch, *J. Pharm. Sci.*, **54**, 891 (1965).
200. A. Matawowski and Z. Stopinska, *Chem. Anal. (Warsaw)*, **9**, 863 (1964).
201. A. A. Zhukhovitskii, S. M. Yanovskii, and V. P. Shvartsman, *J. Chromatogr.*, **119**, 591 (1976).

202. C. Dumazert and C. Ghiglione, *Bull. Soc. Pharm. Mars.*, **11**, 26 (1962).
203. B. R. Warner and L. Z. Raptis, *Anal. Chem.*, **27**, 1783 (1955).
204. F. Wenger and K. O. Kutschke, *Anal. Chim. Acta*, **21**, 296 (1959).
205. H. M. Rosenberger and C. J. Shoemaker, *Anal. Chem.*, **29**, 100 (1957).
206. L. K. Obukhova, *Zh. Anal. Khim.*, **11**, 193 (1956).
207. F. V. Fair and R. J. Friedrich, *Anal. Chem.*, **27**, 1886 (1955).
208. B. N. Murthy, G. N. Bhat, and N. R. Kuloor, *Chem. Age, India*, **14**, 327 (1963).
209. B. Soucek and E. Frankova, *Prac. Lek.*, **4**, 264 (1952).
210. E. Uksila, P. Roine, E. L. Syvaoja, and A. Alivara, *Acta Chem. Scand.*, **17**, 2622 (1963).
211. L. G. Pakhomov and V. A. Mitrofanova, *Zh. Anal. Khim.*, **29**, 406 (1974).
212. K. T. Lee, *Nature*, **182**, 655 (1958).
213. E. R. Biehl and H. M. Li, *Anal. Chem.*, **38**, 1422 (1966).
214. K. Matsuda, N. Yamaoka, M. Nakajima, M. Kusaka, and K. Aso, *J. Agric. Chem. Soc. Jap.*, **38**, 413 (1964).
215. B. H. Clampitt, *Anal. Chem.*, **35**, 577 (1963).
216. A. A. Polyakova, R. A. Khmelnitskii, F. A. Medvedev, and K. I. Zimina, *Khim. Teknol. Topl. Masel*, **1962**, (12), 59.

CHAPTER 5

Separation of Acidic Substances

5.1 SEPARATION BY ION-EXCHANGE REACTIONS

Generally speaking, an organic mixture exhibits acidic behavior when it contains one or more compounds that possess the carboxyl, sulfonyl, or phenolic function.[1] Hence the separation of acidic substances usually involves methods to differentiate phenols, carboxylic acids, and sulfonic acids. For such a purpose, ion-exchange is the method of choice, because the experimental procedure requires only simple equipment and is easy to perform. Combination of two or more ion-exchange columns in series[2] often permits the separation of complex mixtures. When the separated acidic compounds are present in quantities of 0.1 millimole or more, they can be determined by titrimetry. For smaller quantities, the individual constituents can be determined by spectrophotometric methods.

Table 5.1 gives examples of mixtures of acidic compounds that have been separated by ion-exchange reactions. It should be mentioned that some mixtures containing compounds less acidic than phenols also can be separated. For instance, Seki[34] reported the separation of dinitrophenylhydrazides of nine fatty acids using Amberlite IRC-50 with ethyl methyl ketone–acetone–H_2O as eluent; the u.v. absorption of each fraction was measured at 340 nm and recoveries ranged from 97 to 105%.

Elution at elevated temperatures is prescribed in a number of procedures. Larsson et al.[35] investigated the influence of temperature in the anion-exchange separation of organic acids in acetate medium up to 80°C. Results of experiments at two concentrations of sodium acetate (0.05 and 0.1 *M*) for 10 acids show that anions with low distribution coefficients exhibit an increased distribution coefficient at high temperature, whereas anions that are firmly held by the resin show a decreased affinity for the resin. In most

Table 5.1 Separation of Acidic Compounds by Ion-Exchange Resins

Acidic function	Mixture	Ion-exchange resin	Reference
Carboxyl	Acetic, propionic, etc.	Dowex 1-X8	Egashira[3]
	Water-soluble aliphatic acids	Diaion CA 08	Kasai[3a]
	Hydroxy acids	Dowex 1-X8	Samuelson,[4] Gedda,[5] Carlsson[6]
	Lactic, glycolic	Dowex 1	Samuelson[7]
	Lactic, malic, tartaric	Dowex 2-X8	Courtoisier[8]
	Malic, tartaric, citric	Dowex 1	Schenker[9]
	Acrylic, methacrylic	Dowex 1-X10	Gueron[10]
	Mono-, di-, trichloracetic	Amberlite IRA-410	Funasaka[11]
	Levulinic, glyoxylic, glycolic	Dowex 1-X8	Larsson[12]
	Aldonic acids	Dowex 1	Samuelson[13]
	Idonic, gluconic	Dowex 1-X8	Aoki[14]
	Mono-, dicarboxylic	Aminex 50W-X4	Richards[15]
	Di-, tricarboxylic	Dowex 1-X4	Bengstsson[16]
	Dicarboxylic, hydroxy acids	Dowex 1-X8	Ericsson[16a]
	Phthalic, maleic, adipic, etc.	Weakly basic (OH^-)	Toryanik[17]
	Benzoic, phthalic, terephthalic	Dowex W-X4	Czerwinski[17a]
	Fumaric, glutaric, succinic	Amberlite CG-120	Seki[18]
	Trimesic, toluic, terephthalic	Amberlite CG-45	Funasaka[19]
	Maleic, fumaric	Amberlite CG-450	Funasaka[20]
	Maleic, fumaric	Dowex 50W-X4	Patel[21]
	Maleic, fumaric, acrylic, methacrylic	Aminex A27	Lefevre[22]
	Chlorobenzoic isomers	Amberlite CG-50	Funasaka[23]
	D-, L-Aspartic or mandelic	Bio-Rex 70	Gaal[24]
	Acidic amino acids	Amberlite CG-45	Jurahasi[24a]
Sulfonyl	Mono-, disulfonic	Amberlite XAD-2	Scoggins[25]
	o-, *p*-aminobenzenesulfonic	Amberlite IRA-400	Spillane[26]
	Hydroxy-, aminonaphthalenesulfonic	Amberlite CG-50	Funasaka[27]
	Naphthalene disulfonic	Amberlite CG-50	Funasaka[28]
	1-,2- and 1-,4,hydroxynaphthalene-sulfonic	Amberlite CG-50	Funasaka[29]

Table 5.1 (*Contd.*)

Acidic function	Mixture	Ion-exchange resin	Reference
	Anthraquinonesulfonic isomers	Amberlite CS-50	Funasaka[30]
	1-, 2-Naphthalenesulfonic	Amberlite CG-50	Funasaka[31]
Phenolic	Phenol, substituted phenols	Amberlite IR-120	Seki[32]
	Phenolic compounds		Krampitz[33]

systems, the selectivity was lower at higher temperature. It was also found that epimerization occurs with aldonic acids at 60° C, and serious decomposition occurs with uronic acids at this temperature.

The technique and optimal conditions for elution depend on the composition of the mixture. For example, for separations of aldobionic and aldonic acids, Samuelson and Wallenius[36] recommended stepwise elution for systems containing only one simple aldonic acid and several aldobionic acids; with complex mixtures, the aldonic acids were first eluted as a group in acetate medium and then separated with $Na_2B_4O_7$ solution as eluent.

While elution with a single solvent (pure liquid or solvent mixture of fixed proportions) provides the simplest operation, sometimes gradient elution may be employed with advantage for complex systems. Thus Jorysch and Marcus[37] described a method to differentiate grape juices by gradient elution of acids present in the samples through columns of Dowex 1- X10 (formate form) with formic acid in increasing concentration as eluent.

5.2 SEPARATION BY GAS CHROMATOGRAPHY

5.2.1 Carboxylic Acids

It is generally recognized that the gas-chromatographic column is not best suited for the separation of highly polar compounds such as carboxylic acids. Nevertheless, a number of investigators have worked out procedures to separate free fatty acids by gas chromatography. The conditions are shown in Table 5.2. Most columns contain a nonvolatile acid in the stationary phase, H_3PO_4 being used in the majority of cases. The direct separation of aliphatic dicarboxylic acids has not been reported. Hoffman and White[58] analyzed mono- and disubstituted malonic acids by decarboxylating them, in the injection port of the chromatograph, to the corresponding mono-

Table 5.2 Separation of Free Acids by Gas Chromatography

Acid mixture	GC column recommended	Reference
C_1 to C_6 in tobacco smoke	Firebrick, H_3PO_4, silicone oil, stearic acid	De Wet[38]
C_4 to C_{10} in beer	Chromosorb, H_3PO_4	Arkima[39]
C_5 to C_{13}	Stainless steel, Chromosorb, H_3PO_4	Parimskii[40]
C_5 to C_{20}	Glass, silanized beads, H_3PO_4	Maruyama[41]
Short-chain	Porapak, H_3PO_4	Mahadevan[42]
C_2 to C_{10}	Glass, Carbowax, isophthalic acid	Clarke[43]
C_1 to C_4	Chromosorb, isophthalic acid	Lee[44]
C_1 to C_4	Silanized Celite, sebacate–citric acid	Kaplanova[45]
Acids in vegetable oils	Diethyleneglycol polyester, succinic acid	Craig[46]
C_1 to C_6	Celite, behenic acid	Fukui[47]
C_1 to C_4	Fluorinated polymer, behenic, azelaic acids	Parimskii[48]
C_2 to C_{10}	Powdered glass washed with CrO_3, HCl, H_2O	Aminov[49]
C_1 to C_7	Stainless-steel, PEGA	Doelle[50]
C_2 to C_5	Stainless-steel, two-section	Rigaud[51]
C_2 to C_8	Diasolid, Tween	Nakae[52]
C_{17} to C_{20}	Diatomacious earth, silicone	Shcherbakov[53]
Hydroxy, keto acids	Celite, Apiezon	Szczepanska[54]
Acetic, propionic, acrylic	Glass, Chromosorb, dimer acid–silicone	Variali[55]
2-Fluoro-C_1 to C_6	Chromosorb, H_3PO_4	Gershon[56]
Mono-, dichloro-acids	Chromosorb, H_3PO_4	Smith[57]

carboxylic acids, which were then separated on a column containing Carbowax 20M–terephthalic acid on Chromosorb.

Direct gas chromatographic determination of free acids has certain advantages. It simplifies the operation and eliminates the errors when derivatives of the carboxylic function are prepared and then analyzed. For example, Shelley et al.[59] determined C_1 to C_4 acids in foodstuffs by liberating these acids by steam distillation of the sample, collecting the acids as sodium salts, regenerating the acids with dichloroacetic acid in acetone, and injecting the acid solution directly into the gas chromatograph; recoveries for 0.2 to 4.9 mg were 95 to 105%. Similarly, Gehrke and Lamkin[60] analyzed steam-volatile fatty acids using etheral solutions.

The following studies on the gas chromatographic determination of free fatty acids are of interest. Jackson[61] found that columns containing silicone–stearic acid or silicone–behenic acid as the liquid phase rapidly lost their separating power when operated with a dry carrier gas, but the activity was restored by passing the carrier gas through H_2O or by incorporating H_3PO_4 in the liquid phase. In the determination of the fatty acid composition of soya or linseed oil, Tiscornia and Boniforti[62] observed that the nature of the polyester stationary phase had little effect when helium was used as carrier gas, but a marked difference was seen between results obtained with helium as carrier gas and those obtained with hydrogen; the latter gave lower results for the more unsaturated acids and Tiscornia and Boniforti ascribed the discrepancy to the purity of hydrogen and not to hydrogenation. Ackman et al.[63] found that formic acid vapor must be added to the helium carrier gas in the analysis of higher fatty acids using a column of ethanediol succinate on Chromosorb; the formic acid gave no response and did not affect the response of other compounds. Zeman[64] investigated the behavior of fatty acids having triple bonds. A polar column containing 20% of poly(ethylene glycol adipate) on Celite operated at 180°C and a nonpolar column containing 20% of Apiezon L at 220°C were used, with nitrogen as carrier gas. Results confirmed that the high polarity of the triple bond caused the acids to have longer retention times on the polar column than the corresponding acids with two double bonds. With the nonpolar column, the ynoic acids moved faster than the corresponding saturated acids, but not as fast as the enoic acids.

Although it is possible to perform direct separation of certain carboxylic acids as described above, the common practice is to separate carboxylic acids via their derivatives. A summary of the derivatives used for the separation of aliphatic and alicyclic acids is given in Table 5.3. Using these derivatives makes it easy to find suitable columns for separation. When the same column can be employed for both the free acids and their corresponding esters, the column temperature required for the latter is considerably lower. For instance, Eidus et al.[103] determined branched-chain alkanoic acids and their isobutyl esters using a column (3.5 m × 4 mm) packed with Chromosorb W HMDS supporting 18% of diethylene glycol adipate polyester and 4% of H_3PO_4; the column was operated at 190°C for the free acids and at 150°C for the esters. On the other hand, it should be noted that the conversion of the carboxylic acids to their derivatives is rarely complete. Hence, for quantitative analysis, the preparation of the calibration graphs must be carried out in the proportions and under conditions very similar to those of the samples.

Table 5.3 shows that the methyl esters are most commonly employed as derivatives. These esters can be synthesized in several ways. Methylation by means of diazomethane[104] is instantaneous and usually gives high yields.

Table 5.3 Gas Chromatographic Separation of Aliphatic and Alicyclic Acids via Derivatives

Acid mixture	Derivative	References
Volatile acids in tabacco leaves	Methyl esters	Kaburaki[65]
C_{10} to C_{20}	Methyl esters	Sato[66]
Fatty acids in pig fats	Methyl esters	Jaarma[67]
Fatty acids in cacao butter	Methyl esters	Van Wijngaarden[68]
Oleic, elaidic, stearic	Methyl esters	Cartoni[69]
Polyunsaturated	Methyl esters	Gerson[70]
Fatty acids in cocoanut and linseed oils	Methyl esters	Tudisco[71]
Fatty acids in rapeseed oils	Methyl esters	Zeman[72]
cis-, *trans*-fatty acids	Methyl esters	Litchfield[73]
Fruit acids	Methyl esters	Hautala[74]
Krebs-cycle acids	Methyl esters	Kuksis,[75] Estes,[76] Barnett[77]
Oxo acids	Methyl esters	Hagenfeldt,[78] Harrison[79]
Resin acids	Methyl esters	Nestler,[80], Brooks[81]
Tall-oil fatty acids	Methyl esters	Iden,[82] Paylor[83]
Fatty, cyclic acids	Methyl esters	Black[84]
2,4-D, 2,4,5-T	Methyl esters	Pursley[85]
Mono-, dibasic aliphatic	Methyl esters	Zhavnerko,[86] Lindemann[87]
Dibasic aliphatic	Methyl esters	Martynov,[88] Kazinik[89]
Cyclohexanetricarboxylic	Methyl esters	Bendel[90]
cis-, *trans*-Hydroxycyclohexane-carboxylic	Methyl esters	Kaneda[90a]
Volatile fatty acids, C_2 to C_5	Ethyl esters	Kumeno[91]
C_1 to C_7	Butyl esters	Schwarze[92]
C_1 to C_7	Pentyl esters	Langner[93]
C_3 to C_9	Decyl, phenacyl esters	Craig[94]
Succinic, fumaric, malic, citric	Trifluoroethyl esters	Aguggini[94a]
Hydroxy fatty acids C_{14} to C_{26}	Acetoxy acid methyl esters	O'Brien[95]
Cyclopropene fatty acids	Methanethiol addition products	Raju[96]

Table 5.3 (*Contd.*)

Acid mixture	Derivative	References
Pyruvic, 2-oxoglutaric	Quinoxalinones	Hoffman[97]
Fatty acids	Anilides	Umeh[98]
Aldonic acids	Trimethylsilyl derivatives	Sjöström[99]
Krebs-cycle acids	Trimethylsilyl derivatives	Horning[100]
Resin acids	Trimethylsilyl derivatives	Zinkel[101]
2,5-Dimethylmuconic acid isomers	Trimethylsilyl derivatives	Foulds[102]

Esterification of fatty acids using a methanol–BF_3 reagent[105] can be accomplished by boiling the reaction mixture for a few minutes; Hautala and Weaver[106] recommended room temperature and 16 hrs in the case of lactic, pyrivic, fumaric, succinic, malic, and citric acids. Scoggins and Fitzgerald[107] used dimethyl sulfate–methanol (1 : 19) in the presence of anhydrous Na_2SO_4 for the methylation of chlorophenoxyacetic acids. Schneider et al.[108] refluxed cyclopropenoid fatty acids with methanolic sodium methoxide. For the analysis of tall-oil rosin acids, Hetman et al.[109] injected a solution of the tetramethylammonium salts of the acids into the inlet (at 400°C) of the chromatograph; pyrolysis occurred and the resulting methyl esters were conducted into the chromatographic column.

Noguchi and Nakazawa[110] studied the analysis of a mixture of fatty acid methyl esters with different columns and reported that it was necessary for the two stationary phases used to have widely different polarities to achieve complete separation of the components. Buoncristiani et al.[111] investigated the influence of temperature on column efficiency in separating the methyl esters of linolenic, arachidic, and eicosenoic acids. Using a column packed with 20% of polyester succinate supported on Chromsorb and helium as the carrier gas, the best separation was achieved at 218°C; at 245°C the peaks for linolenic and arachidic esters coalesced. Using a column packed with 20% of polyoxyethylene succinate supported on Celite, the arachidic ester was eluted before linolenic, with good separation of the three esters at 224°C; the peaks for linolenic and arachidic esters coalesced when the column was maintained at 204°C, while the peaks for linolenic and eicosenoic esters coalesced at 243°C.

According to Ronkainen,[112] the 2,4-dinitrophenylhydrazones of oxo-monocarboxylic acids were converted into the methyl esters by treatment with diazomethane in ether at 0°C, followed by oxidation with ozone. Hishta and Bomstein[113] prepared the diethylamides from acid chlorides for gas chromatography; this reaction can be applied to the analysis of carboxylic acids.

Mixtures of aromatic acids are given in Table 5.4. Separation of the free acids was performed either on a column of Chromosorb W treated with hexamethyldisilazane,[114] or on a class column containing 15% of LAC446 and 3% of H_3PO_4 on Chromosorb W. Methyl esters are the preferred derivatives. Williams[121] investigated the derivatives of six isomeric dihydroxybenzoic acids and found that separation of other derivatives was not satisfactory. Komers[126a] studied the methyl esters of benzene mono- and dicarboxylic acids on stationary phases of different polarities. On weakly polar phases, the *m*- and *p*-isomers were eluted together after the *o*-compound. Polar phases caused a change in the order of elution, the *p*-isomer being eluted first. With polar phases of the ester type, the *o*- and *m*-compounds were not separated, but with phases containing hydroxyl and cyanoethoxy groups, all isomers were completely separated.

Table 5.4 Gas-Chromatographic Separation of Aromatic Acids

Mixture	Separated as	Reference
Benzoic, *o*-, *m*-, *p*-toluic, 3,5-dimethyl-, 4-*t*-butylbenzoic	Free acids	Hill[114]
Mono-, dicarboxylic	Free acids	Witte[115]
Mono-, di-, tri-, tetra-, penta-, hexacarboxylic	Methyl esters	Kukharenko[116]
15 components	Methyl esters	Jirak[117]
Nitrobenzoic acids	Methyl esters	Smith[118]
Mono-, dicarboxylic	Methyl esters	Korbina[119]
Tri-, tetra-, pentacarboxylic	Methyl esters	Funasaka[120]
Dihydroxybenzoic acids	Methyl esters	Williams[121]
12 components, mono- to hexacarboxylic	Methyl esters	Schnitzer[122]
Dihydroxy acids	Acetylated methyl esters	Williams[123]
Salicylic acid, aspirin	Trimethylsilyl derivatives	Walter[124]
Acids in vanilla	Trimethylsilyl derivatives	Fitelson[125]
Phenolic acids	Trimethylsilyl derivatives	Blakley[126]

5.2.2 Sulfonic Acids

All methods for gas chromatography of sulfonic acids are dependent on their transformation to suitable derivatives. Here again the methyl esters[127,128] are commonly used because of the ease of preparation. Baker et al.[129] separated toluenemonosulfonic and toluenedisulfonic acids by converting them into the ethyl esters, which were prepared by refluxing the acids with a twentyfold excess of triethyl formate in toluene solution.

Himes and Dowbak[130] converted toluenesulfonic acid isomers into the sulfonyl chlorides and separated them on a stainless-steel column packed with 20% of high-vacuum grease on Chromosorb. Subsequently, Parsons[131] recommended the conversion of the sulfonyl chlorides into sulfonyl fluorides for separation; the retention times of the sulfonyl fluorides are shorter than those of the corresponding chlorides and the fluorides are more stable. Spryskov and Kozlov[132] reported on the separation of sulfonyl fluorides derived from toluene, ethylbenzene, cumene, xylenes, disulfonyl fluorides of benzene and toluene, and chloro-, dichloro-, and bromobenzenes. The same workers[133] proposed a method to analyze mixtures of isomeric chloro- or nitrobenzenesulfonic acids that is based on transforming them into the corresponding di-, tri-, or tetrachlorobenzenes through the sulfonyl chlorides.

Parsons[134] obtained volatile derivatives of aminobenzenesulfonic acids by heating the acids with a mixture of PCl_5 and $POCl_3$, whereupon the sulfonic function is converted into $—SO_2Cl$ and the amino function into $—NHPOCl_2$. Siggia et al.[135] determined arylsulfonic acids by converting them into the corresponding phenols through alkali fusion; phenols can be separated by direct gas chromatography[136] (see next section).

5.2.3 Phenolic Compounds and Thiols

Phenolic compounds containing a single hydroxyl group can be conveniently separated by direct gas chromatography. Table 5.5 gives the mixtures that has been separated and the stationary liquid phases employed. One example[148] involves di- and trihydric phenols. The behavior of polyhydric phenols was studied by von Rudloff[160] who found that for all the stationary phases investigated, except SE-30, the temperature required for elution was above the limits of thermal stability, but the use of low-loaded columns (1% of liquid phase on glass beads) overcame this difficulty. However, in such cases, it is better to separate the derivatives instead of the free phenolic compounds. Table 5.6 shows those derivatives of phenolic compounds, that are used in gas-chromatographic separation.

For quantitative analysis of phenols by gas chromatography, Weiss and Kreyenbuhl[174] observed that the specific peak area decreased with retention time. These workers recommended that a rough analysis should first be made with an internal standard, and the sample should then be analyzed with a separate internal standard for each component; each standard should be

Table 5.5 Direct Gas Chromatography of Phenolic Acids

Phenolic mixture	Stationary liquid phase	References
Phenol homologues	Polyoxyethylene glycol	Malz[137]
Phenol, ethylphenol, cresols, xylenols	Di(trimethylcyclohexyl)-phthalate	Sassenberg[138]
Phenol, cresols, xylenols	Dinonyl phthalate	Wrabetz[139]
m-, *p*-Cresols	Dimethyl phthalate	Rudolfi[140]
Alkylphenols	Apiezon	Zakupra[141]
Cresols, xylenols	Disperol–KOH	Anwar[142]
Thymol isomers	Lanolin	Porcaro[143]
t-Butylphenols	Silicone rubber	Krumbholz[144]
t-Butylcresols, 2,6-di-*t*-butylcresol	Carbowax	Kardos[145]
Butylated hydroxyanisole, di-*t*-butylcresol	Silicone gum	Choy[146]
2,6-di, 2,4,6-tri-*t*-butylphenols	Apiezon	Norikov[147]
Phenol, cresols, (poly methyl)-phenols	Phthalate–phosphate	Landault[147a]
Catechol, pyrogallol, phloroglucinol	SE-30 on silanized Diatoport	Mendez[148]
Phenols of wood tar	Poly(methylphenylsiloxane)	Shaposhinikov[149]
Chlorocresols, chloroxylenols	Reoplex	Husain[150]
Halogenated phenols	Versamid	Spiegel[151]
Mono-, di-, trichlorophenols	Polyoxyethylene glycol, H_3PO_4	Simonov[152]
Mono-, dichlorophenols	Carbowax	Kolloff,[153] Nosal[154]
Fluorophenols, chlorophenols	Tris(2,4-xylenyl)phosphate	Smith[155]
Nitrophenols	Polyoxyethyleneglycol, H_3PO_4	Hrivnak[156]
Nitrophenol isomers	Versamid	Ma[157]
Alkyldinitrophenols	LAC-2R, H_3PO_4	Clifford[158]
Phenolic pesticides	Butanediol succinate, H_3PO_4	Hrivnak[159]

similar in chemical type and concentration to the corresponding component and should form a peak close to the component.

According to Tominaga,[175] phenols containing two or more aromatic rings were partly decomposed during gas chromatography on columns of Chromosorb P or W, but could be separated on a column (40 cm × 2.5 mm) containing 10% of Apiezon L grease supported on Celite 545 (<140 mesh), operated at 300°C with helium (30 ml per min) as carrier gas. German and

Table 5.6 Gas Chromatographic Separation of Phenolic Compounds via Derivatives

Phenolic compounds	Derivative	Reference
Phenolic pesticides	Methyl or ethyl ethers	Stanley[161]
Phenolic mixtures from coal	Methyl ethers	Bhattacharyya[162]
Phenols	Methyl ethers	Bergmann[163]
Phenols in re-solution prepolymers	Acetates	Higginbottom[164]
Methylated phenols, naphthols	Trifluoroacetates	Shulgin[165]
C_6 to C_8 monohydric phenols	Trimethylsilyl ethers	Adlard[166]
Low-boiling phenols	Trimethylsilyl ethers	Freedman[167]
Hindered phenols	Trimethylsilyl ethers	Freedman[168]
Close-boiling phenolic isomers	Trimethylsilyl ethers	Oydvin[169]
Phenols from lignin	Trimethylsilyl ethers	Clark[170]
Phenols from beer	Trimethylsilyl ethers	Dallas[171]
Bisphenols	Trimethylsilyl ethers	Brydia[172]
Dichlorodiphenol isomers	Trimethylsilyl ethers	Tomori[173]

Ciot[176] determined mixtures of chlorophenols using a column of Celite impregnated with 20% of dinonyl phthalate; since the peak areas for the components were not strictly proportional to concentration, it was necessary to apply correction factors obtained by running standard prepared mixtures. Higgenbotham et al.[177] described a method for determining tetra- and pentachlorophenols in fats and oils that involves sulfonation and analysis of the resulting derivatives on a column of diethylene glycol succinate plus H_3PO_4.

Thiols are determined through their derivatives by a procedure similar to that for phenols. For instance, Croitoru and Freedman[178] prepared the acetates of toluene thiols and xylene thiols and separated them on a column of poly(propene sebacate) supported on Chromosorb; recoveries were >97%. Kawahara[179] converted thiols into pentafluorobenzyl thioethers, which were easily separated on an aluminum column of 5% of F.S. 1265 and 3% of D.C. 200 on Chromosorb P (60 to 80 mesh) operated at 200°C with nitrogen as carrier gas.

5.3 SEPARATION BY LIQUID CHROMATOGRAPHY

There are two general types of operations for liquid chromatography. One type uses wide (open-bed) columns and atmospheric or slightly higher pres-

Table 5.7 Separation of Carboxylic Acids by Liquid Chromatography

Acid mixture	Column	Mobile phase	Reference
C_2 to C_6 fatty acids	Silica gel	CCl_4–C_4H_9OH	Vioque[180]
C_1 to C_6 fatty acids	Silica gel	$CHCl_3$–*t*-C_4H_9OH	Raveux[181]
Higher fatty acids		$(CH_3)_2CO$–CH_3COOH	Sliwiok[182]
C_{10} to C_{20} fatty acids	Kieselguhr	H_2O–CH_3OH–$(CH_3)_3$-C_5H_9	Chobanov[183]
C_1 to C_4 fatty acids	Silica gel	C_4H_9OH–$CHCl_3$	Stradomskaya[184]
Plant acids	Silica gel	$CHCl_3$–C_4H_9OH	Freeman[185]
Saturated, unsaturated, C_{10} to C_{18}	Silica gel	H_2O–$(CH_3)_2CO$	Savary[186]
Citric, oxalic, malonic, succinic	Silica gel	C_4H_9OH–$CHCl_3$	Pyatnitskii[187]
Polycarboxylic	Silica gel		Frohmann[188]
Adipic, glutaric, succinic	Silica gel	C_4H_9OH–$CHCl_3$	Smith[189]
Saturated dicarboxylic	Silica gel	Benzene	Poddubnyi[190]
Acetic, chloroacetic, succinic	Silica gel	$CHCl_3$–C_4H_9OH	Markova[191]
Urinary acids	Silica gel		Orten[192]
cis-, *trans*-Cyclopentane-1,3-dicarboxylio	Silica gel	$CHCl_3$–*t*-C_4H_9OH	Bossert[193]
Pimaric, isopimaric	Silica gel	$(CH_3)_3C_5H_9$	Bruun[194]
Rosin acids	Silica gel		Bardyshev[195]
Tall-oil acids	Bio-bead gel	Tetrahydrofuran	Chang[196]
Chloro C_1 to C_2 acids	Cellulose	Ether–ligroin	Taylor[197]
Chlorophenoxy acids	Super-Cel	Ether–$CHCl_3$	Haddock[198]
Perfluoro fatty acids	Silica gel	$CHCl_3$–C_4H_9OH	Chelnokova[199]
Bezenecarboxylic acids	Silica gel	$CHCl_3$–C_4H_9OH	Kochloefl[200]
Isomeric phenolic carboxylic acids	Silica gel		Hiller[201]
Phthalic acids	Silica gel	Ether–benzene	Kamneva[202]
Mesitylenic, triesic methylisophthalic	Silica gel	$CHCl_3$	Samodumov[203]
Benzoic, salicylic	Celite	CH_3COOH–ether	Weber[204]
Mono-, dinitrobenzoic acids	Alumina	Benzene–$(CH_3)_2CO$	Oglobin[205]

sure; the equipment is simple and inexpensive, but the separation process is slow. The other type, known as high-performance liquid chromatography, operates at very high pressure with narrow columns; the apparatus is considerably more sophisticated and costly. As shown in Table 5.7, mixtures of free carboxylic acids can be separated in wide columns. After separation, the individual acids are determined generally by titration, or occasionally by spectrophotometry.[195] For high-pressure operations, the reader is referred to the study by Deming and Turoff[206] on optimization.

Less frequently, the carboxylic acids in a mixture are converted into derivatives for separation. For instance, Kuzdzal-Savoie and Raymond[207] separated the methyl esters of fatty acids in butter on a column of silicic acid impregnated with $AgNO_3$, using petroleum–benzene for elution. Coles[207a] separated branched-chain and straight-chain acids as their methyl esters on columns containing urea; elution with light petroleum yielded branched-chain compounds and elution with methanol gave straight-chain compounds. Sagredos[208] separated cyclic and straight-chain fatty acids from linseed or tung oil on a urea column via their methyl esters. Stevens and Lott[209] attempted to separate the racemic mixture of mandelic acid on an optically active polymer prepared by treating Porapak Q with chloromethyl methyl ether and $SnCl_4$ solution in light petroleum and condensing the product with (−)-α-phenethylamine; no resolution was obtained with $CHCl_3$ or ethanol as eluent, but partial resolution was achieved with H_2O and with methanol.

Separation of phenolic compounds by liquid chromatography is sum-

Table 5.8 Separation of Phenolic Compounds by Liquid Chromatography

Phenolic compounds	Column	Mobile phase	Reference
m-, *p*-Cresols	Celite	Cyclohexane	White[210]
Cresols	Alumina	Chlorobenzene	Stoltenberg[211]
Phenols, cresols, xylenols	Silica gel	Cyclohexane	Pearson[212]
Xylenols, ethylphenols, cresols	Silica gel	Cyclohexane–H_2O	Franc[213]
Phenol, cresols	Kieselguhr	$CHCl_3$–H_2O	Lakota[214]
Cresols	Silica gel	Cyclohexane	Majewska[215]
Phenols	Teflon	H_2O–CH_3OH	Fritz[216]
Xylenols, cresols	Firebrick	$(CH_3)_3C_5H_9$	Kogan[217]
Polyphenolic glycosides	Extran gels	H_2O–CH_3OH	Nordström[218]
Phenol–mono-, disulfonic acids	Cellulose	C_3H_7OH–$CHCl_3$–H_2O	Grebenovsky[219]

marized in Table 5.8. After separation the phenols are usually determined by spectrophotometric methods. When the phenolic compound also contains a strong acidic function,[219] determination by titration with alkali is suitable.

High-performance liquid chromatography is a recently developed technique. It is employed for the separation of complex mixtures and is discussed in the various chapters in Part Two of this book.

5.4 SEPARATION BY PAPER CHROMATOGRAPHY

Not so long ago paper chromatography was the most commonly used technique in the analytical and organic research laboratories for the separation of nonvolatile substances. Its prominent position now has been taken over by thin-layer chromatography. For quantitative analysis, however, paper chromatography possesses certain desirable features: (1) Square filter paper measuring 60 cm on each side or round filter paper with a diameter of 60 cm can be handled conveniently. Thus it provides a large area for two-dimensional or circular development of the chromatogram. (2) When time and labor are available, cutting out a well-defined section of the paper chromatogram for extraction of the particular compound to be determined spectrophotometrically gives precise and accurate results. (3) Paper chromatography is by far the least expensive method with respect to equipment, operation, and maintenance among all chromatographic techniques.

Paper chromatography can be carried out in various modifications to achieve the separation of the components in a mixture. For instance, Schmall et al.[220] separated carboxylic acid, barbiturates, and phenols on multibuffered paper utilizing the fact that many acidic compounds form salts at a particular pH level with alkaline buffer solutions that are applied on the paper in order of ascending pH. When the chromatogram is developed with $CHCl_3$ by a descending technique, stronger acids are immobilized at lower pH levels than are the weaker ones.

Glass-fiber paper[221] is employed on some occasions. For example, Hamilton and Dieckert[222] used glass-fiber impregnated with 0.1 M KH_2PO_4 to separate bile acids. Weeks and Deutech[223] reported quantitative separation and assay of ascorbic and isoascorbic acids on a silicic acid impregnated glass-fiber paper that was treated with H_3PO_3–glycerol using ethyl methyl ketone as the developing solvent. Bark and Graham[224] separated nitrophenols on alumina-impregnated paper. Ion-exchange paper is prepared by impregnation with ion-exchange resins, and it separates acidic compounds based on the ion-exchange principle. Thus Clark[225] described the separation of mono-, di-, and trihydric phenols on paper that was loaded with a strongly basic ion-exchange resin (Cl^- form), using cyclohexane–H_2O–acetic acid to develop the chromatogram.

Table 5.9 gives examples of paper-chromatographic separation of mixtures containing carboxylic acids. In these methods the free acids are placed directly on the paper and are separated as such. In other methods the carboxylic acids are converted into derivatives for separation. For instance, Smirnov[250] separated bile acids as methyl esters on paper treated with $Al(OH)_3$ using benzene–ethanol for development, and they separated the methyl esters of fatty acids from castor oil on paper impregnated with petroleum jelly using acetic acid–acetone for development. According to Daniels and Enzell,[251] resin acids could be methylated and separated on glass-fiber paper impregnated with hexadecane, while an attempt to separate the free acids on paper impregnated with dimethyl sulfoxide was not successful. Keto acids have been separated via their 2,4-dinitrophenylhydrazones[252] or via the amino acids obtained by reduction of the hydrazones.[253] Fluoroacetic acid and other straight-chain fatty acids have been separated via their hydroxamates.[254]

Table 5.10 shows direct paper chromatography of phenolic compounds. Sometimes it is necessary to convert the phenolic compounds into derivatives to effect separation. Thus, for the determination of monohydric phenols in effuents, Lure and Nikolaeva[267] separated phenol, *p*-cresol, and *o*- plus *m*-cresol via their compounds with diazotized sulfanilic acid using ethyl methyl ketone–H_2O for development, while phenol, *o*-cresol, and *m*- plus *p*-cresol were separated as such by using cyclohexane–heptane–H_2O as developing solvent. The same derivatives were recommended by Yuditskaya[268] for the analysis of simple phenols in smoked-food products. Bidlo[269] separated monohydric phenols as *p*-nitrophenylazo dyes obtained by reaction with diazotized *p*-nitroaniline; the dyes from phenols having unsubstituted *para*-positions were developed with CCl_4-hexane on paper impregnated with formamide, whereas the dyes from *para*-substituted phenols were separated on paper impregnated with paraffin, with phenol–$HCOOH$–H_2O as developer. Pohloudek-Fabini and Münchow[270] determined phenolic compounds in essential oils via their 4-phenylazobenzoic esters, which were separated on paper impregnated with paraffin, using butyl acetate–pyridine–H_2O as the solvent.

The general techniques for quantitation of the separated spots are presented in Chapter 4, Section 4.6.2. A few special methods are given below. Using $HCOOH$–CH_3COOH–H_2O as the developer, Oster et al.[271] separated cotton-seed-oil fatty acids on paper loaded with EDTA and treated with paraffin; the chromatogram was then treated with 1% $Bi(NO_3)_3$ solution, heated at 80°C, washed with H_2O, and exposed to aq. NH_3 saturated with H_2S; the fatty acids appeared as dark spots and were measured in a densit-

Table 5.9 Separation of Carboxylic Acids by Paper Chromatography

Acid mixture	Developing solvent	Reference
Formic, acetic	C_4H_9OH–cyclohexane–propanediol–$C_2H_5NH_2$–H_2O	Gomoryova[226]
Formic, acetic	C_4H_9OH–hexane–ethanediol–NH_3–$(C_2H_5)_2NH$–H_2O	Pszonka[227]
Straight-chain aliphatic	C_4H_9OH–cyclohexane–propanediol–NH_3–H_2O	Osteux[228]
Iso-acids	$C_6H_5CH_2OH$–NH_3–H_2O	Osteux[228]
Aliphatic	Two-dimensional: C_2H_5OH–NH_3–H_2O; C_4H_9OH–$HCOOH$–H_2O	Wollmann[229]
Higher fatty acids	$(CH_3)_2CO$–CH_3COOH–H_2O	Sliwick[230]
Unsaturated fatty acids	CH_3COOH—H_2O	Jurecka[231]
Succinic, citric, tartaric, lactic, malonic	C_4H_9OH–$HCOOH$–H_2O	Usmanov[232]
Tartaric, oxalic, citric, malic, succinic	C_3H_7OH–H_2O–pyridine	Traiter[233]
Citric, tartaric, malic	C_4H_9OH–CH_3COOH–H_2O	Ma[234]
Malic, tartaric, citric in fruit juice	Ethyl acetate–CH_3COOH–H_2O	Lindner[235]
C_2 to C_9 dibasic acids		Kuvaev[236]
Glucuronic, galacturonic	$(CH_3)_2CO$–C_2H_5OH–*i*–C_3H_7OH–0.05 *M* borate	Mukerjee[237]
Oxogluconic, oxogulonic	Phenol–H_2O	Nakanishi[230]
Tall-oil acids	CH_3COOH–CH_3CN–H_2O	Müller[239]
Bromo and iodo acids	C_3H_7OH–NH_3–H_2O	Hashmi[240]
Mono-, di-, trichloroacetic	C_4H_9OH–NH_3–H_2O	Chittum[241]
2-Halogenated aliphatic		Meikle[242]
Aromatic acids	C_4H_9OH–C_2H_5OH–pyridine–H_2O	Franc[243]
Benzenemono-, benzenedi-, benzene-tetracarboxylic	C_4H_9OH–NH_3–H_2O–pyridine	Abidova[244]
Benzene–polycarboxylic	C_4H_9OH–CH_3COOH–H_2O	Germain[245]
Phenolic acids	*i*-C_3H_7OH–NH_3–H_2O	Saini[246]
Phenolic acids	Benzene–CH_3COOH–H_2O	Metche[247]
Pyridinecarboxylic acids	C_4H_9OH–CH_3COOH–H_2O	Morimoto[248]
Pyrazolecarboxylic acids	*i*-C_3H_7OH–$CHCl_3$–NH_3–H_2O	Tabak[249]

Table 5.10 Separation of Phenolic Compounds by Paper Chromatography

Mixture of phenolic compounds	Developing solvent	Reference
Phenol, cresols	Cyclohexane–$CHCl_3$–C_2H_5OH	Mlodecka[255]
Phenols, alkoxyphenols, coumaranols	C_2H_5OH	Marcinkiewicz[256]
Monohydric phenols in vanilla	*i*-C_3H_7OH–NH_3–H_2O	Anwar[257]
Gallates, hydroxyanisole		Sedlacek[258]
Dinitrophenol, picric acid, herbicides		Henneberg[259]
Dihydroxyphenols in artichoke leaves	0.1 *N* HCl	Nichiforescu[260]
Chlorinated cresols and xylenols	Ligroin–HCOOH	Husain[261]
Monosubstituted phenols	Toluene–H_2O	Gumprecht[262]
Bromonaphthols	H_2O–NH_3	Postawka[263]
Mono- to penta-chlorophenols	Buffer solutions	Vacek[264]
Xylenols	C_2H_5OH–$CHCl_3$	Wojdylo[265]
Phenolic antioxidants	Hexane	Korcek[266]

ometer. Alternatively,[272] the spots were cut out, extracted with dilute HCl, and treated with aq. NH_3–Na_2S, and the intensity of the resulting cinnamon color was measured. Alimova et al.[273] determined saturated high-molecular-weight fatty acids by treating the chromatogram with basic lead acetate solution, locating the zones with $(NH_4)_2S$ solution, eluting the lead salts with 10% HNO_3, extracting the lead ions with dithizone in CCl_4, and measuring the absorbance at 520 nm. Vorobev[274] proposed the following method to determine *m*- and *p*-cresols when both compounds occupy the same area on the chromatogram: Determine *m*-cresol by spraying with ethanolic 2,6-dibromo-*p*-benzoquinonechlorimine followed by ethanolic KOH and then comparing the area of the blue spot with standards, *p*-cresol having no interference; determine the sum of the two cresols by spraying with diazotized sulfanilic acid and then comparing the color intensity with standards prepared with *p*-cresol.

5.5 SEPARATION BY THIN-LAYER CHROMATOGRAPHY

Owing to the popularity of thin-layer chromatography, much effort has been made to utilize it for quantitative analysis. Short courses have been organized for such purpose.[275] At the present time, however, thin-layer chromatography is generally performed for identification and seldom for the determination of organic compounds in a mixture.

Table 5.11 gives examples of thin-layer chromatography of carboxylic acids applied directly on the adsorbent. Unless otherwise indicated, the adsorbent is silica gel. It should be mentioned that two kinds of adsorbents can be mixed to form gradient layers. Thus Schettino and Ferrara[310] have investigated the behavior of 10 fatty acids (C_6 to C_{22}) on gradients of silica gel changing to silica gel plus $AgNO_3$, kieselguhr to silica gel plus $AgNO_3$, silica gel to kieselguhr, and silica gel to polyamide.

Table 5.11 Separation of Carboxylic Acids by Thin-Layer Chromatography

Acid mixture	Solvent system	Reference
C_1 to C_8 fatty acids	$(CH_3)_2CO$–NH_3–H_2O–$CHCl_3$	Brümmer[276]
Saturated and unsaturated fatty acids		Akramov[277]
Acids of plant origin	Ether–HCOOH–H_2O	Ting[278]
Tartaric, malic, citric, lactic, succinic	C_4H_9OH–HCOOH–H_2O, on cellulose	Lehmann[279]
Citric, pyruvic, lactic, fumaric	Two-dimensional: C_2H_5OH–NH_3–H_2O; isopentyl formate–HCOOH–H_2O	Whereat[280]
Acids of biological interest	$C_5H_{11}OH$–HCOOH–H_2O, on cellulose	Dittmann[281]
Keto acids	C_3H_7OH–NH_3–H_2O	Passera[282]
Dibasic, oxalic to sebaccic	$(i\text{-}C_3H_7)_2O$–HCOOH–H_2O, on kieselguhr	Knappe[283]
Di-, tribasic	$(i\text{-}C_3H_7)_2O$–ligroin–CCl_4–HCOOH–H_2O, on kieselguhr	Knappe[284]
Dibasic, C_2 to C_{10}	C_4H_9OH–xylene–phenol–HCOOH–H_2O	Rajagopal[285]
Mono-, dibasic, C_2 to C_{10}	Benzene–CH_3COOH–H_2O, on kieselguhr	Miyazaki[286]
cis-, *trans*-Fumaric, maleic	Benzene–CH_3COOH, on kieselguhr	Pastuska[287]
cis-, *trans*-Angelic, tiglic	$CHCl_3$–CH_3OH	Dutta[288]
Hydroxyricinoleic, undecenoic	Ligroin–$(C_2H_5)_2O$	Subbarao[289]
Epoxy fatty acids	Ligroin–$(C_2H_5)_2O$	Subbaram[290]
Bile acids	$(CH_3)_3C_5H_9$–$(i\text{-}C_3H_7)_2O$–CH_3COOH–$i\text{-}C_3H_7OH$	Panveliwalla[291]
Keto bile acids	Benzene–CH_3COOH, on kieselguhr	Usui[292]
Bile acids, keto bile acids	Benzene–CH_3OH	Matkovics[293]

Table 5.11 (*cont.*)

Acid mixture	Solvent system	Reference
Cholenic acids		Duvivier[294]
Dimethylmuconic acid isomers	C_2H_5OH–H_2O–NH_3, on alumina	Stewart[295]
Sialic acids	C_2H_5OH–CH_3COOH–H_2O	Schettino[296]
Resin acids	Ethyl acetate–hexane	Challen[297]
Ascorbic, dehydroascorbic	C_2H_5OH–H_2O, on polyamide	Hüttenrauch[298]
Triterpenoid acids		Elgamal[299]
Halohydroxy fatty acids	$(C_2H_5)_2O$–ligroin	Roomi[300]
Perfluorinated C_1 to C_4 acids	i-$C_5H_{11}OH$–NH_3–H_2O	Chelnokova[301]
Aromatic acids	Benzene–pyridine	Frankenfeld[302]
Dimethylbenzoic, methylphthalic, trimellitic	Benzene–$CHCl_3$–CCl_4–dioxane–HCOOH	Kulicki[303]
Aromatic dibasic	$(i$-$C_3H_7)_2O$–HCOOH–H_2O	Knappe[304]
Isomeric peroxybenzoic	CCl_4–CH_3COOH	Kavcic[305]
Methoxypiperonylic	Ethyl acetate–hexane–CH_3COOH	Beroza[306]
Mono-, dichlorobenzoic		Stedman[307]
Pyridine–monocarboxylic acids	C_4H_9OH–CH_3COOH	Carunchio[308]
Pyridine–dicarboxylic acids	$NaClO_4$–H_2O–CH_3OH–CH_3COOH, on diethylaminocellulose	Carunchio[308]
Pyrole–carboxylic acids	Two-dimensional: C_4H_9OH–C_2H_5OH–NH_3–H_2O; ethyl acetate–C_2H_5OH–CH_3COOH	Chierici[309]

To facilitate separation, the carboxyl function in fatty acids[311,312] or resin acids[313] can be methylated. Sgoutas and Kummerow[314] carried the derivatization one step further by brominating the methyl esters of fatty acids or converting the hydroxyacid methyl esters into the corresponding unsaturated esters through oxidation. Other derivatives of carboxylic acids include anilides,[315] bromanilides,[316] and 2,4-dinitrophenylhydrazides.[317] Keto acids have been separated via their 2,4-dinitrophenylhydrazones[318,319] or rhodamine complexes.[320] Wagner and Pohl[321] separated polyunsaturated fatty acids via their mercury adducts.

Separation of phenolic compounds by thin-layer chromatography is given in Table 5.12. It should be noted that some phenols are susceptible to oxida-

Table 5.12 Separation of Phenolic Compounds by Thin-Layer Chromatography

Mixture of phenols	Solvent system	Reference
Phenol, resorcinol, catechol	Benzene–CH_3COOH	Seeboth[322]
Phenol, cresols, catechol, pyrogallol, etc.	Benzene–CH_3COOH–H_2O	Lapina[323]
Homologous phenols		Halmekoski[324]
Gallates, *t*-butylcresol antioxidants	CH_3OH–$(CH_3)_2O$–H_2O, on polyamide	Copius-Peereboom[325]
Phenols, phenolic acids	On cellulose	Dittmann[326]
Halogenated phenols	Cyclohexane–dioxane, on alumina	Bark[327]
Chloroxylenols	Xylene	Husain[328]
Di-, hexachlorophane	CH_3OH–ethyl acetate	Daisley[329]
Hydroxyalkylphenols, methylenediphenol		Adorova[330]
Polynitro-, nitrosophenols		Parihar[331]
Phenols from cannabis	Benzene–toluene	de Faubert-Maunder[332]
Phenolic glycosides	Ethyl acetate–CH_3OH	Audette[333]
Diphenols	C_2H_5OH–benzene, on alumina	Sopkina[334]

tion on the adsorbent. For instance, Aly[335] reported that silica gel impregnated with $AgNO_3$ oxidized catechol and quinol, but not resorcinol, to the corresponding quinones.

Crump[336] described the separation of 20 simple alkylphenols as *p*-nitrophenylazo dyes by two-dimensional chromatography; the solvents were acetone–$CHCl_3$ for the first dimension and benzene–dipropylamine for the second. Thielemann[337] separated phenol, cresols, and naphthols as their condensation products with 4-amino-2,3-dimethyl-1-phenylpyrazolin-5-one. Moten and Kottemann[338] gave a method to determine naphtholsulfonic acid via the sulfonylazo derivative.

5.6 SEPARATION BY OTHER METHODS

Vietti-Michelina[339] compared electrophoretic and paper-chromatographic methods for the separation of salicylic acid from sulfosalicylic acid; electrophoresis gave more rapid separation, with sulfosalicylic acid moving the faster, whereas salicylic acid moved more quickly than sulfosalicylic acid by paper chromatography. Halmann and Kugel[340] described electrophoresis of methylphosphonic, methylphosphinic, and dimethylphosphinic acids on paper strips; best results were obtained at 750 V and 3 mA, no cooling being necessary.

Peter[341] determined long-chain fatty acids by multistage partition, with 80% acetic acid as the stationary phase and hexane as the mobile phase. Jones et al.[342] analyzed fatty acids in hydrogenated soya bean oil by countercurrent fractionation of the methyl esters. Wehle[343] determined hexachlorophane, pentachlorophenol, and trichlorophenol in the presence of one another and of other phenols by selective extraction into aq. $NaHCO_3$. From an aqueous solution of *o*-toluic and phthalic acids, Ellingboe and Ruybal[344] extracted the toluic acid into $CHCl_3$ and then determined it either gravimetrically or spectrophotometrically; the phthalic acid in the aqueous phase was then measured spectrophotometrically. According to von Schivizhoffen,[345] phenol in salicylic acid can be determined by coupling with diazotized *p*-nitroaniline at pH 5.5, the resulting dyes being extracted with benzene and measured colorimetrically; coupling of salicylic acid takes place more slowly, but any resulting dye is removed by shaking the benzene extract of the mixture of the two dyes with $NaHCO_3$ solution.

Osipova and Khodzhaev[346] described a procedure for the analysis of a mixture of benzene–mono-, di-, and tricarboxylic acids that involves precipitation of mercury and silver salts. Halmekoski[347] separated catechol from quinol and resorcinol by precipitation with lead acetate. According to Beckering and Fowkes,[348] addition of lead acetate and sodium acetate to a mixture of polyhydric phenols precipitates the lead salts of catechol and 4-alkylcatechols; 3,6-disubstituted catechols are not precipitated, and 3-monosubstituted catechols are precipitated only in the absence of other phenols or organic solvents.

5.7 COMBINATION OF SEPARATION METHODS

For the sake of expediency, two separation methods can be carried out consecutively in analyzing a mixture. Thus, for the determination of volatile fatty acids, steam distillation of the sample may be performed first to separate these acids from nonvolatile components; subsequently, the distillate is subjected to gas chromatography. The precision and accuracy of the analytical results are then related to both separation methods. While gas chromatography gives good separation and quantitation, it should be recognized that steam distillation is not easily reproducible. Riddle[349] reported that the recovery of acetic acid was >50%, whereas the recoveries of C_3 to

C_6 acids averaged 87%. In determining volatile acids in dairy products, Hempenius and Liska[350] found that the percent recovery of acetic acid depended on the fat content of the sample, ranging from 70% from skimmed milk to 85% from cream with 37% fat.

Lawson and Purdie[351] described a procedure combining liquid chromatography and paper chromatography to analyze aliphatic acid mixtures. The sample was first resolved into a number of fractions by gradient elution in a column with Dowex-1 resin as the stationary phase and 0 to 90% HCOOH as the mobile phase. The column was connected by a capillary siphon to a reservoir containing water, which was connected to a second reservoir containing 90% HCOOH. Enough of each fraction was collected to form a spot 1 cm in diameter on a line 3 cm from the edge of the filter paper, dried in a stream of warm air, and respotted twice more; the paper chromatogram was developed by the ascending technique with ethyl acetate–CH_3COOH–H_2O. This procedure was used to separate 200 μg each of 12 acids in 0.2 ml of water.

James[352] surveyed the methods for separating long-chain unsaturated fatty acids and recommended that combining gas chromatography with thin-layer chromatography was superior to all other techniques. White and Powell[353] separated 17 trienoic and 15 tetraenoic acids in the range C_{12} to C_{24} from fish oil in the following manner. The fatty acid methyl esters were prepared by base-catalyzed interesterification of the oil, isolated by thin-layer chromatography with hexane–$(C_2H_5)_2O$–CH_3COOH, and extracted with diethyl ether. The methyl esters were converted into the methoxy-bromomercuri adducts, which were separated by thin-layer chromatography with heptane–dioxane as developer, and then decomposed in the individual portions of adsorbent with HCl–CH_3OH. Finally, the methyl esters were subjected to gas chromatography programmed from 100 to 210°C.

REFERENCES

1. N. D. Cheronis and T. S. Ma, *Organic Functional Group Analysis*, Wiley, New York, 1964, pp. 165, 334, 445.
2. T. S. Ma and V. Horak, *Microscale Manipulations in Chemistry*, Wiley, New York, 1976, p. 213.
3. S. Egashira, *Jap. Anal.*, **15**, 1356 (1966).

3a. Y. Kasai, T. Tanimura, Z. Tamura, and Y. Ozawa, *Anal. Chem.*, **49**, 655 (1977).

4. O. Samuelson, *Svensk. Kem. Tidskr.*, **76**, 635 (1964).
5. L. Gedda and O. Samuelson, *Anal. Chim. Acta*, **27**, 63 (1962).
6. B. Carlsson and O. Samuelson, *Anal. Chim. Acta*, **49**, 247 (1970).
7. O. Samuelson and B. Johnard, *Svensk. Kem. Tidskr.*, **73**, 586 (1961).

8. A. J. Courtoisier and J. Ribereau-Gayon, *Bull. Soc. Chim. Fr.*, **1963**, 350.
9. H. H. Schenker and W. Rieman, III, *Anal. Chem.*, **25**, 1637 (1953).
10. A. Gueron, J. P. Lefevre, M. Caude, and R. Rosset, *Analusis*, **4**, 12 (1976).
11. W. Funasaka, T. Kojima, M. Ishibashi, and H. Kita, *Jap. Analyst*, **7**, 69 (1958).
12. U. B. Larsson, T. Isakasson, and O. Samuelson, *Acta Chem. Scand.*, **20**, 1965 (1966).
13. O. Samuelson, K. J. Lungqvist, and C. Parck, *Svensk. Pappidn.*, **61**, 1043 (1958).
14. I. Aoki, M. Hori, and H. Matsumaru, *Jap. Analyst*, **18**, 346 (1969).
15. M. Richards, *J. Chromatogr.*, **115**, 259 (1975).
16. L. Bengtsson and O. Samuelson, *Anal. Chim. Acta*, **44**, 217 (1969).
16a. T. Ericsson and O. Samuelson, *J. Chromatogr.*, **134**, 337 (1977).
17. V. P. Toryanik and I. A. Ilicheva, *Lakokras, Mater. Primen.*, **1973**, 53.
71a. W. Czerwinski and J. Guberska, *Chem. Anal. (Warsaw)*, **13**, 101 (1968).
18. T. Seki, *J. Chromatogr.*, **22**, 498 (1966).
19. W. Funasaka, T. Kojima, G. Takeshima, and J. Kimoto, *Jap. Analyst*, **11**, 434 (1962).
20. W. Funasaka, T. Kojima, and K. Fujimura, *Jap. Analyst*, **13**, 42 (1964).
21. D. J. Patel, R. A. Bhatt, and S. L. Bafna, *Chem. Ind. (Lond.)*, **1967**, 2110.
22. J. P. Lefevre, M. Caude, and R. Rosset, *Analudis*, **4**, 16 (1976).
23. W. Funasaka, T. Kojima, K. Fujimura, and S. Kushida, *Jap. Analyst*, **15**, 835 (1966).
24. J. Gaal and J. Inczedy, *Talanta*, **23**, 78 (1976).
24a. K. Jurahasi, *J. Chromatogr.*, **13**, 278 (1964).
25. M. W. Scoggins and J. W. Miller, *Anal. Chem.*, **40**, 1155 (1968).
26. W. J. Spillane and F. L. Scott, *Lab. Pract.*, **17**, 352 (1968).
27. W. Funasaka, T. Kojima, and K. Fujimura, *Jap. Anal.*, **10**, 374 (1961).
28. W. Funasaka, T. Kojima, K. Fujimura, and S. Kushida, *Jap. Analyst*, **12**, 1170 (1963).
29. W. Funasaka, T. Kojima, K. Fujimura, and T. Minami, *Jap. Analyst*, **12**, 466 (1963).
30. W. Funasaka, T. Kojima, and K. Fujimura, *Jap. Analyst*, **11**, 936 (1962).
31. W. Funasaka, T. Kojima, and K. Fujimura, *Jap. Analyst*, **9**, 852 (1960).
32. T. Seki, *J. Chem. Soc. Jap., Pure Chem. Sect.*, **75**, 1297 (1954).
33. G. Krampitz and W. Albersmeyer, *Experientia*, **15**, 375 (1959).
34. T. Seki, *J. Chromatogr.*, **3**, 376 (1960).
35. U. B. Larsson, I. Norstedt, and O. Samuelson, *J. Chromatogr.*, **22**, 102 (1966).
36. O. Samuelson and L. O. Wallenius, *J. Chromatogr.*, **12**, 236 (1963).
37. D. Jorysch and S. Marcus, *J. Assoc. Off. Agric. Chem.*, **48**, 534 (1965).
38. W. J. De Wet, *S. Afr. J. Agric. Sci.*, **6**, 535 (1963).
39. V. Arkima, *Mschr. Brau.*, **21**, 247 (1968).

40. A. I. Parimskii, P. M. Petryakov, and Y. P. Panov, *Zh. Anal. Khim.*, **24**, 1109 (1969).
41. M. Maruyama and H. Sakaguchi, *Jap. Analyst*, **15**, 56 (1966).
42. V. Mahadevan and L. Stenroos, *Anal. Chem.*, **39**, 1652 (1967).
43. J. R. P. Clarke and K. M. Fredricks, *J. Gas Chromatogr.*, **5**, 99 (1967).
44. W. K. Lee and R. M. Bethea, *J. Gas Chromatogr.*, **6**, 582 (1968).
45. B. Kaplanova and J. Janak, *Mikrochim. Acta*, **1966**, 119.
46. B. M. Craig and N. L. Murty, *J. Am. Oil Chem. Soc.*, **36**, 549 (1959).
47. K. Fukui, H. Nagatomi, and S. Murata, *Jap. Analyst*, **11**, 432 (1962).
48. A. I. Parimskii and P. M. Petryakov, *Zh. Anal. Khim.*, **24**, 1425 (1969).
49. S. N. Aminov and A. B. Terentev, *Uzb. Khim. Zh.*, **1968**, 35.
50. H. W. Doelle, *J. Chromatogr.*, **39**, 398 (1969).
51. J. Rigaud and M. Journet, *Ann. Biol Anim. Biochem. Biophys.*, **10**, 151 (1970).
52. T. Nakae and T. Nakanishi, *J. Agric. Chem. Soc. Jap.*, **37**, 302 (1963).
53. P. M. Shcherbakov, B. P. Kotelnikov, and Y. V. Ganin, *Maslob. Zhir. Prom.*, **27** (12) 25 (1961).
54. H. Szczepanska, *Chem. Anal. (Warsaw)*, **8**, 621 (1963).
55. G. Variali, *Ann. Chim. (Rome)*, **59**, 583 (1969).
56. H. Gershon and J. A. A. Renwick, *J. Chromatogr.*, **20**, 134 (1965).
57. E. D. Smith and A. B. Gosnell, *Anal. Chem.*, **34**, 438 (1962).
58. N. E. Hoffman and I. R. White, *Anal. Chem.*, **37**, 1541 (1965).
59. R. N. Shelley, H. Salwin, and W. Horwitz, *J. Assoc. Off. Agric. Chem.*, **46**, 486 (1963).
60. C. W. Gehrke and W. M. Lamkin, *J. Agric. Good Chem.*, **9**, 85 (1961).
61. R. B. Jackson, *J. Chromatogr.*, **22**, 251 (1966).
62. E. Tiscornia and L. Boniforti, *Riv. Ital. Sostanze Grasse*, **45**, 140 (1964).
63. R. G. Ackman, R. D. Burgher, and J. C. Sipos, *Nature*, **200**, 777 (1963).
64. I. Zeman, *J. Gas Chromatogr.*, **3**, 18 (1965).
65. Y. Kaburaki and Y. Sato, *J. Agric. Chem. Soc. Jap.*, **36**, 865 (1962).
66. Y. Sato and M. Momotani, *Jap. Anal.*, **10**, 196 (1961).
67. M. Jaarma, *Acta Chem. Scand.*, **11**, 30 (1964).
68. D. van Wijngaarden, L. A. Thyssen, and T. D. Osinga, *Z. Lebens. Forsch.*, **137**, 171 (1968).
69. G. P. Cartoni and L. Liberti, *Riv. Ital. Sostanze Grasse*, **44**, 178 (1967).
70. T. Gerson, J. E. A. McIntosh, and F. B. Shorland, *Biochem. J.*, **91** (2), 11 (1964).
71. R. Tudisco and D. A. Turner, *Riv. Ital. Sostanze Grasse*, **40**, 528 (1962).
72. I. Zeman, *Prum. Potravin*, **15**, 287 (1964).
73. C. Litchfield, R. Reiser, and A. F. Isbell, *J. Am. Oil Chem. Soc.*, **40**, 302 (1963).
74. E. Hautala, *J. Assoc. Off. Anal. Chem.*, **49**, 619 (1966).
75. A. Kuksis and P. Vishwakarma, *Can. J. Biochem. Physiol.*, **41**, 2353 (1963).

76. F. L. Estes and R. C. Bachmann, *Anal. Chem.*, **38**, 1178 (1966).
77. D. Barnett, R. D. Cohen, C. N. Tassopoulos, J. R. Turtle, A. Dimitriadou, and T. R. Traser, *Anal. Biochem.*, **26**, 68 (1968).
78. L. Hagenfeldt, *Ark. Kemi*, **29**, 63 (1968).
79. G. A. F. Harrison and E. Collins, *Proc. Am. Soc. Brew. Chem.*, **1968**, 101.
80. F. H. M. Nestler and D. F. Zinkel, *Anal. Chem.*, **35**, **1747** (1963).
81. T. W. Brooks, G. S. Fisher, and N. M. Joye, Jr., *Anal. Chem.*, **37**, 1063 (1965).
82. R. B. Iden and E. J. Kahler, *J. Am. Oil Chem. Soc.*, **39**, 171 (1962).
83. R. A. L. Paylor, R. Feinland, and N. H. Conroy, *Anal. Chem.*, **40**, 1358 (1968).
84. L. T. Black and R. A. Eisenhauer, *J. Am. Oil. Chem. Soc.*, **40**, 272 (1963).
85. P. L. Pursley and E. D. Schall, *J. Assoc. Off. Agric. Chem.*, **48**, 327 (1965).
86. K. A. Zhavnerko and B. V. Erofeev, *Izv. Akad. Nauk Beloruss SSR, Ser. Khim Nauk*, **1968** (3), 67.
87. L. de Lindemann, *J. Chromatogr.*, **51**, 297 (1970).
88. Y. N. Martynov and V. A. Proskuryakov, *Zh. Prikl Khim., Leningr.*, **41**, 2094 (1968).
89. E. M. Kazinik, N. V. Novorusskaya, L. M. Lvovich, and G. A. Gudkova, *Zh. Anal. Khim.*, **24**, 1592 (1969).
90. E. Bendel, W. Meltzow, and V. Vogt, *J. Chromatogr.*, **38**, 133 (1968).
90a. T. Kaneda, *J. Chromatogr.*, **92**, 191 (1974).
91. F. Kumeno, *J. Agric. Chem. Soc. Jap.*, **36**, 181 (1962).
92. J. W. Schwarze, *Anal. Chem.*, **41**, 1686 (1969).
93. H. J. Langner, *Angew. Chem., Int. Ed.*, **4**, 71 (1965).
94. B. M. Craig, A. P. Tulloch, and N. L. Murty, *J. Am. Chem. Soc.*, **40**, 61 (1963).
94a. G. Aguggini and P. A. Biondi, *Arch. Vet. Ital.*, **24**, 213 (1973).
95. J. S. O'Brien and G. Rouser, *Anal. Biochem.*, **7**, 288 (1964).
96. P. K. Raju and R. Reiser, *Lipids*, **1**, 10 (1966).
97. N. E. Hoffman and T. A. Killinger, *Anal. Chem.*, **41**, 162 (1969).
98. E. O. Umeh, *J. Chromatogr.*, **51**, 147 (1970).
99. E. Sjöström, P. Haglund, and J. Janson, *Acta Chem. Scand.*, **20**, 1718 (1966).
100. M. G. Horning, E. A. Boucher, A. M. Moss, and E. C. Horning, *Anal. Lett.*, **1**, 713 (1968).
101. D. F. Zinkel, M. B. Lathrop, and L. C. Zank, *J. Gas Chromatogr.*, **6**, 158 (1968).
102. G. Foulds and W. W. Wimer, *Anal. Biochem.*, **30**, 477 (1969).
103. Y. T. Eidus, S. D. Pirozhkov, and K. V. Puzitskii, *Zh. Anal. Khim.*, **22**, 1559 (1967).
104. T. S. Ma and R. Roper, *Microchem. J.*, **1**, 245 (1957).
105. L. D. Metcalfe and A. A. Schmitz, *Anal. Chem.*, **33**, 363 (1961).
106. E. Hautala and M. L. Weaver, *Anal. Biochem.*, **3**, 32 (1969).
107. J. E. Scoggins and C. H. Fitzgerald, *J. Agric. Food Chem.*, **17**, 156 (1969).

108. E. L. Schneider, S. P. Loke, and D. T. Hopkins, *J. Am. Oil Chem. Soc.*, **45**, 585 (1968).

109. N. E. Hetman, H. G. Arlt, Jr., R. Paylor, and R. Feinland, *J. Am. Oil Chem. Soc.*, **42**, 255 (1965).

110. S. Noguchi and F. Nakazawa, *Agric. Biol. Chem.*, **37**, 1973 (1973).

111. D. Buoncristiani, G. Toponeco, and R. Salvadorini, *Olearia*, **16**, 99 (1962).

112. P. Ronkainen, *Acta Chem. Fenn.*, **37**, 209 (1964).

113. C. Hishta and J. Bomstein, *Anal. Chem.*, **35**, 65 (1963).

114. J. T. Hill and I. D. Hill, *Anal. Chem.*, **36**, 2504 (1964).

115. W. Witte and H. Rasse, *Chromatographia*, **1968**, 32.

116. T. S. Kukharenko and E. A. Grigoreva, *Neftekhimiya*, **10**, 303 (1970).

117. J. Jirak and J. Dvoracek, *Chem. Listy*, **64**, 75 (1970).

118. A. Smith and E. Vernon, *J. Chromatogr.*, **43**, 503 (1969).

119. V. N. Korbina, V. A. Utkin, and A. G. Khmelnitskii, *Izv. Sib. Otd. Akad. Nauk SSSR, Ser. Khim. Nauk*, **1968** (6), 114.

120. W. Funasaka, T. Kojima, and H. Hirono, *Jap. Analyst*, **13**, 1116 (1964).

121. C. M. Williams, *Anal. Biochem.*, **11**, 224 (1965).

122. M. Schnitzer and J. G. Desjardins, *J. Gas Chromatogr.*, **2**, 270 (1964).

123. C. M. Williams and R. H. Leonard, *Anal. Biochem.*, **5**, 362 (1963).

124. L. J. Walter, D. F. Riggs, and R. T. Coutts, *J. Pharm. Sci.*, **63**, 1754 (1974).

125. J. Fitelson and G. L. Bowden, *J. Assoc. Off. Anal. Chem.*, **51**, 1224 (1968*h*.

126. E. R. Blakley, *Anal. Biochem.*, **15**, 350 (1966).

126a. R. Komers, *Collect. Czech. Chem. Commun.*, **28**, 1549 (1963).

127. J. J. Kirkland, *Anal. Chem.*, **32**, 1388 (1960).

128. W. Funasaka, T. Kojima, and T. Toyota, *Jap. Analyst*, **14**, 815 (1965).

129. K. M. Baker and G. E. Boyce, *J. Chromatogr.*, **117**, 471 (1976).

130. J. B. Himes and I. J. Dowbak, *J. Gas Chromatogr.*, **3**, 194 (1965).

131. J. S. Parsons, *J. Gas Chromatogr.*, **5**, 254 (1967).

132. A. A. Spryskov and V. A. Kozlov, *Izv. Vyssh. Uchebn. Zaved., Khim.*, **11**, 785 (1968).

133. A. Kozlov and A. A. Spryskov, *Izv. Vyssh. Ucheb. Zaved., Khim.*, **13**, 1752 (1971).

134. J. S. Parsons, *J. Chromatogr. Sci.*, **11**, 659 (1973).

135. S. Siggia, L. R. Whitlock, and J. C. Tao, *Anal. Chem.*, **41**, 1387 (1969).

136. T. S. Ma and A. S. Ladas, *Organic Functional Group Analysis by Gas Chromatography*, Academic, London, 1976, p. 112.

137. F. Malz and J. Gorlas, *Z. Anal. Chem.*, **242**, 81 (1968).

138. W. Sassenberg and K. Wrabetz, *Z. Anal. Chem.*, **184**, 423 (1961).

139. K. Wrabetz and W. Sassenberg, *Z. Anal. Chem.*, **179**, 333 (1961).

140. T. A. Rudolfi, R. I. Sharapova, and V. I. Lushchik, *Zh. Anal. Khim.*, **19**, 903 (1964).

141. V. A. Zakupra and V. S. Doborv, *Khim. Tekhnol. Topl. Masel*, **1969**, (10), 54.

142. M. M. Anwar, C. Hanson, and M. W. T. Pratt, *J. Chromatogr.*, **46**, 200 (1970).
143. P. J. Porcaro and V. D. Johnston, *Anal. Chem.*, **34**, 1071 (1962).
144. K. Krumbholz and V. Vollert, *Chem. Tech. (Berl.)*, **17**, 693 (1965).
145. E. Kardos, V. Kosljar, J. Hudec, and P. Ilka, *Chem. Zvesti*, **22**, 786 (1968).
146. T. K. Choy, J. J. Quattrone, Jr., and N. J. Alicino, *J. Chromatogr.*, **12**, 171 (1963).
147. Y. D. Norikov and V. N. Vetchinkina, *Neftekhimiya*, **5**, 284 (1965).
147a. C. Landault and G. Guiochon, *Anal. Chem.*, **39**, 713 (1967).
148. J. Mendez, *J. Gas Chromatogr.*, **6**, 168 (1968).
149. Y. K. Shaposhnikov, L. V. Kosyukova, and E. Z. Druskina, *Zh. Anal. Khim.*, **23**, *Khim.*, **23**, 609 (1968).
150. S. Husain, *Indian J. Technol.*, **6**, 94 (1968).
151. D. Spiegel, Master's Thesis, Brooklyn College of the City University of New York, 1964.
152. V. D. Simonov, T. F. Akhunov, Y. N. Alyamkin, and L. I. Lushchekina, *Dokl. Neftekhim. Sekts. Bashrish. Resp. Pravl. Vses. Khim. Obshch. Mendeleeva*, **1967** (3) 82.
153. R. H. Kolloff, L. J. Breuklander, and L. B. Barkley, *Anal. Chem.*, **35**, 1651 (1963).
154. A. Nosal, A. Pasternak, and S. Witek, *Chem. Anal.* (*Warsaw*), **14**, 1115 (1969).
155. J. R. L. Smith, R. O. C. Norman, and G. K. Radda, *J. Gas Chromatogr.*, **2**, 146 (1964).
156. J. Hrivnak and Z. Stota, *Collect. Czech. Chem. Commun.*, **30**, 2128 (1965).
157. T. S. Ma and D. Spiegel, *Microchem. J.*, **10**, 67 (1966).
158. D. R. Clifford and D. A. M. Watkins, *J. Gas Chromatogr.*, **6**, 191 (1968).
159. J. Hrivnak and Z. Stota, *J. Gas Chromatogr.*, **6**, 9 (1968).
160. E. von Rudloff, *J. Gas Chromatogr.*, **2**, 89 (1964).
161. C. W. Stanley, *J. Agric. Food. Chem.*, **14**, 321 (1966).
162. A. C. Bhattacharyya, A. Bhattacharjee, O. K. Guha, and A. N. Basu, *Anal. Chem.*, **40**, 1873 (1968).
163. G. Bergmann and D. Jentzsch, *Z. Anal. Chem.*, **164**, 10 (1958).
164. H. P. Higginbottom, H. M. Culbertson, and J. C. Woodbrey, *Anal. Chem.*, **37**, 1021 (1965).
165. A. T. Shulgin, *Anal. Chem.*, **37**, 920 (1964).
166. E. R. Adlard and G. W. Roberts, *J. Inst. Petroleum*, **51**, 376 (1965).
167. R. W. Freedman and G. O. Charlier, *Anal. Chem.*, **37**, 1880 (1964).
168. R. W. Freedman and P. P. Croitoru, *Anal. Chem.*, **36**, 1389 (1964).
169. K. Oydvin, *Medc. Norsk. Farm. Selsk.*, **28**, 116 (1966).
170. I. T. Clark, *J. Gas Chromatogr.*, **6**, 53 (1968).
171. F. C. Dallas, A. F. Lautenbach, and D. B. West, *Proc. Am. Soc. Brew. Chem.*, **1967**, 103.
172. L. E. Brydia, *Anal. Chem.*, **40**, 2212 (1968).

173. L. Tomari, *Magy. Kem. Foly.*, **76**, 437 (1970).
174. H. Weiss and A. Kreyenbuhl, *Bull. Soc. Chim. Fr.*, **1961**, 603.
175. S. Tominaga, *Jap. Analyst*, **12**, 137 (1963).
176. I. A. German and N. Ciot, *Revta Chim.*, **17**, 177 (1966).
177. G. R. Higginbotham, J. Ress, and A. Rocke, *J. Assoc. Off. Anal. Chem.*, **53**, 673 (1970).
178. P. P. Croitoru and R. W. Freedman, *Anal. Chem.*, **34**, 1536 (1962).
179. F. K. Kawahara, *Anal. Chem.*, **40**, 1009 (1968).
180. E. Vioque, *Grasas y Aceites*, **7**, 234 (1956).
181. R. Raveux and J. Bove, *Bull. Soc. Chim. Fr.*, **1957**, 369.
182. J. Sliwiok, *Rocz. Chem.*, **37**, 1497 (1963).
183. D. Chobanov, A. Popov, and E. Chuparova, *Fette, Seifen, Anstrichm.*, **68** (2), 85 (1966).
184. A. G. Stradomskaya and I. A. Goncharova, *Gidrokhim. Mater.*, **39**, 169 (1965).
185. G. G. Freeman, *J. Chromatogr.*, **28**, 338 (1967).
186. P. Savary and P. Desnuelle, *Bull. Soc. Chim. Fr.*, **20**, 939 (1953).
187. M. P. Pyatnitskii and L. I. Kiprach, *Uch. Zap. Krasnodarsk. Gos. Ped. Inst.*, **19**, 177 (1957).
188. C. E. Frohmann and J. M. Orten, *J. Biol. Chem.*, **205**, 717 (1953).
189. A. I. Smith, *Anal. Chem.*, **31**, 1621 (1959).
190. V. I. Poddubnyi, *Izv. Vyssh. Uchebn. Zaved. Pishch. Tekhnol.*, **3**, 170 (1959).
191. A. V. Markova and V. A. Smirnov, *Zh. Anal. Khim.*, **24**, 1271 (1969).
192. J. M. Orten, W. Gamble, C. B. Vaughn, and K. C. Shrivastava, *Microchem. J.*, **13**, 182 (1968).
193. E. C. Bossert, A. Donmoyer, R. D. Hinkel, and R. Mainier, *Anal. Chem.*, **36**, 1213 (1964).
194. H. H. Bruun and H. Lehtonen, *Acta Chem. Scand.*, **17**, 853 (1963).
195. I. I. Bardyshev, K. A. Cherches, and L. A. Meerson, *Zh. Anal. Khim.*, **18**, 895 (1963).
196. T. L. Chang, *Anal. Chem.*, **40**, 989 (1968).
197. E. A. Taylor and M. Rhodes, *Analyst*, **88**, 136 (1963).
198. L. A. Haddock and L. G. Phillips, *Analyst*, **84**, 94 (1959).
199. M. N. Chelnokeva and V. G. Gromova, *Zh. Anal. Khim.*, **22**, 418 (1967).
200. K. Kochloefl, E. Grebenovsky, and V. Bazant, *Chem. Prumy.*, **13**, 303 (1963).
201. K. Hiller, *Pharmazie*, **20**, 353 (1965).
202. A. I. Kamneva and E. S. Panfilova, *Zavod. Lab.*, **29**, 666 (1963).
203. S. A. Samodumov and K. I. Matkovskii, *Ukr. Khim. Zh.*, **31**, 534 (1965).
204. J. D. Weber, *J. Assoc. Off. Agric. Chem.*, **48**, 1151 (1965).
205. K. A. Oglobin and G. V. Markina, *Vestn. Leningr. Univ.*, **1959**, *Ser. Fiz. Khim.*, 149.

206. S. N. Deming and M. L. H. Turoff, *Anal. Chem.*, **50**, 546 (1978).
207. S. Kuzdzal-Savoie and J. Raymond, *Ann. Biol. Anim. Biochim.*, **5**, 497 (1965).
207a. L. Coles, *J. Chromatogr.*, **32**, 657 (1968).
208. A. N. Sagredos, *Fette, Seifen, Anstrichm.*, **69**, 707 (1967).
209. T. S. Stevens and J. A. Lott, *J. Chromatogr.*, **34**, 480 (1968).
210. D. White and D. W. Grant, *Nature*, **175**, 513 (1955).
211. H. Stoltenberg, *Z. Anal. Chem.*, **146**, 181 (1955).
212. R. M. Pearson, *Analyst*, **80**, 656 (1955).
213. J. Franc, *Chem. Listy*, **50**, 547 (1956).
214. V. Lakota, *Chem. Listy*, **52**, 1922 (1958).
215. F. Majewska and W. Michalek, *Twoezywa Wielkoczasteczkowe*, **6**, 136 (1961).
216. J. S. Fritz and C. E. Hedrick, *Anal. Chem.*, **37**, 1015 (1965).
217. I. E. Kogan, M. M. Bragilevskaya, and M. E. Neimark, *Zh. Anal. Khim.*, **20**, 253 (1965).
218. C. G. Nordström, *Acta Chem. Scand.*, **21**, 2885 (1967).
219. E. Grebenovsky, *Z. Anal. Chem.*, **213**, 412 (1965).
220. M. Schmall, E. G. Wollish, R. Colarusso, C. W. Keller, and E. G. E. Shafer, *Anal. Chem.*, **29**, 791 (1957).
221. T. S. Ma and A. A. Bededetti-Pichler, *Anal. Chem.*, **25**, 999 (1953).
222. J. G. Hamilton and J. W. Dieckert, *Arch. Biochem. Biophys.*, **82**, 203 (1959).
223. C. E. Weeks and M. J. Deutsch, *J. Assoc. Off. Anal. Chem.*, **50**, 793 (1967).
224. L. S. Bark and R. J. T. Graham, *Talanta*, **11**, 839 (1964).
225. I. T. Clark, *J. Chromatogr.*, **15**, 65 (1964).
226. A. Gomoryova, *Acta Chim. Acad. Sci. Hung.*, **33**, 251 (1962).
227. B. Pszonka, *Chem. Anal. (Warsaw)*, **13**, 83 (1968).
228. R. Osteux, J. Guillaume, and J. Laturaze, *J. Chromatogr.*, **1**, 70 (1958).
229. H. Wollmann and R. Pohloudek-Fabini, *Pharmazie*, **16**, 21 (1961).
230. J. Sliwiok, *Ann. Soc. Chim. Pol.*, **37**, 1497 (1963).
231. B. Jurecka, *Chem. Anal. (Warsaw)*, **11**, 121 (1966).
232. K. U. Usmanov, A. M. Yakubov, and A. B. Kuchkarev, *Tr. Sredneazat. Politekh. Inst.*, **1957** (2), 3.
233. M. Traiter, *Chem. Zvesti*, **11**, 583 (1957).
234. T. S. Ma and R. Roper, *Mikrochim. Acta*, **1968**, 174.
235. K. Lindner and E. W. Jurics, *Z. Lebensmitt. Forsch.*, **128**, 65 (1965).
236. B. E. Kuvaev and N. S. Imyanitov, *Zavod. Lab.*, **33**, 1382 (1967).
237. H. Mukerjee and J. S. Ramn, *J. Chromatogr.*, **14**, 551 (1964).
238. I. Nakanishi, K. Sasajima, M. Isono, and R. Takeda, *Ann. Rep. Takeda Res. Lab.*, **23**, 48 (1964).
239. W. Müller and K. Berger, *Plaste Kautsch.*, **10**, 632 (1963).

240. M. H. Hashmi, *Anal. Chim. Acta*, **17**, 225 (1957).
241. J. W. Chittum, T. A. Gustin, R. L. McGuire, and J. T. Sweeney, *Anal. Chem.*, **30**, 1213 (1958).
242. R. W. Meikle, *Nature*, **196**, 61 (1962).
243. J. Franc, M. Hajkova, and V. Jehlicka, *Chem. Zvesti*, **17**, 542 (1963).
244. Z. K. Abidova, A. M. Yakubov, K. U. Usmanov, and G. K. Khodzhaev, *Dokl. Akad. Nauk Uzb. SSR*, **1957** (6), 29.
245. J. E. Germain, J. Montreuil, and P. Koukos, *Bull. Soc. Chim. Fr.*, **6**, 115 (1959).
246. A. S. Saini, *J. Chromatogr.*, **19**, 619 (1965).
247. M. Metche, F. Jacquin, Q. H. Nguyen, and E. Urion, *Bull. Soc. Chim. Fr.*, **1962**, 1763.
248. I. Morimoto, K. Furuta, and H. Oida, *Jap. Analyst*, **10**, 664 (1961).
249. S. V. Tabak, I. I. Grandberg, and A. N. Kost, *Zh. Obshch. Khim.*, **32**, 1562 (1962).
250. B. P. Smirnov, *Tr. Komiss. Anal. Khim. Akad. Nauk SSSR*, **13**, 435 (1963).
251. P. Daniels and C. Enzell, *Acta Chem. Scand.*, **16**, 1530 (1962).
252. D. Cavallini and N. Frontali, *Biochim. Biophys. Acta*, **13**, 439 (1954).
235. N. S. Gsiko, V. L. Kretovich, B. D. Polkovnikov, A. A. Balandin, and A. M. Taber, *Dokl. Akad. Nauk SSSR*, **153**, 209 (1963).
254. F. Bergmann and R. Segal, *Biochem. J.*, **62**, 542 (1956).
255. J. Mlodecka, *Chem. Anal. (Warsaw)*, **4**, 157 (1959).
256. S. Marcinkiewicz, J. Green, and D. McHale, *J. Chromatogr.*, **10**, 42 (1963).
257. M. H. Anwar, *Anal. Chem.*, **35**, 1974 (1963).
258. B. A. J. Sedlacek, *Fette, Seifen, Anstrichm.*, **65**, 915 (1963).
259. M. Henneberg, *Acta Pol. Pharm.*, **21**, 296 (1964).
260. E. Nichiforescu and V. Coucou, *Ann. Pharm. Fr.*, **23**, 419 (1965).
261. S. Husain, *J. Chromatogr.*, **18**, 197 (1965).
262. D. L. Gumprecht and F. Schwartzenburg, *J. Chromatogr.*, **18**, 336 (1965), **23**, 131 (1966).
263. A. Postawka and L. Prajer-Janczewska, *Chem. Anal. (Warsaw)*, **10**, 977 (1965).
264. Z. Vacek, Z. Stota, and J. Stanek, *J. Chromatogr.*, **19**, 572 (1965).
265. S. Wojdylo and K. Zielenski, *Chem. Anal. (Warsaw)*, **11**, 1127 (1966).
266. S. Korcek, J. Baxa, and V. Vesely, *Chem. Zvesti*, **22**, 947 (1968).
267. Y. Y. Lure and Z. V. Nikolaeva, *Zavod. Lab.*, **30**, 937 (1964).
268. A. I. Yuditskaya, *Zh. Prikl. Khim. Leningr.*, **41**, 656 (1968).
269. Z. Bidlo, *Z. Anal. Chem.*, **214**, 351 (1965).
270. R. Pohloudek-Fabini and P. Munchow, *Pharmazie*, **19**, 591 (1964).
271. N. R. Oster, L. R. Snisarenko, and M. N. Ismailov, *Lab. Delo*, **10**, 457 (1964).
272. L. P. Snisarenko, N. R. Oster, and M. N. Ismailov, *Lab. Delo*, **1968**, 181.
273. E. K. Alimova, A. T. Astvatsaturyan, E. A. Endokova, and V. A. Stepanova, *Zh. Anal. Khim.*, **18**, 769 (1963).

274. V. Vorobev, *Acta Chim. Acad. Sci. Hung.*, **32**, 337 (1962).
275. Kontes Short Course on Quantitative Thin-Layer Chromatography, 1977.
276. J. M. Brümmer, *Brot Gebeck*, **19**, 238 (1965).
277. S. Akramov and A. L. Markman, *Dokl. Akad. Nauk Uzb. SSR*, **1967** (8), 25.
278. I. P. Ting and W. M. Dugger, Jr., *Anal. Biochem.*, **12**, 571 (1965).
279. G. Lehmann and P. Martinod, *Z. Lebensm. Forsch.*, **130**, 269 (1966).
280. A. F. Whereat, D. R. Snydman, and L. A. Barnes, *J. Chromatogr.*, **36**, 390 (1968).
281. J. Dittmann, *J. Chromatogr.*, **34**, 407 (1968).
282. C. Passera, A. Pedrotti, and G. Ferrari, *J. Chromatogr.*, **14**, 289 (1964).
283. E. Knappe and D. Peteri, *Z. Anal. Chem.*, **188**, 184 (1962).
284. E. Knappe and I. Rohdewald, *Z. Anal. Chem.*, **211**, 49 (1965).
285. N. S. Rajagopal, P. K. Saraswathy, M. R. Subbaram, and K. T. Achaya, *J. Chromatogr.*, **24**, 217 (1966).
286. S. Miyazaki, Y. Suhara, and T. Kobayashi, *J. Chromatogr.*, **39**, 88 (1969).
287. G. Pastuska and H. J. Petrowitz, *J. Chromatogr.*, **10**, 517 (1963).
288. S. P. Dutta and A. K. Barua, *J. Chromatogr.*, **29**, 263 (1967).
289. R. Subbarao, M. W. Roomi, M. R. Subbaram, and K. T. Achaya, *J. Chromatogr.*, **9**, 295 (1962).
290. M. R. Subbaran, M. W. Roomi, and K. T. Achaya, *J. Chromatogr.*, **21**, 324 (1966).
291. D. Panveliwalla, B. Lewis, I. D. P. Wootton, and S. Tabaqchali, *J. Clin. Pathol.*, **23**, 309 (1970).
292. T. Usui, *J. Biochem. (Tokyo)*, **54**, 283 (1963).
293. B. Matkovics and Z. Tegyey, *Microchem. J.*, **13**, 174 (1968).
294. J. Duvivier, *J. Chromatogr.*, **19**, 352 (1965).
295. R. G. Stewart, *Anal. Biochem.*, **35**, 321 (1970).
296. O. Schettino, M. I. La Rotonda, and L. Ferrara, *Boll. Soc. Ital. Biol. Sper.*, **46**, 429 (1970).
297. S. B. Challen and M. Kucera, *J. Chromatogr.*, **32**, 53 (1968).
298. R. Hüttenrauch and I. Keiner, *Pharmazie*, **23**, 157 (1968).
299. M. H. A. Elgamal and M. B. E. Fayez, *Z. Anal. Chem.*, **211**, 190 (1965).
300. M. W. Roomi, M. R. Subbaram, and K. T. Achayz, *J. Chromatogr.*, **24**, 93 (1966).
301. M. N. Chelnokova, T. N. Korotkova, and V. G. Gromova, *Zh. Anal. Khim.*, **22**, 1436 (1967).
302. J. W. Frankenfeld, *J. Chromatogr.*, **18**, 179 (1965).
303. J. Kulicka, R. Baranowski, Z. Gregorowicz, and Z. Kulicki, *Chem. Anal. (Warsaw)*, **13**, 169 (1969).
304. E. Knappe and D. Peteri, *Z. Anal. Chem.*, **188**, 352 (1962).
305. R. Kavic and B. Plesnicar, *J. Chromatogr.*, **38**, 515 (1968).
306. M. Beroza and W. A. Jones, *Anal. Chem.*, **34**, 1029 (1962).
307. E. D. Stedman, *Analyst*, **94**, 594 (1969).

308. V. Carunchio and A. Marino, *J. Chromatogr.*, **31**, 601 (1967).
309. L. Chierici and M. Perani, *Ric. Sci., Rc. A.*, **6**, 168 (1964).
310. O. Schettino and L. Ferrara, *Riv. Ital. Sostanze Grasse*, **47**, 450 (1970).
311. L. D. Bergelson, E. V. Dyatlovitskaya, and V. V. Voronokova, *Izv. Akad. Nauk SSSR, Otd. Khim. Nauk*, **1963**, 954.
312. L. J. Morris, *J. Chromatogr.*, **12**, 321 (1963).
313. D. F. Zinkel and J. W. Rowe, *J. Chromatogr.*, **13**, 74 (1964).
314. D. Sgoutas and F. A. Kummerow, *J. Am. Oil Chem. Soc.*, **40** (4), 138 (1963).
315. G. F. Bories, *J. Chromatogr.*, **36**, 377 (1968).
316. J. P. Lebacq, M. Severin, J. Casimir, and M. Renard, *Bull. Rech. Agron. Gemblous*, **4**, 130 (1969).
317. A. C. Thompson and P. A. Hedin, *J. Chromatogr.*, **21**, 13 (1966).
318. H. J. Stan and J. Stormueller, *J. Chromatogr.*, **43**, 103 (1969).
319. P. Ronkainen, *J. Chromatogr.*, **11**, 228 (1963).
320. M. Rink and S. Herrmann, *J. Chromatogr.*, **14**, 523 (1964).
321. H. Wagner and P. Pohl, *Biochem. Z.*, **340**, 337 (1964).
322. H. Seeboth and H. Görsch, *Chem. Tech. Leipz.*, **15**, 294 (1963).
323. T. G. Lapina, Trudy Khim., *Gorkii*, **1962**, 424.
324. J. Halmekoski and H. Hannikainen, *Suom. Kem. B*, **36**, (2) 24 (1963).
325. J. W. Copius-Peereboom, *Nature*, **204**, 748 (1964).
326. J. Dittmann, *J. Chromatogr.*, **32**, 764 (1968).
327. L. S. Bark and R. J. T. Graham, *J. Chromatogr.*, **25**, 347 (1966).
328. S. Husain and P. A. Swaroop, *J. Chromatogr.*, **22**, 180 (1966).
329. R. W. Daisley and C. J. Olliff, *J. Pharm. Pharmacol.*, **22**, 202 (1970).
330. I. V. Adorova, V. Y. Kovner, and M. I. Siling, *Plast. Massy*, **1968**, 60.
331. D. B. Parihar, S. P. Sharma, and K. C. Tewari, *J. Chromatogr.*, **24**, 230 (1966).
332. M. J. de Faubert-Maunder, *J. Pharm. Pharmacol.*, **21**, 334 (1969).
333. R. C. S. Audette, G. Blunden, J. W. Steele, and C. S. C. Wong, *J. Chromatogr.*, **25**, 367 (1966).
334. A. K. Sopkina and V. D. Ryabov, *Zh. Anal. Khim.*, **19**, 615 (1964).
335. O. M. Aly, *Z. Anal. Chem.*, **234**, 251 (1968).
336. G. B. Crump, *Anal. Chem.*, **36**, 2447 (1964).
337. H. Thielemann, *Z. Chem. Leipz.*, **9**, 350 (1969).
338. L. Moten and C. Kottemann, *J. Assoc. Off. Anal. Chem.*, **52**, 31 (1969).
339. M. Vietti-Michelina, *Z. Anal. Chem.*, **157**, 346 (1957).
340. M. Halmann and L. Kugel, *Bull. Res. Counc. Israel, A*, **10**, 124 (1961).
341. K. Peter, *Fette, Seifen, Anstrichm.*, **66**, 265 (1964).
342. E. P. Jones, C. R. Scholfield, V. L. Davison, and H. J. Dutton, *J. Am. Oil Chem. Soc.*, **42**, 727 (1965).
343. H. Wehle, *Pharmazie*, **23** (4), 180 (1968).

344. J. L. Ellingboe and C. N. Ruybal, *Chemist-Analyst*, **56** (4), 98 (1967).
345. E. von Schivizhoffen, *Z. Anal. Chem.*, **145**, 184 (1955).
346. M. I. Osipova and G. K. Khodzhaev, *Uzb. Khim. Zh.*, **1968**, 32.
347. J. Halmekoski, *Suom. Kem.*, **32** (12), 274 (1959).
348. W. Beckering and W. E. Fowkes, *Anal. Chem.*, **30**, 1336 (1958).
349. V. M. Riddle, *Anal. Chem.*, **35**, 853 (1963).
350. W. L. Hempenius and B. J. Liska, *J. Dairy Sci.*, **51**, 221 (1968).
351. G. J. Lawson and J. W. Purdie, *Mikrochim. Acta*, **1961**, 415.
352. A. T. James, *Analyst*, **88**, 572 (1963).
353. H. B. White, Jr., and S. S. Powell, *J. Chromatogr.*, **21**, 213 (1966), **32**, 451 (1968).

CHAPTER 6

Separation of Basic Substances

6.1 INTRODUCTORY REMARKS

It is difficult to define a basic substance. From the standpoint of quantitative organic analysis, we may include in this category all compounds that, by themselves, can be determined by titrimetry using a very strong acid (e.g., perchloric acid) in nonaqueous medium (e.g., acetic acid, glycol–hydrocarbon).[1] Needless to say, aliphatic amines that can be determined by titration with standardized acid in aqueous solution also belong to this category. For convenicnce, when a compound possesses both acidic and basic functions (e.g., amino acids, aminophenols, barbituric acids), it is discussed in this chapter and not in Chapter 5.

Separation of basic compounds becomes necessary when two or more are present in a mixture and cannot be determined simultaneously using the differential titration technique (see Chapter 3, Section 3.3). Understandably, more than one method or principle may be utilized to separate basic substances. Thus Chodakowski[2] reviewed the literature on the separation of phenothiazines and cited 39 references on procedures using paper, thin-layer, column, and gas-chromatographic techniques. Whereas column chromatography is generally preferred in quantitative work, Lepri et al.[3] reported that separations of aromatic amines on carboxymethylcellulose or alginic acid columns were less satisfactory than those by thin-layer chromatography with similar media and solvents. Therefore it is often advisable to compare the methods when a certain mixture is to be separated for routine analysis.

6.2 SEPARATION ON ION-EXCHANGE COLUMNS

Mixtures containing two or more basic compounds having significantly different ionization constants can be conveniently separated on suitable ion-exchange columns. Table 6.1 gives examples of published procedures. The ion-exchange resins indicated in Table 6.1 are commercially available. Some workers prepared their own ion exchangers. For instance, Foster and Murfin[46] treated alginic acid with formaldehyde and used the material as a carboxylic cation-exchange medium for quantitative separation of 14 organic bases.

Ion-exchange columns are frequently employed for the separation of protein hydrolyzates. On the market are several apparatus known as amino acid analyzers.[47] According to Duncan and Mohler,[48] the concentration range must be kept within narrow limits to obtain 100% recoveries for routine operations. Peters et al.[49] observed that a slight change of conditions could alter the order of elution of certain amino acids.

It is possible to use the ion-exchange technique to separate isomeric compounds even though they are present in the mixture in very large ratios. For example, Urbanyi et al.[50] determined pilocarpine and isopilocarpine using a glass column (10 cm × 6 mm) packed with Aminex A-7 cation-exchange resin with a 5% solution of isopropyl alcohol in 0.1 *M* Tris buffer of pH 9 as mobile phase; 0.1 μg of isopilocarpine was separated from 100 μg of pilocarpine.

After separation, the individual components can be determined spectrophotometrically. For compounds containing free primary amino groups, it is a common practice to convert them into the colored dinitrophenyl

Table 6.1 Separation of Basic Compounds on Ion-Exchange Columns

Mixture	Ion-exchange resin used	References
Mono-, di-, triethanolamines	Dowex 50W-X8	Yoshino[4]
Acetamide, ammonium acetate	Amberlite IRA-120	King[5]
Biuret, urea	Amberlite IRA-400AG	Takahashi[6]
Betaine, choline	Lewatit S100	Niemann[7]
Gentamycin antibiotics	Dowex 1-X2	Maehr[8]
Methyl-adrenalines methyl noradrenalines	Amberlite CG-50	Bigelow[9]
Methyl adrenalines, noradrenalines	Dowex 50W-X2	Kahane[10]
Quaternary ammonium bases	Dowex 50	Christianson,[11] Friedman[12]

Table 6.1 (*Contd.*)

Mixture	Ion-exchange resin used	References
Quaternary ammonium bases	Amberlite CG-120	Christianson[13]
Gluco-, galactosamines	E.E.L. 120B	Donald[14]
Gluco-, galactosamines	KU-2	Baev[15]
Glucos-, galactosaminitols	UR-30	Bella[16]
Hexosamines		Steele[17]
Amino acids	Dowex 50-X4	Moore,[18] Dustin,[19] Hirs[20]
Amino acids	Chromobeads	Padieu[21]
Amino acids	Zeokarb 225	Jacobs[22]
Sulfur amino acids		Frimpter[23]
Asparagine, isoasparagine		Tritsch[24]
Ornithine, lysine	Amberlite CG 120	Pfordte[25]
Ephedrine, barbiturates	Dowex 50-X8	Blake[26]
Caffeine, antipyrine	Dowex 50	Sjöström[27]
Caffeine, antipyrine, phenacetin	Dowex 50	Sjöström[28]
Caffeine, phenacetin, amidopyrine	Amberlite IR-120	Reisch[29]
Caffeine, quinine, strychnine	Dowex 50-X2	Kamp[30]
Phenylephrine, codeine, antihistamines	AG 50W-X4	Montgomery[31]
Berberine, atropine, hyoscine	Duolite CS-101	Watanabe[32]
Opium alkaloids	Duolite C10	Büchi[33]
Berberine, palmatine	Duolite C101	Watanabe[34]
Pyridoxol, pyridoxal, pyridoxamine	Amberlite IR-120	Hedin[35]
Indoles	Dowex 50W-X2	Kaleysa[36]
Purines, pyrimidines	AA-15	Bonnelycke[37]
Purines from ribonucleic acids	Dowex 2 X8	Popa[38]
Vitamins of B complex	Wofatit CP300	Klotz[39]
B_6 vitamins	KH 4B	Fukui[40]
Melamines	Amberlite IR-120	Takimoto[41]
Melamine, ammeline, ammelide, cyanuric acid	Amberlite IRA-410	Bacalogu[42]
Aminobenzoic acids	Amberlite CG-120	Funasaka[43]
Aminophenols		Wheaton[44]
Aminocresols	Zipax SCX	Sakurai[45]

derivatives and measure the absorbances in the visible spectrum. On the other hand, if the basic compound has a low boiling point and the quantity separated is at the milligram level,[5,6] the micro-Kjeldahl distillation and titration procedure[51] offers a simple alternative. Quantitation by precipitation was recommended by some workers. For instance, Niemann[7] separated betaine and choline in sugar beet juice on a column packed with Lewatit S100 (H^+ form) and determined the respective compounds with Reinecke's salt gravimetrically. Büchi and von Moos[33] described the separation of morphine, codeine, narceine, papaverine, noscapine, and thebaine from 0.5 g of opium sample by the use of Duolite-ClO(NH_4^+), Duolite-CS101-(NH_4^+), and Dowex 2-X4(OH^-) columns, followed by precipitation of the reineckates and titration with 0.1 *N* $AgNO_3$ and 0.1 *N* NH_4SCN.

6.3 SEPARATION BY GAS CHROMATOGRAPHY

Aliphatic and aromatic compounds that possess free amino groups are usually stable at their boiling points and are therefore amenable to direct gas chromatography. Vanden-Heuvel et al.[52] studied the conditions for the separation and determination of several categories of amines, including long-chain and alicyclic amines, aliphatic diamines, and aromatic amines; for each category, different conditions were used, and the suitability of the methods for application to biological materials was tested. Table 6.2 gives the amine mixtures separated on the specified columns reported by various workers. Besides simple aliphatic and aromatic amines. Table 6.2 also gives a number of pharmaceutical and biological amines. It should be noted that high-boiling amines may decompose in the gas chromatograph. Thus, while amphetamine and its homologues can be determined by direct gas chromatography,[76] Vessman and Schill[80] reported the decomposition of amphetamine, lidocaine, ephedrine, and other high-boiling amines during gas chromatography under various conditions. On the other hand, the decomposition of quaternary ammonium compounds is the basis for their analysis by gas chromatography. The column can be packed with Apiezon on KOH–Chromosorb (Metcalfe[81]) or on glass beads (Grossi and Vece[82]) and the tertiary amines are determined. Laycock and Mulley[83] determined alkyltrimethyl quaternary ammonium antibacterial agents using a column containing 10% of SE-30 on acid-washed Celite and measured both the tertiary amine and the olefin produced by each compound. According to Craig et al.,[84] tertiary-amine oxides undergo elimination reaction instantaneously in a gas chromatograph under moderate conditions; this affords a rapid means for the determination of the decomposition products, and hence of the original amine oxides.

Table 6.2 Separation of Amines by Gas Chromatography

Amine mixture	Column packing	Reference
Methyl-, dimethyl-, tri- trimethylamines	Triethanolamine on diatoms	Petrova[53]
Methyl-, dimethyl-, ethylamines, NH_3	*o*-Toluidine on firebrick	Amell[54]
Homologous primary-, secondary-, tertiary-amines	Apiezon on Teflon	Landault[55]
Allyl-, butyl-, benzylamines	Carbowax–KOH on Chromosorb	Arad[56]
Diethyl-, triethyl-, butylamines	Gas-Chrom with 4% of $AgNO_3$	Sawardeker[57]
Isoaliphatic amines		Golovnya[58]
Lower aliphatic amines	Tetrahydroxyethylene diamine on Chromosorb	Sze[59]
Aliphatic primary C_2 to C_{12} amines	Polyethylenime on glass capillary	Kudryavtsev[60]
Aliphatic diamines	Apiezon–KOH on Celite	Lindsay[61]
Di-, triamines	Apiezon on KOH–Chromosorb	Metcalfe[62]
Mono-, triethylenediamines, diethylene triamine	Carbowax on KOH–silanized Chromosorb	Törnquis[63]
Diamines, tetramines, guanidines	Ucon on NaOH–Celite	Cincotta[64]
Volatile amines in caviar		Golovnya[65]
Cyclohexylamines	Silicone oil on KOH–Hyprose	Feltkamp[66]
Volatile aromatic amines	Paraffin; Lubrol; benzyldiphenyl	James[67]
Toluidines	Ucon; Carbowax	Parsons[68]
Phenyl-, napthylamines, phenylenediamines		Wise[69]
Phenylenediamines	Triton on Chromosorb	Bryan[70]
Toluenediamines	Siponate on KOH–Chromosorb	Willeboordse[71]
Toluenediamines	Bentone, Hyprose on KOH–Chromosorb	Boufford[72]
Aminopropiophenones	Apiezon–KOH on Chromosorb	Beckett[73]
Chloroanilines	Ucon on Chromosorb	Henkel[74]
Chloroanilines	Siponete on Chromosorb	Bombaugh[75]
Amphetamine and related amines	Carbowax–KOH on Chromosorb	Lebish[76]
Lidocaine and related compounds	SE-30 on Chromosorb	Strong[77]
Amino alcohols	Flexol on firebrick	Koehler[78]
Catecholamines	Silicone oil on Gas-Chrom	Brooks[79]

Aliphatic amides were separated by Metcalfe et al.[85] on a column of Chromosorb–KOH with 5% of Apiezon, and by O'Donnell and Mann[86] on a column containing 20% of Dowfax 9N9 and 2.5% of NaOH supported on Chromosorb P. Glutethimide and its metabolite 2-phenylglutarimide in biological fluids were separated (Grosijean and Noirfalise[87]) on a column packed with 8% of XF1112 on Chromosorb W-HMDS; peak area is proportional to concentration for up to 6 μg of either compound per ml. *N*-Alkylnitrosoamines were separated on columns packed with Chromosorb 101 (Foreman et al.[88]) or diethylene glycol adipate polyester (Saxby[89]); interference due to the presence of pyrazines can be removed by adding a precolumn containing CuSCN and Chromosorb W. Alkylated hydrazines were determined using a column containing 25% of Carbowax plus 7% of KOH on Celite (Bighi and Saglietto[90]) or 10% of 2-hydrazinopyridine on Fluoropak (Dee and Web[91]), or a capillary tube coated with *N*,*N*-bis(hydroxyethyl)trimethylenediamine (Heyns et al.[92]). Alkyl-substituted ureas were determined with a column of 0.5% of Carbowax on glass beads (Reiser[93]) or of 5% of methyl silicone on Gas-Chrom (McKone and Hance[94]). *N*-Chloroethyl- and *N*,*N*-bis(chloroethyl)carbamates were separated on a column containing 5% of Dow-11 on Chromosorb (Zielinski and Fishbein[95]).

Numerous mixtures containing heterocyclic nitrogen compounds have been separated by direct gas chromatography. Examples are given in Table 6.3. Comparing this table with Table 6.2 reveals that most separations of *N*-cyclic mixtures do not require the alkali-treated columns prescribed for the aliphatic and aromatic amines. It may be mentioned that Yakerson et al.[126] recommended alkali-treated diatomite brick containing 15 to 20% of polyoxyethyleneglycol adipate to separate derivatives of pyridine and also of thiophen and furan.

Tailing should be prevented in quantitative work. Tranchant[127] found that

Table 6.3 Separation of Heterocyclic Nitrogen Compounds by Gas Chromatography

Mixture	Column packing	Reference
Pyridine homologues	Tween 20 on Embacel	Kirsten[96]
Pyridine, picolines, 2,6-lutidine	Ethanolamine–phenylphenol on Celite	Kametani[97]
Pyridine, picolines, 2,6-lutidine	Glycerol on silica gel	Mitra[98]
15 pyridines		Decora[99]
Pyridines, toluidines, chloranilines	Polyphenyl ether–glycerol on Chromosorb	Anwar[100]
21 pyridines, 13 quinolines	Silicone or diglycerol	Fitzgerald[101]

Table 6.3 (*Contd.*)

Mixture	Column packing	Reference
Coal tar bases	Glycerol on Chromosorb	Bhattacharya [102]
Piperidines	Tris(2-cyanoethoxymethyl)-4-picoline on Chromosorb	Janak[103]
Piperidines	Polyoxyethyleneglycol on KOH–Kieselgel	Moll[104]
Methyl-, ethylpyridine	Triethanolamine on Celite	Vietti-Michelina[105]
Quinolines	Polyoxyethylene glycol	Funasaka[106]
Quinolines	Glycerol on Celite	Rezl[107]
Pyridine aldehydes	Apiezon on Celite	Vietti-Michelina[108]
Pyridine acid amides, nitriles, esters	Polyethylene glycol adipate on diatomite	Trubnikov[109]
Pyrazoles, isopyrazoles	Polyethylene glycol sebacate on Celite	Grandberg[110]
Piperazine, ethylenediamine	Apiezon on Celite	Castiglione[111]
Nicotine, tobacco alkaloids	Versamide on Chromosorb	Weeks[112]
Nicotine, tobacco alkaloid	Ucon–KOH in capillary	Harke[113]
Nicotine, pyridine	Apiezon on Celite	Pilleri[114]
Carbazoles in cigarette smoke	SE-30 on Chromosorb	Hoffmann[115]
Indole alkaloids	SE-30 on Gas-Chrom	van Binst[116]
Ephedrine, codeine, methapyrilene	Silicone on Diatoport	Wesselman[117]
Codaine, codeine, morphine, methadone	SE-30 on silanized Chromosorb	Vessman[118]
25 alkaloids	QF-1 on Epon resin	Massingill[119]
23 alkaloids	Polyester on Gas-Chrom	Brochmann-Haussey[120]
Atropine, scopolamine	SE-30 on silanized Chromosorb	Solomon[121]
Diazepan and related compounds	OV-17 on Diatomite	Greaves[122]
Xanthines, caffeine, theobromine, theophylline	SE-30 on Chromosorb	Reisch[123]
Caffeine, amidopyrine, allobarbitone	SE-30 on Gas-Chrom	Kawaii[124]
Phenothiazines, quinolines, barbiturates	SE-30 on Gas-Chrom	Vanden-Heuvel[125]

amines, whose peaks otherwise tended to tail, were best separated on strongly polar phases supported on an inert material such as Teflon. Lindsay-Smith[128] studied the cause of tailing of aliphatic amines on coated polymers and concluded that tailing arises from two types of active sites on the polymer: simple acidic sites, which can be neutralized by a base, and metal-ion sites, which must be deactivated by a nonvolatile chelating agent such as the polyamines. For the determination of pyridine bases from coal tar, van der Meeren and Verhaar[129] reported that tailing could be prevented by adding triethanolamine to the stationary phase of polyoxyethylene glycol supported on Celite or kieselguhr.

The general approach to prevent tailing of amino compounds is to convert them into derivatives that are less basic and less polar. Varieties of derivatives are shown in Table 6.4. In some cases, such as the acetyl derivatives of methylephedrine[161] and the methylation of phenolic alkaloids by means of trimethylanilinium hydroxide,[164] it is the hydroxyl group that is modified and not the basic nitrogen function. For amino acids, esterification of the carboxyl group to generate the stable free bases is sufficient for gas-chromatographic separation[145]; however, both the amino and the carboxyl groups are generally derivatized for quantitative analysis. Zlatkis et al.[165] described a method to determine amino acids that involves their conversion into the corresponding aldehydes in a precolumn containing ninhydrin. The aldehydes, after being separated in the gas chromatograph, are cracked to methane over a nickel catalyst; the methane derived from each amino acid is then measured by a thermal-conductivity cell.

Table 6.4 Derivatives of Basic Compounds Suitable for Gas Chromatography

Basic compounds	Derivatives	Reference
Lower aliphatic primary amines	Fluorine-containing Schiff bases	Hoshika[130]
C_1 to C_4 primary, secondary amines	Dinitrophenyl	Day[131]
Primary amines	With hexane-2,5-dione	Walle[132]
Secondary amines	Trifluoroacetyl	Corbin[133]
Amines in tobacco smoke	Amides	Pailer[134]
Ethanolamines	Trifluoroacetyl	Brydia[135]
Aniline, toluidines, xylidines	Trifluoroacetyl	Dove[136]
Toluenediamines	*N*,*N'*-Bis(trifluoroacetyl)	Brydia[137]
Naphthylamines	Pentafluoropropionyl	Hoffmann[138]
Pharmaceutical amines	Trifluoroacetyl	Hishta,[139] Hirtz[140]

Table 6.4 (*Contd.*)

Basic compounds	Derivatives	Reference
Optical isomers of amphetamine	*N*-Trifluoroacetyl-L-prolyl	Wells[141]
Catecholamines	Trifluoroacetyl	Kawai[142]
Catecholamines	Trimethylsilyl ethers	Sen[143]
Tyramine, phenethylamine	Heptafluorobutyramide	Wilk[144]
Amino acids	Methyl esters	Nicholls,[145] Saroff[146]
Amino acids	Butyl esters	McBride[147]
Amino acids	Trimethylsilyl	Smith[148]
Amino acids	Trifluoroacetyl	Carlstrom[149]
Amino acids	*N*-Trifluoroacetyl butyl esters	Marucci[150]
Amino acids	*N*-Acetyl esters	Shlyapnikov[151]
Amino acids	*N*-Acetyl butyl esters	Youngs[152]
Amino acids	*N*-Acetyl propyl esters	Coulter[153]
Amino acids	*N*-Dinitrophenyl methyl esters	Landowne,[154] Pisano[155]
Amino sugars	Acetates	Jones[156]
Carbamates	Trimethylsilyl ethers	Zielinski[157]
Pyridinecarboxylic acids	Esters	Liliedahl[158]
Amides of pyridinecarboxylic acids	Nitriles	Trubnikov[159]
Pyridoxines	Isopropylidenes	Korytnyk[160]
Ephedrine, methylephedrine	Acetyl	Hattori[161]
Morphine, mono-, diacetylmorphines	Trimethylsilyl	Grooms[162]
Imidazoles	Trimethylsilyl	de Silva[163]
Xanthines, barbiturates, phenolic alkaloids	Methylated	Brochmann-Hanssen[164]

6.4 SEPARATION BY LIQUID CHROMATOGRAPHY

Table 6.5 illustrates the separation of basic substances by elution from a partition or adsorption column.[184] The components of a mixture generally are eluted in the order of basicity or solubility. Hence the solvent systems may be altered or the pH of the eluent may be gradually changed in order to recover the various constituents of the original sample placed on top of the column.

Table 6.5 Separation of Basic Compounds by Liquid Chromatography

Mixture	Column	Mobile phase	Reference
Mono-, di-, tri-ethanolamines	Starch	*sec*-Butyl alcohol	Vyakhirev[166]
Aromatic amines	Teflon	CH_3OH–H_2O	Hedrick[167]
Aromatic amines	Silica gel(Cu)	CH_3OH–cyclohexane	Chow[168]
o-, *m*-, *p*-Nitroanilines	Rubber	$CHCl_3$–acetone–aq. H_2SO_4	Buchowski[169]
o-, *m*-, *p*-Nitroanilines	Alumina	Benzene	Missbach[170]
Monosubstituted *o*-, *p*-nitroanilines	Alumina	Ethyl acetate–benzene	Larson[171]
Amino acids	Poly(vinyl-pyrrolidine)	H_2O	Dougherty[172]
Amino sugars	Cellulose	1% Cetylpyridinium chloride	Praus[173]
Pyridine bases			Waksmundzki[174]
Quinoline, isoquinoline	Silica gel	Benzene	Sinha[175]
Hydrogenated quinolines	Alumina	CH_2Cl_2–pentane–H_2O	Snyder[176]
Caffeine, codeine, amidopyrine	Poly(styrene–divinylbenzene)	H_2O	Chmil[177]
Atropine, hyoscyamine	Alumina	H_2O–C_2H_5OH	Kamienski[178]
Alkaloids	Silica gel	Buffer solution	Andreeva[179]
Morphine, corytuberine, magnoflorine	Cellulose	Aqueous NH_3	Nijland[180]
Opiates, quinine	Sephadex	0.1 M NaH_2PO_4	Broich[181]
Benzodiazepines	Durapak	Hexane–isopropyl alcohol	Scott[182]
Aminoalkylphosphonic acids			Roop[183]

If the solid support in the liquid column is not completely inert, other principles besides adsorption and partition may be involved and the order of elution of a given mixture may be different. For instance, when Cu(II)-bonded silica gel is used, complexation between the basic compounds and Cu(II) occurs, and the stability constants of the chelates play an important role. Thus Chow and Crushka[168] found that Cu(II) ion bonded to Partisil-10 is an effective stationary phase for the separation of aromatic amines and the retention times seem to be related to the basicities and structures of the amines.

Ion-pair partition chromatography may be considered as a variation of liquid chromatography for separating basic compounds. It is dependent on the formation of amine salts of large molecular weights with different elution characteristics. For example, Rader[185] analyzed a mixture of pyrilamine, methypyrilene, and chlorphenramine in ethyl ether using a column packed with Celite containing *p*-toluenesulfonic acid and KCl; the first two antihistamines were eluted with $CHCl_3$, and the chlorpheniramine was then eluted with 1% acetic acid solution in $CHCl_3$.

Sometimes partition columns are employed to separate amino compounds in the form of hydrochlorides. For instance, Clasper et al.[186] separated the hydrochlorides of hexamethylenediamine, *p,p'*-diaminodicyclohexylmethane, and ε-aminocaproic acid in the following manner. The aqueous solution of the mixed hydrochlorides was applied to a cellulose column. With *sec*-butyl alcohol–formic acid–H_2O as eluting solvent, ε-aminocaproic acid hydrochloride and the *p,p'*-diaminodicyclohexylmethane dihydrochloride were recovered. The column was then eluted with ethanol–H_2O, and hexamethylenediamine dihydrochloride was recovered.

Preparation of derivatives for liquid chromatography of basic substances is rare. Maitlis[187] determined quinoline and isoquinoline in the presence of one another by nitrating the mixture with an excess of HNO_3 and separated the nitro derivatives on a column of alumina. Nitroquinoline was eluted rapidly with ethyl ether, while nitroisoquinoline remained almost stationary. The latter compound was eluted separately with $CHCl_3$.

6.5 SEPARATION BY PAPER CHROMATOGRAPHY

The use of paper chromatography is not warranted when mixtures of simple aliphatic and aromatic amines are analyzed, because the high volatility of these compounds makes it difficult to control the conditions of separation and determination. Paper chromatography, however, provides an economical technique for the analysis of nonvolatile basic substances. Many biological and medicinal compounds can be determined in complex mixtures using two-dimensional paper chromatography, while the one-dimensional procedure is adequate for less complex mixtures. Some examples are given in Table 6.6. The basic compounds can be separated without alteration, although derivatives are sometimes prepared. For instance, for the determination of amino acids, Habeeb[239] converted them into the dinitrophenylamino acids and carried out two-dimensional chromatography with ethylbenzene–*t*-amyl alcohol–1.6 *N* aq. NH_3 (1 :3 :2) as solvent in one direction, followed by 1.5 *M* phosphate buffer in the other direction. Tulus and Guran[240] recommended conversion of sulfonamides to the corresponding

Table 6.6 Separation of Basic Substances by Paper Chromatography

Mixture	Developing solvent	Reference
Amino acids	H_2O–phenol; butanol–CH_3COOH–H_2O	Lukyanov[188]
Amino acids from plants		Thompson[189]
Amino acids, by ascending technique		Tsukamoto[190]
Amino acids	Butanol–CH_3COOH–H_2O	Szczepaniak[191]
Amino acids, by circular technique	Phenol–butanol–CH_3COOH–H_2O	Barbiroli[192]
Sulfur-containing amino acids	C_2H_5OH–$CHCl_3$–H_2O	Fondarai[193]
Amino acid isomers	*t*-Pentylalcohol–CH_3COOH–H_2O	Gray[194]
3-(Aminonaphthyl)alanine enantiomers	Butanol–pyridine–H_2O	Zaltzman-Nirenberg[195]
Ring-substituted hyppuric acids	Butanol–CH_3COOH–H_2O	Haberland[196]
Aminopolycarboxylates	Acetone–H_2O	Sykora[197]
Aminophenols	*t*-Pentyl alcohol–CH_3NH_2–H_2O; $C_2H_5NO_2$–CH_3COOH–H_2O	Smith[198]
Adrenaline, noradrenaline	CH_3OH–tartaric acid–H_2O	Carassiti[199]
Dicyandiamide, biguanides, etc.	$CHCl_3$–C_2H_5OH–H_2O	Moza[200]
ε-Caprolactone, methyl-ε-caprolactone	Cyclohexane–CH_3COOH–C_3H_7OH	Franc[201]
Ureas, thioureas	Butanol–CH_3COOH–H_2O	Fishbein[202]
Thiopentone and related compounds	H_2O–$CHCl_3$–heptanol–$(CH_3)_2CHOH$–CH_3OH–NH_3	Narbutt-Maring[203]
Urea herbicides	$(CH_3)_2CHOH$–H_2O–CH_3COOH	Oswiecimska[204]
Guanidine compounds		Pant[205]
Alkylpyridines	C_2H_5OH–HCl	Neuhauser[206]
Quinolines	Ethyl acetate–H_2O	Danilovic[207]
Aminoacridines	Decalin–benzene–$CHCl_3$	Waksmundzki[208]
Purines, pyrimidines		Blazsek[209]
Caffeine, 8-methylcaffeine	0.05 *M* Sodium-3-hydroxy-naphthalene-2,7-disulfonate	Stuchlik[210]
Opium alkaloids	Dioxane–$HCOOH$–H_2O	Micheel[211]
Opium alkaloids, by reversed-phase technique	Sodium salt of arylsulfonic acid	Stuchlik[212]

Table 6.6 (*Contd.*)

Mixture	Developing solvent	Reference
Morphine, codeine, thebaine in poppy latex	Butanol–CH_3COOH–H_2O	Fairbairn[213]
Morphine, normorphine, nalorphine	Pyridine–ethyl acetate–H_2O	Penna-Herreros[214]
Cocaine, benzoylecgonine, ecgonine	$CH_3COC_2H_5$–diethylformamide–H_2O	Majlat[215]
Morphine, strychnine	Butanol–HCOOH–H_2O	Schultz[216]
Strychnine, brucine	Butanol–CH_3COOH–H_2O	Dusinsky[217]
Rauwolfia alkaloids, by circular technique	Benzene–cyclohexane–$CHCl_3$–formamide	Ma[218]
Tropane alkaloids	Butanol–CH_3COOH–H_2O	Pfordte[219]
Tropane alkaloids	H_2O–NH_3–octanol	Horak[220]
Pyrrolizidine alkaloids	Butanol–CH_3COOH–H_2O	Chalmers[221]
Alkaloids and their *N*-oxides	Acetone–NH_3–H_2O	Munier[222]
Bis(hydroxymethyl)-quinoxaline *N*-oxide, *N*,*N*′-dioxide	Butanol–buffer solution	Granik[223]
Azauridine, azauracil, orotic acid	Butanol–CH_3COOH–H_2O	Mironov[224]
Colchicine, demecolcine	Benzene–$CHCl_3$	Dusinsky[225]
Riboflavines	5% $NaHPO_4$	Travis[226]
Tetracyclines	$CH_3COCH(CH_3)_2$–ethyl acetate–butanol	Canterelli[227]
Antimalarials	Butanol–CH_3COOH–H_2O	Mshvidobadze[228]
Thalidomide and 12 hydrolysis products	Pyridine, butanol	Schumbacher[229]
Ethionamide and decomposition products	*sec*-Butyl alcohol–HCOOH–H_2O	Pawelczyk[230]
Sulfonamides	Butanol–C_2H_5OH–$(C_2H_5)_2NH$–H_2O	Jakubec[231]
Sulfonamides		Maienthal[232]
Sulfonamides	CH_2Cl_2	Kunze[233]
Sulfonamides	Butanol–H_2O–dimethylformamide	Gräfe[234]
Sulfonamides	H_2O–NH_3–*sec*-butyl alcohol–$(CH_3)_2CHOH$	Ritschel[235]
Sulfonamides	0.2 *N* sodium succinate	Garber[236]

Table 6.6 (*Contd.*)

Mixture	Developing solvent	Reference
Sulfonamides	Butanol–H_2O–NH_3	Luise[237]
Phosphoethanolamine, aminoethylphosphonic acid	0.02 *M* buffer (pH 9)	Neuzil[238]

2,4-dinitrobenzenesulfonyl derivatives and developed the chromatogram with toluene–acetone–methanol–10% aq. NH_3 (75 : 25 : 25 : 1). In this way the yellow color of the dinitrophenyl group serves as a means of quantitation by spectrophotometric measurement.

The chromatographic paper may be modified to improve, or to achieve, the analytical results. Separation of amino acids on ion-exchange paper has been described by several groups of workers (Soczewinski and Rojowska,[241] Hartel et al.,[242] Baerhiem-Svendsen and Brochmann-Hanssen,[243] Roberts and Kolor[244]). Moses[245] reported that pretreatment of Whatman No. 4 paper with oxalic acid reduced the amount of decomposition of nine amino acids by an average of 30%. Tabak et al.[246] used acetylated paper to separate pyrazoles. Pöhm[247] impregnated the paper with formamide for the determination of ergot alkaloids, while Bohinc[248] treated the paper with ethanolic salicylic acid for xanthine alkaloids. Chumakov and Filippovich[249] impregnated the paper with $CuCl_2$ to separate quinoline from isoquinoline; the latter compound gave a blue zone whereas the former gave a violet zone.

6.6 SEPARATION BY THIN-LAYER CHROMATOGRAPHY

There are numerous reports on the separation of basic substances by thin-layer chromatography. The large varieties of compounds separated can be seen in Table 6.7. It should be mentioned that most workers performed the separation for qualitative analysis. However, once the separation procedure has been established, some ways may be found to quantify the results.

In contrast to separation by paper chromatography, which requires many hours to complete, separation on thin-layer plates usually is accomplished in a relatively short time. For this reason, high-boiling amines[250–254] can be separated by thin-layer chromatography by applying them directly on the plates without the danger of loss by vaporization. Some thin-layer methods for the determination of basic mixtures are remarkably fast. For instance, Franc and Hajkova[266] separated and determined mono- and diamino-

Table 6.7 Separation of Basic Substances by Thin-Layer Chromatography

Mixture	Adsorbent	Solvent system	Reference
C_8 to C_{16} fatty amines	Kieselgel	Cyclohexane–CH_3COOH	Lauckner[250]
Primary, secondary, tertiary amines	Al_2O_3	Isobutyl acetate–CH_3COOH	Lane[251]
Tertiary amines, amine oxides	Silica gel	$CHCl_3$–aq. NH_3–CH_3OH	Pelka[252]
Polyamines	Silica gel	Aq. NH_3–C_2H_5OH	Parrish[253]
Cyclohexylamines	Kieselgel	Conc. NH_3–acetone–petroleum	Feltkamp[254]
Tetra C_4 to C_7 ammonium salts	Silica gel	Acetone–*N*-methylacetamide	Gordon[255]
Quaternary ammonium salts	Kieselgel	Cyclohexane–$CHCl_3$–CH_3COOH	Waldi[256]
Amino acids, two-dimensional	Kieselgel	$CHCl_3$–CH_3OH; butanol–CH_3COOH–H_2O	Rossetti[257]
Amino acids	Silica gel		Frodyma[258]
Amino acids	Cellulose powder		Clark[259]
Choline and derivatives	Al_2O_3	Butanol–H_2O–HCOOH	Taylor[259a]
Aromatic diamines	Silica gel	Benzene–CH_3OH	Pinter[260]
Nitrogen compounds	Kieselgel	Benzene–dioxane–CH_3COOH	Pastuska[261]
Nitroanilines	Silica gel–gypsum	Dioxane–CCl_4	Waksmundzki[262]
p-Aminoazo compounds	Silica gel		Hashimoto[263]
Aminophenol, aminosalicylic acid	Al_2O_3	Ethanol	Kinze[264]
Mono-, dihydroxy-phenylamines	Silica gel		Alessandro[265]
Mono-, diamino-anthraquinones	Al_2O_3	Cyclohexane–$(C_2H_5)_2O$	Franc[266]
Hydroxylamines, nitroxides	Silica gel	Acetone–CH_3OH	Weil[267]
Aliphatic nitramines	Silica gel		Bell[268]
Ureas, thioureas, biurets	Silica gel	Ethyl acetate–CH_3OH	Di Bello[269]
Substituted ureas, urethanes	Silica gel	CCl_4–CH_2Cl_2–ethyl acetate–HCOOH	Knappe[270]

Table 6.7 (*Contd.*)

Mixture	Adsorbent	Solvent system	Reference
N-Methylcarbamates	Silanized cellulose	H_2O–C_2H_5OH–$CHCl_3$	Menn[271]
Methylenedioxy-phenyl carbamates	Silica gel	Toluene–ethyl acetate	Fishbein[272]
Guanidines	Cellulose		di Teso[273]
Streptomycin, neomycin, etc.	Kieselgel	C_3H_7OH–ethyl acetate–H_2O	Hüttenrauch[274]
Pyridines, *N*-oxides			Grandberg[275]
Bipyridyls	Kieselgel–gypsum	Benzene–CH_3OH–HCOOH	Kulicka[276]
Alkylpyridines	Kieselgel		Barsch[277]
Quinolines, isoquinolines	Kieselgel	Benzene–ethyl acetate	Troszkiewicz[278]
Nicotinic acid, amide	Cellulose	Butanol–aq. NH_3	Frei[279]
Diazepines	Kieselgel		Roeder[280]
Aminotriazoles	Silica gel	Pyridine–$C_5H_{11}OH$–aq. NH_3	Adamek[281]
Azines	Al_2O_3	Benzene–$CHCl_3$	Klemm[282]
Fluoropyrimidines	Silica gel	Ethyl acetate–acetone–H_2O	Hawrylshyn[283]
Indoles	Silica gel		Pillay[284]
Indoles	Kieselgel	$CHCl_3$–CCl_4–CH_3OH	Ballin[285]
Imidazoles	Silica gel	$CHCl_3$–CH_3OH–aq. NH_3	Grimmett[286]
Amidopyrine, phenacetin, etc.	Silica gel	Acetone–aq. NH_3–benzene	Bachrata[287]
Phenacetin, phenaz-one, caffeine	Kieselgel	Cyclohexane–acetone	Pfandl[288]
Phenacetin impurities			Turi[289]
Caffeine, phenacetin, propyphenazone	Silica gel	$(C_2H_5)_2O$	Soeterboek[290]
Caffeine, amido-pyrine, phenacetin	Kieselgel	Cyclohexane–acetone	Gänshirt[291]
Cynarin, chlorogenic, caffeic acids	Silica gel	Ethyl acetate	Colombo[292]
Caffeine, theo-bromine, theophylline	Al_2O_3	C_4H_9OH–$C_5H_{11}OH$–HCl–H_2O	Li[293]
Caffeine, theobromine, theophylline	Kieselgel	$CHCl_3$–C_2H_5OH	Szasz[294]

Table 6.7 (*Contd.*)

Mixture	Adsorbent	Solvent system	Reference
Caffeine, theobromine, theophylline	Kieselgel	$CHCl_3$–CCl_4–CH_3OH	Senanayake[295]
Theobromine, theophylline	Kieselgel	$CHCl_3$–C_2H_5OH	Franzke[296]
Xanthine compounds	Kieselgel	Benzene–acetone	Schunack[297]
Pyrimidines	Cellulose	H_2O–dimethylformamide	Chou[298]
Uracil herbicides	Silica gel	Benzene–hexane–acetone	von Stryk[299]
Isocarboxazide antidepressives	Silica gel	$CHCl_3$–CH_3OH	Schmid[300]
Isocarboxazide antidepressives	Silica gel	$CHCl_3$–aq. NH_3–CH_3OH	Alessandro[301]
Amphetamines, barbiturates	Silica gel	Dioxane–benzene–aq. NH_3	Morrison[302]
Diphenhydramine, promethazine antihistamines	Silica gel	CH_3COOH–H_2O	Morrison[303]
Tetracyclines	Kieselguhr–EDTA	Ethyl acetate–0.1 *M* EDTA	Ascione[304]
Tetracyclines	Kieselguhr	Ethyl acetate–acetone–H_2O	Alvarez-Fernandez[305]
Tetracyclines	Kieselguhr	CH_2Cl_2–C_2H_5OH–H_2O	Dijkhuis[306]
Erythromycins	Kieselguhr	C_2H_5OH–CH_2Cl_2–ethyl acetate–hexane	Banaszek[307]
Thiamine, riboflavine, etc	Supergel–gypsum	CH_3COOH–H_2O	Ludwig[308]
Thiamine, pyrimidine carboxylic acids	Silica gel	CH_3CN–H_2O–$HCOOH$	Waring[309]
Hydroxocobalamin, cyanocobalamin	Al_2O_3	iso C_4H_9OH–*i*-C_3H_7OH-H_2O	Popova[310]
Lysergiamide-related compounds	Silica gel	CH_3CCl_3–CH_3OH	Dal Cortivo[311]
Lysergiamide-related compounds	Silica gel	Acetone–piperidine	Genest[312]
Berberine, palmatine, etc.	Silica gel	Butanol–CH_3COOH–H_2O	Kurono[313]
Ephedrine related compounds	Kieselgel		Paulus[314]

Table 6.7 (*Contd.*)

Mixture	Adsorbent	Solvent system	Reference
Ephedrine, norpseudoephedrine	Kieselgel		Palitzsch[315]
Ephedrine, codeine	Al_2O_3	Benzene–C_2H_5OH	Sarsunova[316]
Codeine, ethylmorphine	Cellulose	Butanol–CH_3COOH–H_2O	Lang[317]
Tropine, pseudo-tropine, tropinone	Kieselgel	C_2H_5OH–NH_3–H_2O	Neumann[318]
58 alkaloids	Kieselgel		Gendi[319]
Ergot alkaloids	Al_2O_3	Benzene–$CHCl_3$–C_2H_5OH	Li[320]
Ergot alkaloids	Silica gel	Benzene–$CHCl_3$–C_2H_5OH	Wichlinska[321]
Ergot alkaloids	Kieselgel		Roeder[322]
Ergot alkaloids	Kieselgel	$CHCl_3$–C_2H_5OH–H_2O	Sahli[323]
Cinchona alkaloids	Kieselgel	$CHCl_3$–CH_3OH–$(C_2H_5)_2NH$	Suszko[324]
Cinchona alkaloids	Silica gel gypsum	CH_2Cl_2–$(C_2H_5)_2O$ $(C_2H_5)_2NH$	Alamski[325]
Tropine alkaloids	Silica gel	$CH_3COC_2H_5$–CH_3OH–aq. NH_3	Oswald[326]
Genista alkaloids	Kieselgel	Cyclohexane–$(C_2H_5)_2NH$	Bernasconi[327]
Bisbenzylisoquinoline alkaloids	Kieselgel	CH_3OH–$CHCl_3$, etc.	Bhatnagar[328]
Pyrrolizidine alkaloids	Silica gel–$CaSO_4$	$CHCl_3$–CH_3OH–aq. NH_3	Sharma[329]
Quinalzoline alkaloids	Kieselgel–KOH	$CHCl_3$–CH_3OH	Gröger[330]
48 alkaloids	Cellulose	5 systems	Giacopello[331]
Opium alkaloids	Kieselgel	Xylene–$CH_3COC_2H_5$–CH_3OH–$(C_2H_5)_2NH$	Bayer[332]
Opium alkaloids	Polyamide	Cyclohexane–ethyl acetate–C_3H_7OH–$(C_2H_5)_2NH$	Huang[333]
Opium alkaloids		9 systems	Paris[334]
Opium alkaloids	Kieselgel–KOH	Benzene–CH_3OH	Danos[335]
26 opiates	Silica gel	8 systems	Steele[336]
Codeine, ethylmorphine	Silica gel	$CHCl_3$–$(C_2H_5)_2O$–CH_3OH–aq. NH_3	Wullen[337]
Codeine, thebaine, papaverine, narcotine	Al_2O_3	Benzene–CH_3COOH	Poethke[338]

Table 6.7 (*Contd.*)

Mixture	Adsorbent	Solvent system	Reference
35 narcotics	Silica gel	i-C_3H_7OH–$CHCl_3$–aq. NH_3	Uhlmann[339]
Tobacco alkaloids, two-dimensional	Silica gel	$CHCl_3$–CH_3OH–aq. NH_3; $CHCl_3$–CH_3OH–CH_3COOH	Hodgson[340]
Strychnine, brucine	Silica gel	$CHCl_3$–acetone–$(C_2H_5)_2NH$	Noirfalise[341]
Strychnine, securinine	Al_2O_3	$CHCl_3$–benzene–C_2H_5OH	Lapina[342]
Anabasis alkaloids	Al_2O_3	Benzene–$CHCl_3$–CH_3OH	Forostyan[343]
24 tropeines	Silica gel	Ethyl acetate–HCOOH–H_2O	Szendey[344]
Tropane alkaloids	Silica gel	C_2H_5OH–H_2O–$(C_2H_5)_3N$	Levorato[345]
Tropane alkaloids	Silica gel	$CHCl_3$–CH_3OH–aq. NH_3	Fiebig[346]
Atropine group	Cellulose	i-C_4H_9OH–HCl–H_2O	Saint-Firmin[347]
Aminosulfones, benzenesulfonamides	Al_2O_3	Ethyl acetate–benzene	Finley[348]
Sulfonamides	Talc	$CHCl_3$–CH_3OH–cyclohexaxe	Butkiewicz[349]
Sulfonamides	Silica gel	Petroleum–$CHCl_3$–butanol	Pao[350]
Sulfonamides	Al_2O_3	Butanol–H_2O	Wagner[351]
Sulfonamides	Polyamide	3 systems	Lin[352]
Sulfonamides	Al_2O_3–gypsum	Ethyl acetate–H_2O	Margasinski[353]
Sulfonamides	Kieselgel	3 systems	Bican-Fister[354]
N^4-Substituted sulfonamides	Silica gel	CH_3OH–C_2H_5OH	Kho[355]
Sulfonamides, two-dimensional	Al_2O_3	$CHCl_3$–CH_3OH; benzene–CH_3COOH–CH_3OH	Poethke[356]
Sulfonamides, two-dimensional		$CHCl_3$–CH_3OH; $CHCl_3$–acetone–CH_3OH–aq. NH_3	Kamp[357]
Phenothiazines	Silica gel	3 systems	Dobrecky[358]
Phenothiazines	Silica gel		Moza[359]
Phenothiazines	Silica gel	2 systems	Vignoli[360]
Phenothiazines	Silica gel	3 systems	Pastor[361]
Phenothiazines	Cellulose	i-C_4H_9OH	Noirfalise[362]
Phenothiazines	Al_2O_3–$CaSO_4$	Petroleum–acetone–benzene	Margasinski[363]
Benzothiadiazines	Silica gel	Benzene–ethyl acetate	Smith[364]

Table 6.7 (*Contd.*)

Mixture	Adsorbent	Solvent system	Reference
Sulfadimethoxine, chloramphenicol-tetracycline	Silica gel	Ethyl acetate–C_2H_5OH–aq. NH_3	Lombardi[36]
Cyclamate, saccharin, dulcin	Silica gel	Acetone–aq. NH_3	Kamp[366]

anthraquinones in the following manner. The sample, dissolved in dimethylformamide, was applied to a layer of alumina and developed with cyclohexane–ethyl ether (1 : 1). After development, the solvent was allowed to evaporate, and the individual components were determined densitometrically. A complete analysis took about 13 min, and the accuracies were within ±6%.

According to a number of investigators, thin-layer chromatography is the preferred technique for the separation of many basic substances. For example, Barsch et al.[277] reported that gas-chromatographic separation of alkylpyridines is often incomplete and recommended the use of Kieselgel (VEB Greiz-Doelau) layers (0.25 mm) prepared with 0.01 *M* sodium acetate. Gröger and Erge[367] tested eight methods to separate ergot alkaloids and found that thin-layer chromatography was much superior to paper chromatography for the separation of clavines and simple lysergic acid derivatives. Thus a mixture of chanoclavine, penniclavine, ergometrine and elymoclavine required 25 hr to separate on paper, but only 1 hr to separate by the thin-layer method; the alkaloids were eluted from the relevant zones with CH_2Cl_2–CH_3OH (9 : 1) until the eluate gave a negative reaction with Van Urk's reagent, the recoveries being close to 100%.

For the determination of mixtures of amines, Gnehm et al.[368] recommended the conversion of the amines into the corresponding hydrochlorides, which gave clearer spots. The amine sample was treated with 2 *N* HCl and the solution was evaporated to dryness. The residue was then dissolved in 70% ethanol; aliquots were applied to the thin-layer plate coated with a 0.25 mm layer of Kieselgel G, dried overnight at room temperature, and activated by heating at 105°C for 30 min. The plate was developed with $CHCl_3$–CH_3OH–17% aq. NH_3 (2 : 2 : 1), dried, and sprayed with a solution containing 0.5% each of $KMnO_4$ and $K_2S_2O_8$. After 30 min, the areas of the yellow spots were measured. There is a rectilinear relationship between the square root of the spot area and the logarithm of the concentration of the amine.

It is obvious that the lower homologues of aliphatic amines cannot be applied directly to the thin-layer plates. However, separation of the hydrochlorides of C_1 to C_4 amines can be achieved, as demonstrated by Gerlach and Senf[369]; these workers also found that the separation of aromatic amines is improved by first converting the amines into the corresponding *p*-toluenesulfonamides, benzamides, or 3,5-dinitrobenzamides. Schwartz et al.[370] separated normal primary aliphatic amines and symmetrical secondary aliphatic amines via the corresponding dinitrophenyl derivatives; these derivatives were prepared on the thin-layer plate by first applying 1-fluoro-2,4-dinitrobenzene solution in benzene and then the amines dissolved in benzene. The same reagent can be used to prepare derivatives of amino acids. The *N*-thiobenzoyl (Barrett and Khokhar[371]) and nitro-2-pyridyl (Celon et al.[372]) derivatives of various amino acids also are suitable for thin-layer chromatography.

Modification of the one-dimensional and two-dimensional development procedure is sometimes recommended to achieve separation of certain mixtures. For instance, Mefferd et al.[373] described the multiple development in one dimension for the analysis of urinary amino acids: Various volumes of a standard solution of the amino acids in 0.1 *N* HCl, together with 5 and 10 μl portions of sample, were applied to a cellulose layer. The plate was developed by the ascending technique with butanol–CH_3COOH–H_2O (3 : 1 : 1) for 3.5 hr, air-dried in the horizontal position, and then oven-dried at 60°C for 15 min. After cooling, it was redeveloped, redried, and sprayed with 0.2% ninhydrin solution in ethanol containing 1% of anhydrous acetic acid. After the plate was dried for 45 min, the colors were produced by heating at 60°C for 1.5 hr. Twenty ninhydrin-positive spots were obtained from the urine sample.

Reagents can be incorporated into the inert adsorbent to accomplish specific purposes. For example, Shimomura and Walton[374] mixed silica gel G with twice its weight of 10% $Zn(NO_3)_2$ or $Cd(NO_3)_2$ solution, spread the mixture on glass plates to a thickness of 2.5 mm, dried it at 110°C, and used these plates for chromatography of aromatic amines; because of ligand exchange, zinc and cadmium retarded the movement of the amines, but also greatly increased the differences in R_F values among the amines. Paris et al.[375] added optically active reagents to the adsorbent for resolving racemic alkaloids; thus the optical isomers of ephedrine were separated on Kieselgel impregnated with *M* aqueous D-galacturonic acid, using isopropyl alcohol–*M* aqueous D-galaturonic acid (94 : 3) as the developing solvent.

While there are general rules for the selection of solvent systems for thin-layer chromatography, the ultimate choice is still an empirical one depending on the analyst and the specific mixture submitted for analysis. The presence and concentration of a certain component in the solvent system can be crucial. Thus Klein and Kho[376] reported the separation of eight sulfonamides and emphasized that the concentration of H_2O in the solvent system used ($CHCl_3$–C_2H_5OH–heptane) must be between 1.0 and 1.8%. Mistryukov[377] found that the separation of primary and secondary amines on alumina is difficult because of the strong adsorption of the —NH group; addition of H_2O or NH_3 in the solvent system can ameliorate the situation. In the analysis of pharmaceutical mixtures on layers of alumina, Sarsunova and Schwarz[378] observed that the position of caffeine was distorted when amidopyrine was present in the sample. It should be noted also that alumina can exert some catalytic effect[379] on certain reactions that take place on the thin layer. Generally speaking, precision and accuracy are better on inert adsorbents with simple solvent systems. Jeney and Walther[380] described a procedure for the separation and determination of two new sulfonamides, azoseptyl and reseptyl, in a mixture containing six other sulfur drugs using a solvent system that comprises 18 liquids (dichloromethane–1,2-dichloroethane–chloroform–carbon tetrachloride–dioxane–ethyl ether–isopropyl ether–ethyl acetate–propyl acetate–butyl acetate–methyl acetate–methanol–ethanol–propanol–butanol–dimethylformamide–dimethylacetamide–dimethyl sulfoxide); such a system is extremely difficult to reproduce and its composition is practically impossible to maintain for repetitive determinations.

The accuracy required in quantitation after thin-layer chromatography depends on the nature of the problem. For the determination of the alkaloid contents in plant materials,[327] for example, visual estimation of the colored spots may suffice. When the spot areas were measured by the planimeter, Oswald and Flück[326] reported precision of $\pm 6\%$ for tropine alkaloids (30 to 40 μg) and $\pm 8\%$ for cinchona alkaloids (3 to 4 μg). On the other hand, an accuracy of better than $\pm 5\%$ is demanded in the determination of pharmaceutical mixtures. Such accuracy can be achieved by extracting the compounds from the thin layer, followed by spectrophotometric measurements.[259,260,305] Frodyma and Frei[258] measured the total reflectance of amino acids sprayed with ninhydrin–$Cu(NO_3)_2$ and reported that the accuracy was comparable to that obtained by the direct transmission technique in paper chromatography. A densitometer based on diffuse reflectance using fiber optics as described by Beroza et al.[381] is commercially available.[382]

6.7 SEPARATION BY ELECTROPHORESIS

Electrophoresis has been utilized to separate amino acids in various ways. Jirgl[383] carried out quantitative separation of basic amino acids (up to

20 μg of each compound) at 7 V per cm for 2 hr with 0.08 *M* $NaHCO_3$ (pH 7.4 to 7.5) as electrolyte; the electropherogram was neutralized and treated with ninhydrin, and the spots were extracted into methanol and measured at 500 nm. A similar method was described by Braun[384] using a buffer solution of pH 1.9 and 60 to 70 V. For the analysis of mixtures containing up to 18 amino acids, Ansorge[385] first performed the electrophoresis on FN-13 paper at −1 to −3°C in a solution of pH 1.9, which separated lysine, histidine plus arginine, alanine, and glycine, and then in a solution of pH 3.6, which separated arginine, lysine plus histidine, glutamic acid, and aspartic acid. Subsequently, the still unresolved neutral amino acids were separated by electrophoresis in a solution of pH 1.9 followed by ascending paper chromatography at right angles to the direction of electrophoresis, with butanol–HCOOH–H_2O (15 :3 :2) as developing solvent.

Combination of chromatography with electrophoresis was employed by Munier and Sarrazin[386] to improve the separation of amino acids via their dinitrophenyl derivatives. Derivatives of the acids with a single amino group were prepared by means of 1-chloro-2,4-dinitrobenzene, and those of others were prepared by means of 1-fluoro-2,4-dinitrobenzene. Separations of the derivatives were made by column chromatography on Sephadex G25 or by two-dimensional paper chromatography. In some instances, chromatography in one direction was followed by electrophoresis in the other direction.

Biserte et al.[387] described multiple electrophoresis for the separation of amino acids from complex biological media. A preliminary paper electrophoresis at pH 3.9 and 400 V for 4 hr separated the compounds into three groups. The first group containing acidic amino acids was further separated by a two-dimensional combination of electrophoresis in one direction at pH 3.9 and 600 V for 8 hr followed by chromatography in the second direction with butanol–CH_3COOH–H_2O (4 :1 :5). The second group, containing neutral amino acids, was eluted from the electropherogram and further separated by electrophoresis at pH 2.9 and 400 V for 18 hr followed by chromatography in the second direction with the above solvent. The third group, consisting of basic amino acids and amines, was also eluted from the electropherogram and further separated by electrophoresis at 400 V for 15 hr at pH 11.7 and 6.5, respectively.

Hara[388] separated amino acids in an electrodialysis apparatus that was divided into five compartments by ion-exchange membranes. The solution of amino acids at pH 5.6 was placed in the center compartment, with 0.1 to 0.2% NaCl solution in the intermediate compartments and 3 to 6% Na_2SO_4 solution circulating through the electrode compartments. Conditions were established whereby the acid mixture could be separated into anionic and cationic functions.

For the determination of glucosamine and galactosamine, Graham and Neuberger[389] prepared the condensation products of the amino sugars with isothiocyanatonaphthalene and separated them by electrophoresis in tungstate buffer solution. Simmonds[390] separated urinary purines, pyrimidines, and pyrazolopyrimidines by high-voltage electrophoresis on thin-layer plates. Yang and Su[391] determined adenosine, guanosine, and inosine by electrophoretic separation on filter paper, elution of each compound, and spectrophotometry. Stuchlik et al.[392] separated caffeine, hydroxyethyltheophylline, and aminophylline by electrophoresis in alkaline buffered solution at 20 V per cm. Winek et al.[393] separated mixtures containing berberine and quinine or hydrastine by electrophoresis in a starch gel with *N* acetic acid as electrolyte. Kornhauser and Perpar[394] separated ergot alkaloids from the pigments by electrophoresis at pH 2.2 on a thin layer of Kieselgel G; the alkaloids were then separated on the same thin layer by chromatography with benzene–$CHCl_3$–C_2H_5OH (4 :4 :1) as developer.

6.8 SEPARATION BY OTHER TECHNIQUES

Separation by extraction is practical when the components in a mixture differ significantly in their solubilities toward a certain solvent. For example, Klera and Dudzik[395] separated ephedrine from aminophylline and ethylmorphine in aqueous solution by adding Na_2CO_3, saturating the solution with NaCl and extracting the ephedrine into benzene; similarly, ethylmorphine was separated from the other two compounds by extraction into chloroform after the solution was made alkaline with KOH. Brzezinska and Maciaszek[396] also used benzene to separate ephedrine from theophylline and guaiacosulfonate. For the determination of papaverine, phenobarbitone, and theobromine in a mixture, Chodakowski and Wojciechowska[397] extracted papaverine with chloroform and then extracted phenobarbitone with ethyl ether. Bose et al.[398] determined 4-aminosalicylic acid in the presence of isonicotinic hydrazide by extracting the former into ethyl ether. Pernarowski and Padval[399] proposed a method to analyze binary mixtures (e.g., phenacetin and caffeine) by partial extraction and specialized spectrophotometric techniques. The mixture is partitioned between two solvents and the ratio of the absorbances of the two phases is determined at an isoabsorptive wavelength at which the molar extinctions of the two components are the same. For the determination of khellin, phenobarbitone, papaverine and, theobromine in mixtures Machek[400] described a rapid spectrophotometric method that is dependent on the differences in solubility of the four compounds.

Bellen et al.[401] reported the partial separation of ethylenediamine, diethylenetriamine, and triethylenetetramine by fractional distillation on a Vigreux column. The first fraction (boiling range 100 to 140°C) contains ethylenediamine and water, the second (140 to 210°C) contains ethylenediamine and diethylenetriamine, and the residue contains diethylenetriamine and triethylenetetramine. The composition of each fraction can be determined refractometrically, the results being compared with those for standard mixtures, and the composition of the sample is then calculated.

REFERENCES

1. N. D. Cheronis and T. S. Ma, *Organic Functional Group Analysis*, Wiley, New York, 1964, p. 436.
2. A. M. Chodakowski, *Farm. Pol.*, **25**, 637 (1969).
3. L. Lepri, P. G. Desideri, V. Coas, and D. Cozzi, *J. Chromatogr.*, **49**, 239 (1970).
4. Y. Yoshino, H. Kinoshita, and H. Sugiyama, *J. Chem. Soc. Jap., Pure Chem. Sect.*, **86**, 405 (1965).
5. P. King and J. R. Simmler, *Anal. Chem.*, **36**, 1837 (1964).
6. T. Takahashi and D. Yoshida, *Soil Plant Food, Tokyo*, **3**, 142 (1958).
7. A. Niemann, *J. Chromatogr.*, **9**, 117 (1962).
8. H. Maehr and C. P. Schaffner, *J. Chromatogr.*, **30**, 572 (1967).
9. L. B. Bigelow and H. Weil-Malherbe, *Anal. Biochem.*, **26**, 92 (1968).
10. Z. Kahane and P. Vestergaard, *Clin. Chim. Acta*, **25**, 453 (1969).
11. D. D. Christianson, J. S. Wall, R. J. Dimler, and F. R. Senti, *Anal. Chem.*, **32**, 874 (1960).
12. S. Friedman, J. E. McFarlane, P. K. Bhattacharyya, and G. Fraenkel, *Arch. Biochem. Biophys.*, **59**, 484 (1955).
13. D. D. Christianson, J. S. Wall, J. F. Cavins, and R. J. Dimler, *J. Chromatogr.*, **10**, 432 (1963).
14. A. S. R. Donald, *J. Chromatogr.*, **35**, 106 (1968).
15. A. A. Baev, L. F. Nikiforovskaya, and G. I. Melnikova, *Vopr. Med. Khim.*, **14**, 221 (1968).
16. A. M. Bella, Jr. and Y. S. Kim, *J. Chromatogr.*, **51**, 314 (1970).
17. R. S. Steele, K. Brendel, E. Scheer, and R. W. Wheat, *Anal. Biochem.*, **34**, 206 (1970).
18. S. Moore and W. H. Stein, *J. Biol Chem.*, **211**, 893 (1954).
19. J. P. Dustin, C. Czajkowska, S. Moore, and E. J. Bigwood, *Anal. Chim. Acta*, **9**, 256 (1953).
20. C. H. W. Hirs, S. Moore, and W. H. Stein, *J. Am. Chem. Soc.*, **76**, 6063 (1954).
21. P. Padieu, N. Maleknia, and A. M. Thireau, *Bull. Soc. Chim. Fr.*, **1963**, 2960.
22. S. Jacobs, *Glas. Khem. Drus., Beograd*, **25**, 21 (1960).

23. G. W. Frimpter, S. Ohmori, and S. Mizuhara, *J. Chromatogr.*, **10**, 439 (1963).
24. G. L. Tritsch and C. L. Moriarty, *J. Chromatogr.*, **44**, 425 (1969).
25. K. Pfordte, *J. Chromatogr.*, **39**, 506 (1969).
26. M. I. Blake and D. A. Nona, *J. Pharm. Sci.*, **53**, 570 (1964).
27. E. Sjöström, *Anal. Chim. Acta*, **16**, 428 (1957).
28. E. Sjöström and L. Nykänen, *J. Am. Pharm. Assoc., Sci. Ed.*, **47**, 248 (1958).
29. J. Reisch, H. Bornfleth, and J. Rheinbay, *Pharm. Ztg. Ver. Apoth.-Ztg.*, **108**, 1182 (1963).
30. W. Kamp, *Pharm. Weekbl. Ned.*, **99**, 1092 (1964).
31. K. O. Montgomery, P. V. Jennings, and M. H. Weinswid, *J. Pharm. Sci.*, **56**, 141 (1967).
32. H. Watanabe, *Jap. Analyst*, **11**, 233 (1962).
33. J. Büchi and R. von Moos, *Pharm. Acta Helv.*, **41**, 142 (1966).
34. H. Watanabe, *J. Chem. Soc. Jap., Pure Chem., Sect.*, **83**, 51 (1962).
35. P. A. Hedin, *J. Agric. Food Chem.*, **11**, 343 (1963).
36. R. K. Raj and O. Hutzinger, *Anal. Biochem.*, **33**, 43 (1970).
37. B. E. Bonnelycke, K. Dus, and S. L. Miller, *Anal. Biochem.*, **27**, 262 (1969).
38. L. Popa, A. Cruceanu, and R. Portocala, *Rev. Roum. Biochim.*, **4**, 297 (1967).
39. L. Klotz, *Pharm. Zentrahalle*, **104**, 393 (1965).
50. S. Fukui, *Anal. Chem.*, **25**, 1884 (1953).
41. M. Takimoto, *J. Chem. Soc. Jap., Ind. Chem. Sect.*, **64**, 1234 (1961).
42. R. Bacalogu, *Z. Anal. Chem.*, **240**, 244 (1968).
43. W. Funasaka, T. Kojima, K. Fujimura, and S. Kuriyama, *Jap. Analyst*, **14**, 820 (1965), **19**, 104 (1970).
44. T. A. Wheaton and I. Stewart, *Anal. Biochem.*, **12**, 585 (1965).
45. H. Sakurai and M. Kito, *Talanta*, **23**, 842 (1976).
46. J. S. Foster and J. W. Murfin, *Analyst*, **86**, 32 (1961).
47. *Amino-acid Analyzer*, see "*Science Guide to Scientific Instruments,*" *Science*, September 20, 1977, p. 12.
48. E. L. Duncan and H. Mohler, *Mitt. Geb. Lebensmittunters Hyg.*, **53**, 399 (1962).
49. J. H. Peters, B. J. Berridge, Jr., J. G. Cummings, and S. C. Lin, *Anal. Biochem.*, **23**, 459 (1968).
50. T. Urbaņyi, A. Piedmont, E. Willis, and G. Manning, *J. Pharm. Sci.*, **65**, 257 (1976).
51. T. S. Ma and R. C. Rittner, *Modern Organic Elemental Analysis*, Dekker, New York, 1979.
52. W. J. A. Vanden-Heuvel, W. L. Gardiner, and E. C. Horning, *Anal. Chem.*, **36**, 1550 (1964).
53. M. P. Petrova and A. I. Dolgina, *Zh. Anal. Khim.*, **19**, 239 (1964).
54. A. R. Amell, P. S. Lamprey, and R. C. Schiek, *Anal. Chem.*, **33**, 1805 (1961).
55. C. Landault and G. Guichon, *J. Chromatogr.*, **13**, 327 (1964).

56. Y. Arad, M. Levy, and D. Vofsi, *J. Chromatogr.*, **13**, 565 (1964).
57. J. S. Sawardeker and J. L. Lach, *J. Pharm. Sci.*, **52**, 1109 (1963).
58. R. V. Golovnya, G. A. Mironov, and I. L. Zhuravleva, *Zh. Anal. Khim.*, **22**, 797 (1967).
59. Y. L. Sze, M. L. Borke, and D. M. Ottenstein, *Anal. Chem.*, **35**, 240 (1963).
60. R. V. Kudryavtsev and V. M. Shirochenkova, *Zh. Anal. Khim.*, **24**, 1431 (1969).
61. J. R. Lindsay-Smith and D. J. Waddington, *J. Chromatogr.*, **42**, 195 (1969).
62. L. D. Metcalfe and A. A. Schmitz, *J. Gas Chromatogr.*, **2**, 15 (1964).
63. J. Törnquist, *Acta Chem. Scand.*, **19**, 777 (1965).
64. J. J. Cincotta and R. Feinland, *Anal. Chem.*, **34**, 774 (1962).
65. R. V. Golovnya, G. A. Mironov, and I. L. Zhuravleva, *Zh. Anal. Khim.*, **22**, 612, 956 (1967).
66. H. Feltkamp and K. D. Thomas, *J. Chromatogr.*, **10**, 9 (1963).
67. A. T. James, *Anal. Chem.*, **28**, 1564 (1956).
68. J. S. Parsons and J. C. Morath, *Anal. Chem.*, **36**, 237 (1964).
69. R. W. Wise and A. B. Sullivan, *Rubber Age, N.Y.*, **91**, 773 (1962).
70. W. H. Bryan, *Anal. Chem.*, **36**, 2025 (1964).
71. F. Willeboordse, Q. Quick, and E. T. Bishop, *Anal. Chem.*, **40**, 1455 (1968).
72. C. E. Boufford, *J. Gas Chromatogr.*, **6**, 438 (1968).
73. A. H. Beckett and R. D. Hossie, *J. Pharm. Pharmacol.*, **21**, 610 (1969).
74. H. G. Henkel, *J. Gas Chromatogr.*, **3**, 320 (1965).
75. K. J. Bombaugh, *Anal. Chem.*, **37**, 72 (1965).
76. P. Lebish, B. S. Finkle, and J. W. Brackett, Jr., *Clin. Chem.*, **16**, 195 (1970).
77. J. M. Strong and A. J. Atkinson, Jr., *Anal. Chem.*, **44**, 2287 (1972).
78. H. M. Koehler and J. J. Hefferren, *J. Pharm. Sci.*, **53**, 745 (1964).
79. C. J. W. Brooks and E. C. Horning, *Anal. Chem.*, **36**, 1540 (1964).
80. J. Vessman and G. Schill, *Svensk. Farm. Tidskr.*, **66**, 601 (1962).
81. L. D. Metcalfe, *J. Am. Oil Chem. Soc.*, **40**, 25 (1963).
82. G. Grossi and R. Vece, *J. Gas Chromatogr*, **3**, 170 (1965).
83. H. H. Laycock and B. A. Mulley, *J. Pharm. Pharmacol.*, **18**, 9S (1966).
84. J. C. Craig, N. Y. Mary, and S. K. Roy, *Anal. Chem.*, **36**, 1142 (1964).
85. L. D. Metcalfe, G. A. Germanos, and A. A. Schmidt, *J. Gas Chromatogr.*, **1**, 32 (1963).
86. J. F. O'Donnell and C. K. Mann, *Anal. Chem.*, **36**, 2097 (1964).
87. M. H. Grosjean and A. Noirfalise, *J. Chromatogr.*, **36**, 347 (1968).
88. J. K. Foreman, J. F. Palframan, and E. A. Walker, *Nature*, **225**, 554 (1970).
89. M. J. Saxby, *Anal. Lett.*, **3**, 397 (1970).
90. C. Bighi and G. Saglietto, *J. Chromatogr.*, **18**, 297 (1965).
91. L. A. Dee and A. K. Webb, *Anal. Chem.*, **39**, 1165 (1967).

92. K. Heyns, R. Stute, and J. Winkler, *J. Chromatogr.*, **21**, 302 (1966).
93. R. W. Reiser, *Anal. Chem.*, **36**, 96 (1964).
94. C. E. McKone and R. J. Hance, *J. Chromatogr.*, **36**, 234 (1968).
95. W. L. Zielinski, Jr., and L. Fishbein, *J. Chromatogr.*, **23**, 175 (1966).
96. W. J. Kirsten and R. G. G. Andren, *J. Chromatogr.*, **8**, 531 (1962).
97. F. Kametani and S. Kubota, *J. Pharm. Soc. Jap.*, **82**, 659 (1962).
98. G. D. Mitra, S. K. Ghosh, N. C. Saha, and A. Sinha, *Technology*, *Sindri*, **4**, 105 (1967).
99. A. W. Decora and G. U. Dinneen, *Anal. Chem.*, **32**, 164 (1960).
100. M. Anwar, C. Hanson, and A. N. Patel, *J. Chromatogr.*, **34**, 529 (1968).
101. J. S. Fitzgerald, *Aust. J. Appl. Sci.*, **12**, 51 (1961).
102. R. N. Bhattacharya, *Indian J. Technol.*, **6**, 279 (1968).
103. J. Janak, M. Holik, and M. Ferles, *Collect. Czech. Chem. Commun.*, **31**, 1273 (1966).
104. F. Moll, *Naturwissenschaften*, **49**, 450 (1962).
105. M. Vietti-Michelina, *Z. Anal. Chem.*, **204**, 110 (1964).
106. W. Funasaka and T. Kojima, *J. Chem. Soc. Jap.*, *Ind. Chem. Sect.*, **64**, 769 (1961).
107. V. Rezl, *Chem. Prum.*, **12**, 246 (1962).
108. M. Vietti-Michelina, *Rass. Chim.*, **13**, 23 (1961).
109. V. I. Trubnikov, L. M. Malakhova, and E. S. Zhdanovich, *Zh. Anal. Khim.*, **23**, 1546 (1968).
110. I. I. Grandberg, A. P. Krasnoshchek, and L. B. Dimitriev, *Izv. Timiryazevsk. sekl. Akad.*, **1969**, 224.
111. A. Castiglioni, *Z. Anal. Chem.*, **182**, 428 (1961).
112. W. W. Weeks, D. L. Davis, and L. P. Bush, *J. Chromatogr.*, **43**, 506 (1969).
113. H. P. Harke and C. J. Drews, *Z. Anal. Chem.*, **242**, 248 (1968).
154. R. Pilleri and M. Vietti-Michelina, *Z. Anal. Chem.*, **174**, 172 (1960).
115. D. Hoffmann, G. Rathkamp, and H. Woziwodzki, *Beitr. Tabakforsch.*, **4**, 253 (1968).
116. G. Van Binst, L. Denolin-Dewaerseggar, and R. H. Martin, *J. Chromatogr.*, **16**, 34 (1964).
117. H. J. Wesselman and W. L. Koch, *J. Pharm. Sci.*, **57**, 845 (1968).
118. J. Vessman, *Acta Pharm. Suec.*, **1**, 183 (1964).
119. J. L. Massingill, Jr., and J. E. Hodgkins, *Anal. Chem.*, **37**, 952 (1965).
120. E. Brochmann-Hanssen and C. R. Fontan, *J. Chromatogr.*, **20**, 394 (1965).
121. M. J. Solomon, F. A. Crane, B. L. W. Chu, and E. S. Mika, *J. Pharm. Sci.*, **58**, 264 (1969).
122. M. S. Greaves, *Clin. Chem.*, **20**, 141 (1974).
123. J. Reisch and H. Walker, *Pharmazie*, **21**, 467 (1966).
124. S. Kawai, T. Hattori, and M. Kotaki, *Jap. Analyst*, **13**, 872 (1964).

125. W. J. A. Vanden-Heuvel, E. O. A. Haahti, and E. C. Horning, *Clin. Chem.*, **8**, 351 (1962).
126. V. I. Yakerson, L. I. Lafer, S. Z. Taits, F. M. Stoyanovich, V. P. Litvinov, Y. A. Danyushevskii, and Y. L. Goldfarb, *J. Chromatogr.*, **23**, 67 (1966).
127. J. Tranchant, *Bull. Soc. Chim. Fr.*, **1963**, 365.
128. J. R. Lindsay-Smith and D. J. Waddington, *Anal. Chem.*, **40**, 522 (1968).
129. A. A. F. van der Meeren and A. L. T. Verhaar, *Anal. Chim. Acta*, **40**, 343 (1968).
130. Y. Hoshika, *Anal. Chem.*, **49**, 541 (1977).
131. E. W. Day, Jr., T. Golab, and J. R. Koons, *Anal. Chem.*, **38**, 1053 (1966).
132. T. Walle, *Acta Pharm. Suec.*, **5**, 353 (1968).
133. J. A. Corbin and L. B. Rogers, *Anal. Chem.*, **42**, 974 (1970).
134. M. Pailer and W. J. Hübsch, *Mh. Chem.*, **97**, 1541 (1966).
135. L. E. Brydia and H. E. Persinger, *Anal. Chem.*, **39**, 1318 (1967).
136. R. A. Dove, *Anal. Chem.*, **39**, 1188 (1967).
137. L. E. Brydia and F. Willeboordse, *Anal. Chem.*, **40**, 110 (1968).
138. D. Hoffmann, Y. Masuda, and E. L. Wynder, *Nature*, **221**, 254 (1969).
139. C. Hishta and R. F. Lauback, *J. Pharm. Sci.*, **58**, 745 (1969).
140. J. Hirtz and A. Gerardin, *Ann. Pharm. Fr.*, **17**, 581 (1969).
141. C. E. Wells, *J. Assoc. Off. Anal. Chem.*, **53**, 113 (1970).
142. S. Kawai and Z. Tamura, *Chem. Pharm. Bull. (Tokyo)*, **16**, 699 (1968).
143. N. P. Sen and P. L. McGeer, *Biochem. Biophys. Res. Commun.*, **13**, 390 (1963).
144. S. Wilk, S. E. Gitlow, M. J. Franklin, and H. E. Carr, *Clin. Chim. Acta*, **10**, 193 (1964).
145. C. H. Nicholls, S. Makisumi, and H. A. Saroff, *J. Chromatogr.*, **11**, 327 (1963).
146. H. A. Saroff, A. Karmen, and J. W. Heally, *J. Chromatogr.*, **9**, 122 (1962).
147. W. J. McBride, Jr. and J. D. Klingman, *Anal. Biochem.*, **25**, 109 (1968).
148. E. D. Smith and H. Sheppard, Jr., *Nature*, **208**, 878 (1965).
149. G. Carlstrom, *Acta Vet. Scand.*, **9**, 71 (1968).
150. F. Marcucci, E. Mussini, F. Poy, and P. Gagliardi, *J. Chromatogr.*, **18**, 487 (1965).
151. S. V. Shlyapnikov and M. Y. Karpeiskii, *Biokhimiya*, **29**, 1076 (1964).
152. C. G. Youngs, *Anal. Chem.*, **31**, 1019 (1959).
153. J. R. Coulter and C. S. Hann, *J. Chromatogr.*, **36**, 42 (1968).
154. R. A. Landowne, *Chim. Anal. (Paris)*, **47**, 589 (1965).
155. J. J. Pisano, W. J. A. Vanden-Heuvel, and E. C. Horning, *Biochem. Biophys. Res. Commun.*, **7**, 82 (1962).
156. H. G. Jones, J. K. N. Jones, and M. B. Perry, *Can. J. Chem.*, **40**, 1559 (1962).
157. W. L. Zielinski, Jr. and L. Fishbein, *J. Chromatogr.*, **25**, 475 (1966).
158. H. Liliedahl, *Acta Chem. Scand.*, **20**, 95 (1966).
159. V. I. Trubnikov, L. M. Malakhova, E. S. Zhdanovich, and N. A. Preobrazhenski, *Khim.-Farm. Zh.*, **1968** (12, 14.

160. W. Korytnyk, G. Fricke, and B. Paul, *Anal. Biochem.*, **17**, 66 (1966).
161. T. Hattori, S. Kawai, and M. Nishiumi, *Jap. Analyst*, **14**, 586 (1965).
162. J. O. Grooms, *J. Assoc. Off. Anal. Chem.*, **51**, 1010 (1968).
163. J. A. F. de Silva, N. Munno, and N. Strojny, *J. Pharm. Sci.*, **59**, 201 (1970).
164. E. Brochmann-Hanssen and T. O. Oke, *J. Pharm. Sci.*, **58**, 370 (1969).
165. A. Zlatkis, J. F. Oro, and A. P. Kimball, *Anal. Chem.*, **32**, 612 (1960).
166. D. A. Vyakhirev and Y. N. Suloev, *Zh. Anal. Khim.*, **11**, 739 (1956).
167. C. E. Hedrick, *Anal. Chem.*, **37**, 1044 (1965).
168. F. K. Chow and E. Grushka, *Anal. Chem.*, **49**, 1756 (1977).
169. H. Buchowski and W. Pawlowski, *Chem. Anal. (Warsaw)*, **4**, 135 (1959).
170. D. Missbach and H. Schütze, *J. Prakt. Chem.*, **22**, 225 (1963).
171. J. E. Larson and S. H. Harvey, *Chem. Ind. (Lond.)*, **1954**, (2), 45.
172. T. M. Dougherty and A. I. Schepartz, *J. Chromatogr.*, **42**, 415 (1969).
173. R. Praus, *J. Chromatogr.*, **48**, 535 (1970).
174. A. Waksmundzki and J. Oscik, *Roczn. Chem.*, **28**, 239 (1954).
175. A. Sinha and G. D. Mitra, *Technology, Sindri*, **2**, 193 (1965).
176. L. R. Snyder, *J. Chromatogr., Sci.*, **7**, 595 (1969).
177. V. D. Chmil and Y. V. Shostenko, *Zh. Anal. Khim.*, **20**, 1122 (1965).
178. B. Kamienski and K. Puchalka, *Acad. Pol. Sci.*, **1**, 305 (1953).
179. L. G. Andreeva, *Pharmazie*, **20**, 95 (1965).
180. M. M. Nijland, *Pharm. Weekbl. Ned.*, **100**, 88 (1965).
181. J. R. Broich, M. M. de Mayo, and L. A. Dal, *J. Chromatogr.*, **33**, 526 (1968).
182. C. G. Scott and P. Bommer, *J. Chromatogr. Sci.*, **8**, 446 (1970).
183. B. L. Roop and W. E. Roop, *Anal. Biochem.*, **25**, 260 (1968).
184. T. S. Ma and V. Horak, *Microscale Manipulations in Chemistry*, Wiley, New York, 1976, p. 174.
185. B. R. Rader, *J. Pharm. Sci.*, **58**, 1535 (1969).
186. M. Clasper, J. Haslam, and E. F. Mooney, *Analyst*, **81**, 587 (1956).
187. P. M. Maitlis, *Analyst*, **82**, 135 (1957).
188. V. B. Lukyanov and E. F. Simonov, *Vestn. Mosk. Gos. Univ., Ser. Khim.*, **1966**, 36.
189. J. F. Thompson and C. J. Morris, *Anal. Chem.*, **31**, 1031 (1959).
190. T. Tsukamoto, T. Komori, and Y. Inoue, *J. Pharm. Soc. Jap.*, **81**, 146 (1961).
191. S. Szczepaniak, I. Krzeczkowska, and E. Nowicka, *Chem. Anal. (Warsaw)*, **13**, 155 (1968).
192. G. Barbioroli, *Rass. Chim.*, **16**, 79 (1964).
193. J. Fondarai and C. Richert, *J. Chromatogr.*, **9**, 262 (1962).
194. D. O. Gray, J. Blake, D. H. Brown, and L. Fowden, *J. Chromatogr.*, **13**, 277 (1964).
195. P. Zaltzman-Nirenberg, *Anal. Biochem.*, **15**, 517 (1966).
196. G. L. Haberland, F. Bruns, and K. I. Altman, *Biochim. Biophys. Acta*, **15**, 578 (1954).

197. J. Sykora and V. Eybl, *Cesk. Farm.*, **17**, 91 (1968).
198. P. Smith, *Nature*, **195**, 174 (1962).
199. V. Carassiti and A. M. Ferrero, *Boll. Sci. Fac. Chim. Ind. Bologna*, **13** (2), 37 (1955).
200. B. K. Moza, P. C. Das, and U. P. Basu, *Indian J. Chem.*, **6**, 407 (1968).
201. J. Franc, *Chem. Listy*, **50**, 1246 (1956).
202. L. Fishbein, *Rec. Trav. Chim. Pays-Bas Belg.*, **84**, 465 (1965).
203. A. B. Narbutt-Maring and W. Weglowska, *Acta Pol. Pharm.*, **22**, 13 (1965).
204. M. Oswiecimska and L. Golcz, *Farm. Pol.*, **25**, 123 (1969).
205. R. Pant and H. C. Agrawal, *Z. Physiol. Chem.*, **335**, 203 (1964).
206. S. Neuhaeuser and F. Wolf, *J. Chromatogr.*, **39**, 53 (1969).
207. M. Danilovic and K. Nikolic, *Acta Pharm. Jugosl.*, **15**, 107 (1965).
208. A. Waksmundzki and D. Ratajewicz, *Chem. Anal. (Warsaw)*, **10**, 1129 (1965).
209. V. Blazsek, *Naturwissenschaften*, **45**, 42 (1958).
210. M. Stuchlik and L. Krasnec, *Cesk. Farm.*, **16**, 123 (1967).
211. F. Micheel and W. Leifels, *Mikrochim. Acta*, **1961**, 444.
212. M. Stuchlik and L. Krasnec, *Cesk. Farm.*, **16**, 70 (1967).
213. J. W. Fairbairn and G. Wassel, *J. Pharm. Pharmacol.*, **15**, 216T (1963).
214. A. Penna-Herreros, *J. Chromatogr.*, **14**, 536 (1964).
215. P. Majlat and I. Bayer, *J. Chromatogr.*, **20**, 187 (1965).
216. O. E. Schultz and D. Strauss, *Dtsch. Apoth. Ztg.*, **95**, 642 (1955).
217. G. Dusinsky and M. Tyllova, *Nature*, **181**, 1335 (1958).
218. T. S. Ma and W. J. McGahren, unpublished work; see W. J. McGahren, Master's Thesis, Brooklyn College of the City University of New York, 1957, p. 47.
219. K. Pfordte, *J. Chromatogr.*, **21**, 495 (1966).
220. P. Horak and J. Zyka, *Cesk. Farm.*, **12**, 394 (1963).
221. A. H. Chalmers, C. C. J. Culvenor, and L. W. Smith, *J. Chromatogr.*, **20**, 270 (1965).
222. R. Munier, *Bull. Soc. Chim. Biol., Paris*, **35**, 1225 (1953).
223. E. M. Granik and A. A. Chemerisskaya, *Khim.-Farm. Zh.*, **2**, 36 (1968).
224. A. F. Mironov, R. P. Evstigneeva, I. V. Ponomarev, N. M. Mishima, K. I. Savenkova, and N. A. Preobrazhenskii, *Med. Prom. SSSR*, **1964** (5), 40.
225. G. Dusinsky, F. Machovicova, and M. Tyllova, *Farm. Obz*, **36**, 397 (1967).
226. J. Travis and A. D. Robinson, *Can. J. Biochem. Physiol.*, **40**, 1251 (1962).
227. G. Canterelli, *Farmaco, Ed. Prat.*, **17**, 728 (1962).
228. A. E. Mshvidobadze, B. I. Chumburidze, and O. V. Sardzhveladze, *Tr. Nauchno-Med. Obshch. Gruz. SSR*, **3**, 73 (1963).
229. H. Schumacher, R. L. Smith, R. B. L. Stagg, and R. T. Williams, *Pharm. Acta Helv.*, **39**, 394 (1964).
230. E. Pawelczyk and S. Domeracki, *Dissnes Pharm. (Warsaw)*, **18**, 651 (1966).

231. I. Jakubec and M. Zahradnicek, *Cesk. Farm.*, **5**, 400 (1956).
232. M. Maienthal, J. Carol, and F. M. Kunze, *J. Assoc. Off. Agric. Chem.*, **44**, 313 (1961).
233. F. M. Kunze and L. Espinoza, *J. Assoc. Off. Agric. Chem.*, **46**, 899 (1963).
234. G. Gräfe, *Dtsch. Apoth. Ztg.*, **104**, 1412 (1964).
235. G. Ritschel-Beurlin, *Arzneim.-Forsch.*, **15**, 1247 (1965).
236. C. Garbner, E. M. Aseem, and L. M. Pinol, *Safybi*, **8**, 33 (1968).
237. M. Luise, *Boll. Chim. Farm.*, **108**, 223 (1969).
238. E. Neuzil, H. Jensen, and J. Le Pogam, *J. Chromatogr.*, **39**, 238 (1969).
239. A. F. S. A. Habeeb, *J. Pharm. Pharmacol.*, **10**, 591 (1958).
240. R. Tulus and A. Guran, *Bull. Fak. Med. Istanbul, Ser. C.*, **28**, 114 (1963).
241. E. Soczewinski and M. Rojowska, *J. Chromatogr.*, **32**, 364 (1968).
242. J. Hartel, J. A. Copper, and C. van Bochove, *Rec. Trav. Chim. Pays-Bas*, **82**, 264 (1963).
243. A. Baerhiem-Svendsen and E. Brochmann-Hanssen, *J. Pharm. Sci.*, **51**, 514 (1962).
244. H. R. Roberts and M. G. Kolor, *Anal. Chem.*, **31**, 565 (1959).
245. V. Moses, *J. Chromatogr.*, **9**, 241 (1962).
246. S. Tabak, I. I. Grandberg, and A. N. Kost, *Zh. Anal. Khim.*, **20**, 869 (1965).
247. M. Pöhm, *Arch. Pharm.*, **291**, 468 (1958).
248. P. Bohinc, *Acta Pharm. Jugosl.*, **15**, 3 (1965).
249. Y. I. Chumakov and M. N. Filippovich, *Zh. Anal. Khim.*, **20**, 856 (1965).
250. J. Lauckner, E. Helm, and H. Fürst, *Chem. Tech. (Berl.)*, **18**, 372 (1966).
251. E. S. Lane, *J. Chromatogr.*, **18**, 426 (1965).
252. J. R. Pelka and L. D. Metcalfe, *Anal. Chem.*, **37**, 603 (1965).
253. J. R. Parrish, *J. Chromatogr.*, **18**, 535 (1965).
254. H. Feltkamp and F. Koch, *J. Chromatogr.*, **15**, 314 (1964).
255. J. E. Gordon, *J. Chromatogr.*, **20**, 38 (1965).
256. D. Waldi, *Naturwissenschaften*, **50**, 614 (1963).
257. V. Rossetti, *Biochim. Appl.*, **11**, 225 (1964).
258. M. M. Frodyma and R. W. Frei, *J. Chromatogr.*, **15**, 501 (1964).
259. M. E. Clark, *Analyst*, **93**, 810 (1968).
259a. E. H. Taylor, *Lloydia*, **27**, 96 (1964).
260. I. Pinter and M. Kramer, *Elelmiszervizsgalati Kozl.*, **12**, 193 (1966).
261. G. Pastuska and H. J. Petrowitz, *Chem. Ztg.*, **88**, 311 (1964).
262. A. Waksmundzki, J. Rozylo and J. Oscik, *Chem. Anal. (Warsaw)*, **8**, 965 (1963).
263. Y. Hashimoto and K. Samejima, *J. Pharm. Soc. Jap.*, **86**, 451 (1966).
264. W. Kinze, *Pharm. Zentralhalle.*, **105**, 365 (1966).
265. A. Alessandro and F. De Sio, *Boll. Chim.-Farm.*, **104**, 498 (1965).
266. J. Franc and M. Hajkova, *J. Chromatogr.*, **16**, 345 (1964).

267. J. T. Weil, *J. Chromatogr.*, **37**, 331 (1968).
268. J. A. Bell and I. Dunstan, *J. Chromatogr.*, **24**, 253 (1966).
269. C. Di Bello and E. Celon, *J. Chromatogr.*, **31**, 77 (1967).
270. E. Knappe and I. Rohdewald, *Z. Anal. Chem.*, **217**, 110 (1966).
271. J. J. Menn and J. B. McBain, *Nature*, **209**, 1351 (1966).
272. L. Fishbein and J. Fawkes, *J. Chromatogr.*, **20**, 521 (1965).
273. F. di Ieso, *J. Chromatogr.*, **32**, 269 (1968).
274. R. Hüttenrauch and J. Schulze, *Pharmazie*, **19**, 334 (1964).
275. I. I. Grandberg, G. K. Faizova, and A. N. Kost, *Zh. Anal. Khim.*, **20**, 268 (1965).
276. J. Kulicka, Z. Greporowicz, and W. Karminski, *Chem. Anal. (Warsaw)*, **10**, 1347 (1965).
277. H. Barsch, K. K. Moll, and U. Thomanek, *J. Prakt. Chem.*, **311**, 159 (1969).
278. C. Troszkiewicz, J. Suwinski, and W. Zielinski, *Chem. Anal. (Warsaw)*, **13** (3), 3 (1968).
279. R. W. Frei, A. Kunz, G. Pataki, T. Prims, and H. Zuercher, *Anal. Chim. Acta*, **49**, 527 (1970).
280. E. Roeder, E. Mutschler, and H. Rochelmeyer, *Z. Anal. Chem.*, **244**, 45 (1960).
281. M. Adamek and A. Cee, *Chem. Prum.*, **18**, 332 (1968).
282. L. H. Klemm, C. E. Klopfenstein, and H. P. Kelley, *J. Chromatogr.*, **23**, 428 (1966).
283. M. Hawrylyshyn, B. Z. Senkowski, and E. G. Wollish, *Microchem. J.*, **8**, 15 (1964).
284. D. T. N. Pillay and R. Mehdi, *J. Chromatogr.*, **32**, 592 (1968).
285. G. Ballin, *J. Chromatogr.*, **16**, 152 (1964).
286. M. R. Grimmett and E. L. Richards, *J. Chromatogr.*, **18**, 605 (1965).
287. M. Bachrata, J. Cerna, and S. Szucsova, *Cesk. Farm.*, **18**, 18 (1969).
288. A. Pfandl, *Dtsch. Apot.-Ztg.*, **108**, 568 (1968).
289. P. Turi and J. Polesuk, *J. Pharm. Sci.*, **57**, 2180 (1968).
290. A. M. Soeterboek and M. can Thiel, *Pharm. Weekl.*, **109**, 962 (1974).
291. H. Gänshirt, *Arch. Pharm. (Berl.)*, **296**, 129 (1963).
292. E. Colombo, *Farmaco, Ed. Prat.*, **23**, 43 (1968).
293. C. Li, *Acta Chim. Sinica*, **31**, 518 (1965).
294. G. Szasz, M. Szasz, and V. Polankay, *Acta Pharm. Hung.*, **35**, 207 (1965).
295. U. M. Senanayake, *J. Chromatogr.*, **32**, 75 (1968).
296. C. Franzke, K. S. Grunert, U. Hilderbrandt, and H. Griehl, *Pharmazie*, **23**, 502 (1968).
297. W. Schunack, E. Mutschler, and H. Rochelmeyer, *Dt. Apoth. Ztg.*, **105**, 1551 (1965).
298. T. C. Chou and H. H. Lin, *J. Chromatogr.*, **27**, 307 (1967).
299. F. G. von Stryk and G. F. Zajacz, *J. Chromatogr.*, **41**, 125 (1969).
300. E. Schmid, E. Hoppe, C. Meythaler, Jr., and L. Zicha, *Arzneim.-Forsch.*, **13**, 969 (1963).

301. A. Alessandro, F. Mari, and S. Settecase, *Farmaco, Ed. Prat.*, **22**, 437 (1967).
302. J. C. Morrison and J. M. Orr, *J. Pharm. Sci.*, **55**, 936 (1966).
303. J. C. Morrison and L. G. Chatten, *J. Pharm. Sci.*, **53**, 1205 (1964).
304. P. P. Ascione, J. B. Zagar, and G. P. Chrekian, *J. Pharm. Sci.*, **56**, 1393 (1967).
305. A. Alvarez-Fernandez, V. Torre-Noceda, and E. Sanchez-Carrera, *J. Pharm. Sci.*, **58**, 443 (1969).
306. I. C. Dijkhuis and M. R. Brommet, *J. Pharm. Sci.*, **59**, 558 (1970).
307. A. Banaszek, K. Krowicki, and A. Zamojski, *J. Chromatogr.*, **32**, 581 (1968).
308. E. Ludwig and U. Freimuth, *Nahrung*, **9**, 41 (1965).
309. P. P. Waring, W. C. Goad, and Z. Z. Ziporin, *Anal. Biochem.*, **24**, 185 (1968).
310. Y. G. Popova, K. Popov, and M. Ilieva, *J. Chromatogr.*, **24**, 263 (1966).
311. L. A. Dal Cortivo, J. R. Broich, A. Dihrberg, and B. Newman, *Anal. Chem.*, **38**, 1959 (1966).
312. K. Genest, *J. Chromatogr.*, **19**, 531 (1965).
313. G. Kurono, K. Ogura, and K. Sasaki, *J. Pharm. Soc. Jap.*, **85**, 262 (1965).
314. W. Paulus, S. Goenechea, and V. Wienert, *Arch. Kriminol.*, **135**, (3), 84 (1965).
315. R. Palitzsch, T. Beyrich, and R. Pohloudek-Fabini, *Pharmazie*, **23**, 246 (1968).
316. M. Sarsunova and N. T. K. Chi, *Cesk. Farm.*, **15**, 474 (1966).
317. E. Lang, *Pharmazie*, **25**, 493 (1970).
318. D. Neumann and H. B. Schröter, *J. Chromatogr.*, **16**, 414 (1964).
319. S. El Gendi, W. Kisser, and G. Machata, *Mikrochim. Acta*, **1965**, 120.
320. L. N. Li and C. C. Fang, *Acta Pharm. Sinica*, **11**, 189 (1964).
321. L. Wichlinski and Z. Skibinski, *Farm. Pol.*, **22**, 194 (1966).
322. E. Roeder, E. Mutschler, and H. Rochelmeyer, *Z. Anal. Chem.*, **244**, 46 (1969).
323. M. Sahli and M. Oesch, *Pharm. Acta Helv.*, **40**, 25 (1965).
324. A. Suszko-Purzycka and W. Trzebny, *J. Chromatogr.*, **16**, 239 (1964), **17**, 114 (1965); *Chem. Anal. (Warsaw)*, **9**, 1103 (1964).
325. R. Adamski and J. Bitner, *Farm. Pol.*, **24**, 17 (1968).
326. N. Oswald and H. Flück, *Pharm. Acta Helv.*, **39**, 293 (1964).
327. R. Bernasconi, S. Gill, and E. Steinegger, *Pharm. Acta Helv.*, **40**, 246 (1965).
328. A. K. Bhatnagar and S. Bhattacharji, *Indian J. Chem.*, **3**, 43 (1965).
329. R. K. Sharma, G. S. Khajuria, and C. K. Atal, *J. Chromatogr.*, **10**, 433 (1965).
330. D. Gröger and S. Johne, *Pharmazie*, **20**, 456 (1965).
331. D. Giacopello, *J. Chromatogr.*, **19**, 172 (1965).
332. I. Bayer, *J. Chromatogr.*, **16**, 237 (1964).
333. J. T. Huang, H. C. Hsiu, and K. T. Wang, *J. Chromatogr.*, **29**, 391 (1967).
334. R. Paris and M. Sarsunova, *Pharmazie*, **22**, 483 (1967).
335. B. Danos, *Acta Pharm. Hung.*, **34**, 221 (1964).
336. J. A. Steele, *J. Chromatogr.*, **19**, 300 (1965).

337. H. Wullen and H. Thielemans, *J. Pharm. Belg.*, **23**, 307 (1968).
338. W. Poethke and W. Kinze, *Arch. Pharm.*, **297**, 593 (1964).
339. H. J. Uhlmann, *Pharm. Ztg. Ver. Apoth.-Ztg.*, **109**, 1998 (1964).
340. E. Hodgson, E. Smith, and F. E. Guthrie, *J. Chromatogr.*, **20**, 176 (1965).
341. A. Noirfalise and G. Mees, *J. Chromatogr.*, **31**, 594 (1967).
342. K. P. Lapina, *Farmatsiya Mosk.*, **17** (5), 54 (1968).
343. Y. N. Forostyan and V. I. Novikov, *Zh. Obshch. Khim.*, **38**, 1222 (1968).
344. G. L. Szendey, *Z. Anal. Chem.*, **244**, 257 (1969).
345. C. Levorato, *Boll. Chim.-Farm.*, **107**, 574 (1968).
346. A. Fiebig, J. Felczak, and S. Janicki, *Farm. Pol.*, **25**, 971 (1969).
347. A. R. Saint-Firmin and R. R. Paris, *J. Chromatogr.*, **31**, 252 (1967).
348. K. T. Finley and R. S. Kaiser, *J. Chromatogr.*, **39**, 195 (1969).
349. K. Butkiewicz and E. Kowalczyk, *Chem. Anal. (Warsaw)*, **11**, 1209 (1966).
350. C. H. Pao, *Acta Pharm. Sin.*, **13**, 67 (1966).
351. G. Wagner and J. Wandel, *Pharmazie*, **21**, 105 (1966).
352. Y. T. Lin, K. T. Wang, and T. I. Yang, *J. Chromatogr.*, **20**, 610 (1965).
353. Z. Margasinski, R. Danielak, and H. Rafalowska, *Acta Pol. Pharm.*, **22**, 423 (1965).
354. T. Bican-Fister and V. Kajganovic, *J. Chromatogr.*, **16**, 503 (1964).
355. B. T. Kho and S. Klein, *J. Pharm. Sci.*, **5** , 494 (1963).
356. W. Poethke and W. Kinze, *Pharm. Zentralhalle*, **10**, 95 (1964).
357. W. Kamp, *Pharm. Weekbl. Ned.*, **99**, 1309 (1964).
358. J. Dobrecky and C. Garber, *Pharma Int.*, **4**, 47 (1968).
359. P. N. Moza and G. S. Khajuria, *J. Chromatogr.*, **24**, 261 (1966).
360. L. Vignoli, B. Cristau, F. Gouezo, and J. M. Vassalo, *Bull. Trav. Soc. Pharm Lyon*, **9**, 277 (1965).
361. J. Pastor and V. Valimamod, *Bull. Trav. Soc. Pharm. Lyon*, **9**, 312 (1965).
362. A. Noirfalise and M. H. Grosjean, *J. Chromatogr.*, **16**, 236 (1964).
363. Z. Margasinski, R. Danielak, T. Pomazanska, and H. Rafalowska, *Acta Pol. Pharm.*, **21**, 253 (1964).
364. P. J. Smith and T. S. Hermann, *Anal. Biochem.*, **22**, 134 (1968).
365. N. M. Lombardi, J. Dobrecky, and R. C. d'A. de Carnevale-Bonino, *Revta Asoc. Bioquim. Argent.*, **33**, 20 (1968).
366. W. Kamp, *Pharm. Weekbl. Ned.*, **101**, (3) 7 (1966).
367. D. Gröger and D. Erge, *Pharmazie*, **18**, 346 (1963).
368. R. Gnehm, H. U. Reich, and P. Guyer, *Chimia*, **19**, 585 (1965).
369. H. Gerlach and H. J. Senf, *Pharm. Zentralhalle*, **105**, 93 (1966).
370. D. P. Schwartz, R. Brewington, and O. W. Parks, *Microchem. J.*, **8**, 402 (1964).
371. G. C. Barrett and A. R. Khokhar, *J. Chromatogr.*, **39**, 47 (1969).
372. E. Celon, L. Biondi, and E. Bordignon, *J. Chromatogr.*, **35**, 47 (1968).

373. R. B. Mefferd, Jr., R. M. Summers, and J. G. Fernandez, *Anal. Lett.*, **1**, 279 (1968).
374. K. Shimomura and H. F. Walton, *Sep. Sci.*, **3**, 493 (1968).
375. R. R. Paris, M. Sarsunova, and M. Semonsky, *Ann. Pharm. Fr.*, **25**, 177 (1967).
376. S. Klein and B. T. Kho, *J. Pharm. Sci.*, **51**, 966 (1962).
377. E. A. Mistryukov, *J. Chromatogr.*, **9**, 314 (1962).
378. M. Sarsunova and V. Schwarz, *Pharmazie*, **18**, 34 (1963).
379. G. H. Posner and D. Z. Rogers, *J. Am. Chem. Soc.*, **99**, 8208 (1977).
380. E. Jeney and J. Walther, *Acta Pharm. Hung.*, **38**, 28 (1968).
381. M. Beroza, K. R. Hill, and K. H. Norris, *Anal. Chem.*, **40**, 1608 (1968); Dr. M. Beroza, personal communication, 1978.
382. Kontes Densitometer, U.S. Patents 3,562,539 and 3.924,948.
383. V. Jirgl, *Anal. Biochem.*, **13**, 381 (1965).
384. L. Braun, *Biochem. Z.*, **339**, 8 (1963).
385. S. Ansorge, *Pharmazie*, **23**, 16 (1968).
386. R. L. Munier and G. Sarrazin, *Bull. Soc. Chim. Fr.*, **1963**, 2939.
387. G. Biserte, T. Plaquet-Schoonaert, P. Boulanger, and P. Paysant, *J. Chromatogr.*, **3**, 25 (1960).
388. Y. Hara, *Bull. Chem. Soc. Jap.*, **36**, 1373 (1963).
389. E. R. B. Graham and A. Neuberger, *Biochem. J.*, **109**, 645 (1968).
390. H. A. Simmonds, *Clin. Chim. Acta*, **23**, 319 (1969).
391. T. I. Yang and J. C. Su, *Chin. Agric. Chem. J.*, **5**, 71 (1967).
392. M. Stuchlik, I. Csiba, and L. Krasnec, *Cesk. Farm.*, **16**, 187 (1967).
393. C. L. Winek, J. L. Beal, and M. P. Cava, *J. Chromatogr.*, **10**, 246 (1963).
394. A. Kornhauser and M. Perpar, *Arch. Pharm.*, **298**, 321 (1965).
395. M. Klera and Z. Dudzik, *Farm. Pol.*, **23**, 863 (1967).
396. D. Brzezinska and L. Maciaszek, *Farm. Pol.*, **1**, 353 (1965).
397. A. M. Chodakowski and J. Wojciechowska, *Farm. Pol.*, **21**, 255 (1965).
398. P. C. Bose, T. Sen, and G. K. Ray, *Indian J. Pharm.*, **20**, 324 (1958).
399. M. Pernarowski and V. A. Padval, *J. Pharm. Sci.*, **52**, 218 (1963).
400. G. Machek, *Sci. Pharm.*, **28**, 252 (1960).
401. N. Bellen, Z. Bellen, and Z. Lada, *Chem. Anal. (Warsaw)*, **11**, 273 (1966).

CHAPTER 7

Separation of Neutral Substances

7.1 GENERAL REMARKS

In this chapter we discuss methods for the separation of organic compounds that do not contain acidic or basic functions. Mixtures of these compounds are called neutral substances, since they do not respond to an acid or alkali titrant in aqueous or nonaqueous medium to produce meaningful analytical results. Inspection of the list of organic functional groups[1] reveals that a vast majority of functions have neutral characteristics, although the practicing analyst frequently encounters samples that are either acidic or basic, or are comprised of components possessing both neutral and acidic (or basic) functions in the molecule.

When a mixture contains compounds possessing different functional groups, it is possible to determine the individual compounds (e.g., alcohol, ketone, ester, etc.) by chemical reactions without separation.[2] Owing to the popularity of chromatography, however, the current practice is to carry out a preliminary separation step, as can be seen from the many published methods described in Part Two of this book. On the other hand, if the sample contains two or more compounds possessing the same functional group, a separation scheme must be chosen to obtain the percent of each component in the mixture. In the following sections, we discuss suitable separation methods for compounds possessing a particular functional group. It should be noted that the examples cited are taken from the literature, so the reader can obtain the detailed experimental procedure if desired. Understandably, there may be better methods that have not been published or have escaped the writer's attention.

7.2 SEPARATION OF COMPOUNDS POSSESSING OXYGEN FUNCTIONS

7.2.1 Alkoxy Compounds (Ethers)

Since alkyl ethers are generally stable at their boiling points, gas chromatography is suitable for separation. Examples may be cited for glycol ethers (Singliar and Dykyk,[3] Kosenko et al.[4]) and alkyl ether derivatives of *p*-hydroxybenzoates (Wilcox[5]). Mixtures of cyclic ethers, such as furan, methylfuran, tetrahydrofuran, tetrahydropyran, and dihydropyran, also have been separated by gas chromatography (Anderson et al.,[6] Shuikin et al.,[7] Polyakova et al.[8]). For a homologous series, a linear relationship was found[7] between the logarithm of the retention volume and the number of carbon atoms.

Thin-layer and paper chromatography are used for separating nonvolatile ethers. For instance, Nealey[9] separated polyphenyl ethers by multiple development on a layer of silica gel with benzene–cyclohexane as solvent. Hänsel and Rimper[10] separated 18 natural and synthetic substituted 4-methoxy-α-pyrones on poly(vinyl cyanide), with ethyl acetate–cyclohexane as developer. Giebelmann[11] separated phenoxy compounds (e.g., methacetin, phenacetin, mephensein) on filter paper using butanol–formic acid or methanol–acetone–triethanolamine as the developing solvent.

If the compounds in the mixture contain different alkoxyl groups, they can be separated indirectly via gas chromatography of the alkyl iodides obtained by reacting the sample with hydroiodic acid (Schachter and Ma,[12] Mitsui and Kitamura,[13] Gutenmann and List[13a]). Isomeric monoethers of glycerol have been converted to the corresponding diacetyl (Albro and Dr Amer[14]), trifluoroacetyl, or trimethylsilyl (Wood and Snyder[15]) derivatives for quantitative gas-chromatographic separation.

7.2.2 Carbonyl Compounds (Aldehydes, Ketones)

Some examples of the published methods for the direct separation of carbonyl compounds are given in Table 7.1. As is to be expected, gas chromatography predominates. Using columns of Celite supporting Apiezon, dinonyl phthalate polyethylene glycol adipate, or 3,3′-oxydipropionitrile to separate aldehydes and methyl ketones, Golovnya and Uralets[22] found that there was a rectilinear relationship between the retention indices and the number of carbon atoms and also between the former and the boiling points of the compounds. When the carbonyl compounds also contain a hydroxyl group, the latter is acetylated in the case of hydroxyaldehydes (Andronov and Norikov[44]) or oxidized to carbonyl in the case of α-hydroxyketones (Ron-

Table 7.1 Direct Separation of Carbonyl Compounds

Separation method	Carbonyl compounds	Reference
Gas chromatography	Acetaldehyde, acetone, crotonaldehyde	Lucchesi[16]
	Acetone, ethyl methyl ketone	UKAEA[17]
	C_3 compounds	Andrev[18]
	C_2 to C_4 compounds	Mizuno[19]
	Methyl ketones	Berridge[20]
	Alkanals, alkan-2-ones, alk-2-enals	Golovnya[21]
	Alkanals, methyl ketones	Golovnya[22]
	Long-chain aldehydes	Wood[23]
	Long-chain ketones	Yakerson[24]
	Acetone, mesityl oxide, diacetone alcohol	Fore[25]
	Cyclic ketones	Komers[26]
	Furaldehyde, methylfuraldehyde	Bagaev[27]
	Aromatic aldehydes	Prabucki[28]
	Vanillin, ethylvanillin	Martin[29]
	Vanillin, *p*-hydroxybenzaldehyde, syrinaldehyde	Pepper[30]
Thin-layer chromatography	Aliphatic, alicyclic, aromatic ketones	Buzlanova[31]
	Cyclohexanone, methyl-, dimethylcyclohexanones	Franc[32]
	Diketones	Gudrinietse[33]
	Ionones, methylionones	Dhont[34]
	Aromatic aldehydes	Klouwen[35]
	Vanillin, ethylvanillin	Blanc[36]
	Hydroxybenzophenones	Durisinova[36a]
Paper chromatography	Aldehydes, ketones	Schulte[37]
	Vanillin, ethylvanillin	Fitelson[38]
Column elution	Polynuclear cyclic ketones	Sawicki[39]
	Pyrenediones	Fatiadi[40]
Ion exchange	Formaldehyde, acetaldehyde, pyruvaldehyde	Christofferson[41]
	Lactaldehyde, acetoxypropanone	Sandman[42]
	High-molecular-weight ketones	Sherma[43]

kainen and Brummer[45]) prior to gas-chromatographic separation. Fediel and Cirimele[46] converted C_1 to C_{12} straight-chain aldehydes to the corresponding phenylhydrazones, which were separated on columns of SE 30 on Chromosorb at 120 to 190°C with a flame ionization detector and nitrogen as carrier gas; the relationship between the logarith of the retention time and number of carbon atoms was shown to be rectilinear.

Derivitization to form the 2,4-dinitrophenylhydrazones is frequently recommended for the separation of carbonyl compounds. Table 7.2 shows examples in which the separation is carried out by thin-layer, paper, or liquid chromatography. Severin[50] studied the separation of formaldehyde, acetaldehyde, butyraldehyde, isobutyraldehyde, hexanal, and octanal using kieselguhr or alumina as adsorbent with various solvent systems, including benzene–cyclohexanol containing methyl acetate, ethyl acetate, dichloromethane, cyclohexene, tetrahydrofuran, pyridine, or dimethylformamide; kieselguhr was preferable to alumina, but no one solvent was superior to the others.

Franc and Celikovska[58] described the separation of 50 aldehydes and ketones via their cyanoacetic acid hydrazones, which were prepared by condensing the carbonyl compounds with cyanoacetohydrazide. The hydrazones were subjected to descending chromatography on paper impregnated with formamide, dimethylformamide, or acetamide, with chloroform as the mobile phase; separation at 20°C required 2 hr.

Table 7.2 Chromatographic Separation of Carbonyl Compounds via Their 2,4-Dinitrophenylhydrazones

Chromatography	Original carbonyl compounds	Reference
Thin-layer	C_2 and C_3 compounds	Nano[47]
	C_2 to C_{14} aliphatic aldehydes	Takeuchi[48]
	n-Alkanals, *n*-alkan-2-ones	Libbey[49]
	Aliphatic aldehydes	Severin[50]
	Aldehydes in leather	Bloem[51]
	Vicinal dicarbonyls	Cobb[52]
Paper	Saturated aliphatic aldehydes	Ellis[53]
	Aldehydes in wine	Maruta[54]
Liquid column	Aliphatic carbonyl compounds	Monty[55]
	Carbonyl compounds in butter	Parsons[56]
	Dicarbonyls	Corbin[57]

Whereas 2,4-dinitrophenylhydrazones are prepared for the quantitative analysis of carbonyl compounds by gas chromatography, it should be noted that these derivatives are not separated as such in the gas chromatograph. Actually, the original carbonyl compounds are regenerated by pyrolysis in the presence of oxoglutaric acid (Jones[59]), sodium bicarbonate (Ralls[60]), or phthalic acid (Nishi[61]) using a technique known as flash-exchange, and the aldehydes or ketones are then separated.

7.2.3 Epoxy Compounds (Olefin Oxides)

Kaliberdo and Vaabel[62] described the separation of ethylene oxide, propylene oxide, and butylene oxide by gas chromatography at 45°C on a column (2 m×4 mm) containing lubricating oil–diisoamyl phthalate (2:1) on a diatomaceous support; 1 to 2 mg of sample was determined by measurement of peak areas. Venetten and Bottini[63] reported the separation of alkyl-substituted oxirans, including *cis*- and *trans*-isomers.

For the analysis of epoxy alkanes, Glowacki et al.[64] recommended converting these compounds either into their methyl ethers (for C_{12} to C_{16} epoxides) using methanolic BF_3, or into secondary alcohols (for cyclic or tertiary epoxides), obtained by $LiAlH_4$ reduction; a stainless-steel column packed with 5% FFAp on Gas-Chrom Q was used, the temperature being programmed from 100 to 250°C at 6°C per min, with helium as carrier gas and flame ionization detection. Fioriti et al.[65] determined epoxyglycerides by preparing the dioxolane derivatives with cyclopentanone.

Ethylene oxide adducts have been analyzed in several ways. Puschmann[66] treated the adducts of secondary alcohols with acetic anhydride and then separated the acetates of the alcohols and the diesters of the glycols, respectively, by gas chromatography. Wickbold[67] used liquid chromatography to separate ethylene oxide adducts according to the length of the ethylene oxide chain on a column (1 m×1.6 cm) of silica gel with ethyl methyl ketone as eluent. Skelly and Crummett[68] analyzed ethylene oxide adducts of nonylphenol by thin-layer chromatography on alumina with water-saturated butan-2-one as the solvent.

7.2.4 Esters

Mixtures of esters are found in nature, obtained in industrial processes, or specifically prepared for the analysis of carboxylic acids, anhydrides, and acyl halides. In the last case, the methyl esters are most commonly synthesized and are then separated by gas chromatography. Table 7.3 gives examples of the separation methods for methyl or ethyl esters. Numerous studies have been performed to elucidate the chromatographic behavior of these esters

Table 7.3 Separation of Methyl or Ethyl Esters

Separation method	Methyl or ethyl esters of	Reference
Gas chromatography	C_{16} to C_{24} acids	Hadorn[69]
	C_{16} to C_{20} acids	Adlard[70]
	Fatty acids of vegetable oils	Jernejcic[71]
	Acids of whale oil	Ito[72]
	Acids of goose fat	Pupin[73]
	Acids from lipids	Metcalfe[74]
	Plant acids	Kellogg[75]
	Resin acids	Chang[76]
	Acids from oil of turpentine	Brus[77]
	Octadecadienoic acids	Christie[78]
	Isomeric fatty acids	Allen[79]
	Hexanoic acids	Fell[80]
	Linoleaic acids	Litchfield[81]
	Alkylmalonic acids	Tepe,[82] Prochazkova[83]
	Dibasic alkanoic, phthalic acids	Mori[84]
	Aliphatic dibasic, cyclopentanecarboxylic acids	Rudenko[85]
	Perfluoroalkanoic acids	Genkin[86]
	Toluic, phthalic acids	Kazinik[87]
	Phthalic acids	Ohashi[88]
Thin-layer chromatography	Benzene–di-, tri-, tetracarboxylic acids	Olafsson[89]
	Hydroxybenzoic acids	Cavello[90]
	Acids in butter fat	Simal-Lozano[91]
	Oxygenated fatty acids in seed oils	Vioque[92]
	Cyclopentyl, cyclopentenyl fatty acids	Mani[93]
Liquid chromatography	Hydroxystearic acids	Dolev[94]
Ion exchange	Formic, acetic acids	Sherma[95]
Countercurrent distribution	Mono-, di-, trienoic acids	Scholfield[96]

and to improve the accuracy of the analysis. For instance, Haken and Allen.[97] found that the changes in retention times with structural variations of the methyl esters of fatty acids are related by a simple expression and can be described by a nomogram[98]; plots of the logarithm of relative retention versus the number of carbon atoms in the alcohol are rectilinear for normal and isoesters; the slopes decrease as the number of carbon atoms in the acid chain is increased up to 3 and are then parallel when this number is 4 to 6; branching in the acid chain reduces the retention, but the effect becomes less as the number of carbon atoms increases.[99] For normal aliphatic ester isomers, Connell[100] reported that the logarithm of the retention time is directly proportional to the boiling point; thus, for any group of isomers, if sufficient boiling points or retention times are known, the remainder can be found graphically. Stern et al.[101] studied the separation of diastereoisomeric esters of 2-haloalkanoic acids and offered explanations for the better separation of 2-bromoesters compared with 2-chloro esters. Sheppard et al.[102] undertook calibration studies of gas-chromatographic systems for quantitative analysis of fatty acid methyl esters with respect to molar and weight responses. Caster et al.[103] described a digital-computer program for quantitation of gas chromatography that accepts peak heights and retention times as input and provides tables of sample composition and information on component identification as output. A five-parameter exponential function is used to correct for systematic nonlinearities of the gas chromatography. Corrections are introduced for incomplete peak resolution. A nonlinear function is used to correct carbon number values for the systematic errors related to the fact that each component delays the appearance of all the latter components. The corrections increase the accuracy by a factor of up to 10. Vandenheuvel[104] demonstrated that 5 to 15 μg of esters (from C_{14} to C_{22}) could be resolved and determined to within $\pm 1\%$ using short columns under conditions designed to ensure stability at low signal attenuation with relatively high temperature and flow rate.

Examples of the methods for separating esters of higher alcohols, glycols, and polyols are shown in Table 7.4. Generally speaking, when the esters contain double bonds, they should be converted into saturated esters for gas chromatography. Mounts and Dutton[123] described a vapor-phase hydrogenation accessory for the analysis of complex mixtures of polyunsaturated fatty acid esters of natural origin. When the esters contain free hydroxyl groups, the common practice is to convert the latter into nonpolar groups to facilitate gas-chromatographic separation. For instance, the trimethylsilyl derivatives were employed for lactates (Brandt et al.[124]), as well as for mono- and diglycerides (Wood et al.,[125] Watts and Dils[126]). Comparing the trifluoroacetyl and trimethylsilyl derivatives of hydroxy fatty acid esters, Freeman[127] reported that the trifluoroacetyl derivatives had

Table 7.4 Separation of Esters Derived from Alcohols, Glycols, Polyols

Esters	Separation method	Reference
Acetates of primary and secondary alcohols	Gas chromatography	Volodina[105]
Lactates of pentanols, hexanols	Ion-exchange column	Leitch[106]
Lactate and mandelate of *sec*-butyl alcohol	Ion-exchange column	Spitz[107]
Benzyl esters of C_1 to C_3 acids	Gas chromatography	Watson[108]
Pentafluorobenzyl esters of C_2 to C_{18} fatty acids	Gas chromatography	Kawahara[109]
Natural occurring lactones	Polyamide-layer chromatography	Wang[110]
Esters of butane-2,3-diol	Gas chromatography	Nurok[111]
Propane-1,2-diol fatty acid esters	Silica gel column	Sahasrabudhe[112]
Natural triglycerides	Kieselguhr layer + $AgNO_3$	Wessels[113]
Triglycerides	Silica gel layer + $AgNO_3$	Jurriens[114]
Acetylated fats	Alumina layer	Pokorny[115]
Triglycerides in *Cuphea llavia* seed	Consecutive chromatography	Litchfield[116]
Glycerides in olive oil	Paper chromatography	Sliwiok[117]
Mono-, di-, triglycerides	Liquid chromatography	Distler[118]
Glycerides in cacao	Liquid chromatography	Black[119]
Triglycerides in milk	Gas chromatography	Smith[120]
Acetates of polyols	Silica gel layer	Dumazert[121]
Esters of sorbitol	Liquid chromatography	Sahasrabudhe[122]

shorter retention times and gave better resolution of saturated and unsaturated hydroxy esters than did the corresponding trimethylsilyl derivatives. Kuksis and Gordon[128] studied the separation of acetyl and trifluoroacetyl derivatives of bile acid methyl esters in narrow-bore columns (0.125 in. × 3 or 5 ft) coated with silicone polymers and operated at 195 to 220°C. Satisfactory results were obtained only with the trifluoroacetyl derivatives on columns of 1% of QF-1 fluoroalkylsilicone on Gas-Chrom P (100 to 200 mesh). The acetyl derivatives were completely separated on columns of 2.25% of SE-30 silicone on Chromosorb W (60 to 80 mesh), but the separation was not quantitative because of losses on the column. Columns containing 2% of SE-60 silicone nitrile polymers on Anachrom ABS (110 to 120 mesh) did not give satisfactory separations.

7.2.5 Hydroxy Compounds (Alcohols, Glycols, Polyols)

Even though the hydroxyl function is moderately polar, compounds containing hydroxyl groups only are amenable to gas-chromatographic separation. As shown in Table 7.5, direct chromatography has been worked out for the analysis of many alcohol mixtures. For instance, Kamibayashi[129] investigated various liquids as stationary phases for the analysis of 13 alcohols from C_1 to C_5 and reported that, with a few exceptions, a linear relationship was observed between carbon number or boiling point and logarithm of retention volume for homologous series, columns treated with tetraethylene glycol dimethyl ether or polyoxyethylene glycol being the most satisfactory.

Table 7.5 Separation of Alcohols

Separation method	Alcohols and conditions	Reference
Gas chromatography	13 C_1 to C_5 alcohols	Kamibayashi[129]
	C_1 to C_5 alcohols	Bighi[130]
	C_1 to C_5 alcohols, 75°C	Bluestein[131]
	8 C_1 to C_5 alcohols, 78°C	Sokolov[132]
	C_1 to C_5 alcohols, 85, 100°C	Kolesnikova[133]
	C_1 to C_4 alcohols	Urone[134]
	C_1 to C_6 alcohols	Rogozinski[135]
	C_1 to C_7 alcohols	Doelle[136]
	C_1 to C_{10} alcohols, 135°C	Struppe[137]
	Fusel oils	Kahn,[138] Maurel[139]
	C_5 to C_{24} fatty alcohols, 150, 180, 220°C	Ionescu[140]
	C_1 to C_{20} alcohols, 50–300°C	Robinson[141]
	C_6 to C_{18} fatty alcohols, 220°C	Raccagni[141a]
	Long-chain fatty alcohols, 190°C	Wood[142]
	Isomeric aliphatic alcohols to C_5	Drews[143]
	Isomeric alcohols to C_5, 70–100°C	Schlunegger[144]
	Primary C_5 alcohols	Porcaro[145]
	Isomeric normal heptanols	Kolesnikova[146]
	C_9 alcohols, 124°C	Bendel[147]
	Enantiomers of C_4 to C_{19} *n*-alkan-2-ols	Gil-Av[148]
	CH_3OH, C^3H_3OH, C_2H_5OH, $C_2{}^3H_5OH$	Cartoni[149]

Table 7.5 (*Contd.*)

Separation method	Alcohols and conditions	Reference
	Aryl–alkyl alcohols	Arkima[150]
	Cyclohexanols	Casselman[151]
	Stereoisomeric alkylcyclohexanols	Komers[152]
	Hydroxyethyl compounds	Fishbein[153]
	Phenethyl, benzyl alcohols in wine	Gomes[154]
	Menthos, 165°C	Terada[155]
	Triterpene alcohols, 230°C	Capella[156]
Liquid chromatography	Fatty alcohols on Al_2O_3 column	Pollerberg[157]
Thin-layer chromatography	5 alcohols on silica gel	Singh[158]
	C_8 to C_{22} fatty alcohols on kieselguhr	Bandyopadhyay[159]
	C_{10}, C_{15}, C_{20} terpene alcohols	McSweeney[160]
	Terpene, sesquiterpene alcohols	Stahl[161]
Paper chromatography	C_1 to C_{18} alcohols	Syper[162]
	C_1 to C_5 alcohols	Ruzicka[163]
Ion exchange	C_7 to C_{12} alcohols	Locke[164]

When C_1 to C_5 alcohols were separated on Carbowax, Bighi et al.[130] found that the components in the mixture were eluted in an order different from that predicted from their boiling points, but the gas-chromatographic behavior of alcohols can be predicted from vapor pressures and calculated activity coefficients. Bluestein and Posmanter[131] performed quantitative analysis of aqueous C_1 to C_5 alcohol mixtures on a copper column (12 ft × 0.25 in.) packed with 5% of Triton X-305 on Teflon at 75°C using helium as carrier gas and thermal-conductivity detection; tailing was negligible even at high concentrations of H_2O.

Table 7.6 gives examples of separation of mixtures of glycols and polyhydroxy compounds. When diols are separated by gas chromatography (Novoselov et al.[165]), the retention times are related to the boiling points. Since glycols and polyols have low vapor pressures, separation by paper or thin-layer chromatography is suitable, the choice of technique depending on the mixture. For instance, Naff et al.[178] compared the separation of but-2-yne-1,4-diol from its mono- and diesters on glass-fiber paper and on silica gel plates and recommended the thin-layer method because it gave better reproducibility of R_F values and did not cause tailing. On the other hand, Fischer and Koch[186] preferred paper chromatography to the thin-layer technique for separating 1,2-diols because paper supports gave better

Table 7.6 Separation of Glycols and Polyhydroxy Compounds

Separation method	Hydroxy compounds and conditions	Reference
Gas chromatography	Glycols	Novoselov[165]
	C_2 to C_3 glycols	Naddeau[165a]
	Glycols, glycerol, 100–200°C	Maksimenko[166]
	Ethanediol, diethylene glycol, 190°C	Dorlini[167]
	Di-, triethylene glycols, 200°C	Spencer[168]
	Di- to octaethylene glycols, 260°C	Ruschmann[169]
	Dipropylene glycol isomers, 175°C	Gross[170]
	Glycols, polyhydric alcohols, 138–145°C	Balakhontseva[171]
	Polyhydric alcohols, 80–185°C	Vaver[172]
	Polyhydric alcohols, 150–290°C	Assmann[173]
	Glycerol, polyhydric alcohols, 250°C	Dooms[174]
	Polyhydric alcohols in alkyd resins, 68–225°C	de la Court[175]
	Polyglycerols, 100–275°C	Sen[176]
Liquid chromatography	Polyoxyethylene glycols, reversed-phase partition	Konishi[177]
Thin-layer chromatography	1,4-Diols	Naff[178]
	Sorbitol, mannitol in cherry pulps	Haeseler[179]
	Mannitol, sorbitol	Castagnola[180]
	Sugar alcohols, on kieselguhr	Waldi[181]
	Sugar alcohols, on magnesium silicate	Grasshof[182]
	Polyhydric alcohols, on alumina	Borisovich[183]
	Polyhydric alcohols, on cellulose	Dyatlovitskaya[184]
	Polyhydric alcohols, on silica gel +Pb(II)	de Simone[185]
Paper chromatography	1,2-Diols	Fischer[186]
	Glycols	Adelberg[187]
	Polyglycerols	Zajic[188]
Ion exchange	Glycols	Sargent[189]
	Ethanediol, glycerol, xylitol	Dabagov[190]
	Polyhydric compounds	Samuelson[191]
	Alditols	Samuelson[192]
	Anhydro alditols, alditols	Matsui[193]
	Isomeric sugar alcohols	Barker[194]

chromatograms for quantitative measurements. Separation of glycols by ion exchange is based on the formation of complexes such as those with boron.[189,195]

For mixtures containing hydroxy compounds only, it is generally unnecessary to convert the hydroxyl group into derivatives prior to separation. Derivatization, however, is sometimes advantageous, especially when certain other functional groups are present in some of the compounds. For gas chromatography, conversion of the hydroxyl function into trimethylsilyl ether is frequently employed, for example, for alcohols (Berezkin and Kruglikova[196]), dialkylglycerols (Wood et al.[197]), pentaerythritols (Smith and Tullberg[198]), terpene alcohols (Seidenstücker[199]), sterols, and tocopherols (Prevot and Barbati[200]). According to Smith and Carlsson,[201] quantitative silylation of polyhydric compounds is achieved by mixing the sample (10 to 50 mg) with trimethylchlorosilane and anhydrous pyridine in test tubes at the temperature of solid CO_2. Sato et al.[202] used ketene to prepare the acetyl derivatives for the analysis of C_6 to C_{22} fatty alcohols, because normal acylation methods do not give complete conversion. Jung et al.[203] recommended the conversion of polyols into trifluoroacetates, while Sanz-Burata et al.[204] prepared the 2-acetoxypropionates for the resolution of racemic alcohols. Celades and Paquot[205] treated polyoxyethylene glycols with dimethyl sulfate; Hefendehl[206] reacted terpene alcohols with metaboric acid in a precolumn. For the separation of alcohols and glycols from complex mixtures by liquid chromatography, Wekell et al.[207] converted the hydroxy compounds into nitrates with a solution of pentyl nitrate in acetic anhydride–acetic acid; the nitrates were eluted from a column of silicic acid using light petroleum. Hurd and Miles[208] separated 38 acetylated sugars in a column containing Magnesol and Celite, with ethanol–benzene as the developing solvent. Acyl derivatives of sugars are also employed in thin-layer chromatography (Deferrari et al.[209]). The 3,5-dinitrobenzoates of terpene alcohols (von Schantz et al.[210]) and the 2,4-dinitrobenzenesulfinates of methyl, ethyl, phenethyl, and furfuryl alcohols (Lefebure et al.[211]) are suitable for separation on silica gel plates. For paper chromatography of lower aliphatic alcohols, the corresponding 2,4-dinitrobenzyl ethers (Churacek et al.[212]) or 2,4-dinitrobenzenesulfinates (Tulus and Güran[213]) have been suggested.

Some special techniques for quantitation may be mentioned. Berezkin et al.[214] determined alcohols in the gas chromatograph by recording the hydrogen obtained when each component reacted with $NaAlH_4$ in a tube inserted before the detector. Katz and Keeney[215] converted the individual alcohols separated by gas chromatography into their *p*-phenylazobenzoates and measured the absorbances at 322 nm. Gil-Av et al.[148] determined optical isomers of C_4 to C_{19} alkan-2-ols either by polarimetry or by measurement of peak areas.

7.2.6 Methylenedioxy Compounds

Although the methylenedioxy function is cleaved from the molecule in the presence of strong acids to yield a mole of formaldehyde,[1] it is possible to separate methylenedioxy compounds by direct gas chromatography under suitable conditions. Zielinski and Fishbein[216] studied the behavior of 18 aromatic compounds containing a methylenedioxy function attached to the benzene ring; these compounds are synergists for pyrethrins as insecticides. Separation was achieved either at 160°C on a column (6 ft × 0.25 in.) of 3% of Carbowax 20M on silanized Chromosorb W, or at 100°C on a similar column of 4% of QF-1 on silanized chromosorb G. If the compounds also possess a carboxyl function, conversion of the latter to the corresponding methyl ester function is better than conversion to the trimethylsilyl derivative. Thus, for 3,4-methylenedioxybenzoic, 3,4-methylenedioxyphenylacetic, and 3,4-methylenedioxycinnamic acids, the respective methyl esters gave an elution ratio of 4.6 : 1.4 : 1.0, whereas the trimethylsilyl derivatives gave a ratio of 3.1 : 1.0 : 1.0 following 3,4-methylenedioxybenzaldehyde.

7.2.7 Peroxy Compounds (Peroxides, Hydroperoxides)

In spite of the thermal instability of peroxides, attempts have been made to separate them by gas chromatography. Bukata et al.[217] claimed that organic peroxides can be determined with a coefficient of variation in the range from 0.3 to 1.0% on a column of didecyl phthalate at 80 to 100°C, or on silicone grease at 138°C. Vlodavets et al.[218] separated hydroperoxides at 60°C using a column of tritolyl phosphate on NaCl.

Kucher et al.[219] employed liquid chromatography to separate mono-, di-, and hydroxy hydroperoxides of diisopropylbenzene in the following manner. The peroxides were adsorbed on a column of silica gel. After elution of the monohydroperoxide with heptane, the hydroxy and dihydroperoxides were eluted with ethanol. The ethanolic solution was evaporated at 40°C and the residue was dissolved in 1 ml of toluene–ethanol (5 : 1) and placed on a silica gel column. The dihydroperoxide was eluted with toluene, and the hydroxy hydroperoxide was then eluted with ethanol. Ivanenko and Volga[220] separated hydroperoxides on a diatomite column and determined each component in the eluate iodometrically.

Thin-layer chromatography is the preferred technique for separating peroxides. Bernhardt[221] used Kieselgel plates with benzene–ether as solvent and aq. KI–starch as spray reagent. Fijolka and Gnauck[222] used silica gel plates and a solution of *N*,*N*-dimethyl-*p*-phenylenediamine hydrochloride to reveal red-purple spots; the color was stable for only a short time. Hayano et al.[223] studied one- and two-dimensional developments on silica gel and found that the solvent used for the first dimension affected the R_F values in the second development. Knappe and Peteri[224] described the separation of a number of technically important peroxides and hydroperoxides on silica gel with toluene–CCl_4 and toluene–CH_3COOH as mobile phases. Separation of cumene hydroperoxide and *t*-butyl peracetate was achieved only with light petroleum–ethyl acetate (49 : 1). Buzlanova et al.[225] separated cyclohexyl peroxides on silica gel bound with gypsum; it was observed that alumina caused partial decomposition of all the peroxides.

7.2.8 Quinones

Paluch[226] separated benzoquinone and its mono- and dichloro compounds by gas chromatography using a column of silicone rubber supported on silica gel, operated at 160°C, with nitrogen as carrier gas. Furuya et al.[227] reported the retention times of many naturally occurring and synthetic anthraquinones; for quinones that also possess the hydroxy function, retention usually increases with the number of hydroxyl groups present and decreases with intramolecular bonding. Dugan and Lundgren[228] used gas chromatography to study quinoid compounds in biological materials. Vitamins of the K group have been analyzed on a glass column (1.2 m × 6 mm) packed with Apiezon on Celite, with argon as carrier gas and argon ionization detection (Yanotovskii et al.[229]), and also in a Teflon tube (1.75 m × 1.5 mm) filled with superfine Sephadex G-25 using CH_3OH—$CHCl_3$–heptane (1 : 1 : 2) as eluent and a platinum-chain flame detector (Nyström-Sjovall[230]).

Pettersson[231] separated benzoquinones on silica gel plates and determined the individual compounds by quantitative u.v. absorptiometry at 270 nm; the molar extinction coefficients range from 5×10^3 to 5×10^4. Kotakemori and Okada[232] employed Kieselgel G to separate naphthoquinones using $CHCl_3$ as solvent for unsubstituted quinones and $CHCl_3$–CH_3COOH–ethyl acetate (10:1:10) for hydroxy- or aminoquinones. Anthraquinones have been separated on layers of silica gel with light petroleum–ethyl acetate as solvent (Danilovic and Neumovic-Stevanovic[233]), or on Kieselgel with isopropyl ether for development (Poethke and Behrendt[234]). Labadie[235] separated naturally occurring hydroxyanthraquinones, hydroxyanthrones, and hydroxydianthrones on Kieselgel; the sample was applied in the dark in a nitrogen atmosphere and the chromatogram was developed with ethyl acetate. Nikolaeva et al.[236] separated tetrahydroanthraquinone, 2-ethylanthraquinone, and 2-ethyltetrahydroanthraquinone on layers of alumina and determined each compound polarographically.

7.3 SEPARATION OF NEUTRAL COMPOUNDS POSSESSING NITROGEN FUNCTIONS

7.3.1 Nitro Compounds, Nitrites, Nitrates

Mixtures of nitro compounds have been separated by gas chromatography. As shown in Table 7.7, the examples include both the aliphatic and aromatic series. Courtier et al.[239] performed quantitative analysis of mono- and di-nitrotoluenes resulting from nitration of toluene using columns operated at 175 or 220°C; collaborative study by three laboratories gave a precision of ±0.6%. It should be recognized, however, that there is the potential danger of explosion due to the presence of polynitro aromatics. Aliphatic nitro compounds also may decompose upon heating; Crawforth and Waddington described the analysis of gases formed from the pyrolysis of nitroalkanes. It seems desirable, therefore, to avoid the use of gas chromatography for the separation of nitro compounds. Kemula et al.[243] recommended separation by liquid chromatography on columns packed with Ni(4-picoline)$_4$(SCN)$_2$-4-picoline clathrate, elution with solvent containing alcohol, picoline, and NH_4SCN, followed by polarographic determination.

Table 7.7 Separation of Nitro Compounds

Separation method	Nitro compounds	Reference
Gas chromatography	C_1 to C_3 nitroparaffins	Starshov[237]
	C_1 to C_4 nitroparaffins	Biernacki[238]
	Mono-, dinitrotoluenes	Courtier[239]
	Nitrotoluenes, ethylbenzenes	Etienne[240]
	Trinitrotoluenes	Gehring[241]
Liquid chromatography	C_1 to C_4 nitroparaffins	Kemula[242]
	Nitrotoluenes	Kemula[243]
	Nitronaphthalenes	Kemula[244]
Thin-layer chromatography	Dinitrobenzenes	Pejkovic-Tadic[245]
	Mono-, di-, trinitrotoluenes	Beider[246]
	Nitro compounds	Trachman[247]
	Polynitro compounds	Hoffsommer[248]
	Nitrofurans	Bortoletti[249]
Paper chromatography	14 substituted trinitrobenzenes	Colman[250]
	Nitronaphthalenes	Kitahara[251]
	Nitrodianthrimides	Cherkasskii[252]

Hoffsommer and McCullough[248] described quantitative analysis of polynitro aromatic compounds in complex mixtures by thin-layer chromatography and visible spectrophotometry. Colman[250] utilized paper chromatography to separate explosives.

Alkyl nitrites can be separated in copper tubes (10 ft × 0.125 in.) containing 20% of 3,3′-oxydipropionitrile on Diatoport and operated at 20 to 25°C; when N_2, N_2O, and NO are also present, a 30 ft tube and 35°C are required for complete separation (Harrison and Stevenson[254]). Paraskevopoulos and Cvetanovic[255] recorded the relative retention times and hydrogen flame detector responses of C_1 to C_4 alkyl nitrites, nitrates, and nitroparaffins separated in a stainless-steel capillary column (300 ft × 0.015 in.) coated with dinonyl phthalate and operated at 0°C; the relationship between the logarithm of relative retention time and boiling point was rectilinear for the nitroalkanes and alkyl nitrates, but not for the nitrites.

Pimkin et al.[256] separated C_5 to C_{10} alkyl nitrates on a stainless-steel column (2 m × 4 mm) of polytetrafluoroethylene powder supporting 10% of silicone at 164°C. Litchfield[257] determined mono- and dinitrates of ethanediol and propane-1,2-diol in blood using a column (5 ft × 0.125 in.) of 30% of E301 on Celite, operated at 100°C. Camera and Pravisani[258] analyzed mixtures of ethylene nitrate, nitroglycerin, diethylene glycol dinitrate, dinitropropanediol, and dinitropentanediol using a 35 or 50 cm column containing 10% of ethanediol succinate polyester supported on Celite at 145°C. Rao et al.[259] described the separation of blasting explosives (ethanediol dinitrate, glycerol trinitrate, diethylene glycol dinitrate, and diglycerol tetranitrate) on silica gel–plaster of Paris thin layers, with benzene–light petroleum for development.

7.3.2 N—N Compounds

Hydrazones are the N—N compounds most frequently encountered. In spite of high melting points, attempts were made to separate phenyl- and 2,4-dinitrophenylhydrazones by gas chromatography. For example, Fedeli et al.[260] separated these derivatives of aldehydes and ketones on a 2.5 m column of SE-30 on Chromosorb operated at 150 to 220°C with injection at 260 to 280°C, and a flame ionization detector; compounds derived from C_1 to C_6 aldehydes were determined with an accuracy of ±0.5%. Other chromatographic techniques, however, are more commonly employed. For liquid chromatography, columns packed with polyethylene (Freytag[261]), Celite (Schwartz[262]), or silica gel (de Jong et al.[263]) can be used. Thin-layer chromatography of 2,4-dinitrophenylhydrazones derived from hydroxycarbonyl compounds can be separated on alumina or silica gel (Anet[264]). Stroh and Schüler[265] recommended alumina–gypsum layers for the separation of sugar phenylhydrazones, while Haas and Seeliger[266] specified polyamide layers for phenylosazones. Hunt[267] described the preparation of 3-methylbenzothiazolin-2-one hydrazone derivatives of C_1 to C_8 aliphatic aldehydes and their separation on silica gel plates. Rasmussen[268] determined oxo acid dinitrophenylhydrazones by photometric measurement of spot areas on Kieselgel thin layers. Secor and White[269] utilized reversed-phase paper chromatography to separate sugar hydrazones and osazones; the paper was impregnated with formamide, with formamide–ethyl acetate–H_2O (1 : 20 : 1, upper phase) as the developing solvent.

Camp and O'Brien[270] separated the semicarbazones of 10 aldehydes on silica gel thin layers; spots were located by spraying with molybdophosphoric acid and heating at 105°C. Paper electrophoresis or thin-layer chromatography can be used to separate acyl hydrazides; visualization is achieved by spraying with ethanolic solution of picryl chloride (Russell[271]).

According to Gritter,[272] diazonium salts can be separated on a layer of silica gel G, with dimethyl sulfoxide–HCOOH–CH_2Cl_2–dioxane (25 : 1 : 5 : 20) as solvent. Because of the sensitivity of these compounds to heat and light, the chromatograms are developed at ambient temperature and in yellow light; spots are revealed by spraying with naphthalene-2,3-diol. Bowins et al.[273] determined isomeric nitrophenyldiazonium salts by coupling with 2-naphthol and separation on an alumina column with ethyl acetate–benzene as eluent.

7.3.3 Other Nitrogen Compounds

Oximes of C_2 to C_{10} alkanals were separated by eluting from a silica gel column with light petroleum–ethyl acetate (Pejkovic-Tadic[274]); all oximes exist in two forms—liquid and crystalline—and the liquid form is eluted first. The two isomeric forms of benzaldoxime, benzoin oxime, and anisoin oxime were resolved on a thin layer of silica gel, with anhydrous methanol as solvent (Hranisavljevic-Jakovljevic et al.[275]). Furil dioxime isomers were separated on paper impregnated with 30% ethanolic formamide, using ethyl ether as solvent (Toul and Okac[276]).

Mixtures of C_7 to C_{20} fatty acid nitriles can be analyzed by gas chromatography on a column (2.8 × 6 mm) of silicone oil supported on diatomaceous brick; Parimskii and Shelomov[277] gave three methods for quantitative measurement of peak areas. Aliphatic nitriles in aqueous solution are extracted into *o*-dichlorobenzene prior to gas chromatography (Arad-Talmi et al.[278]). Separations of isomeric C_4 to C_1 nitriles (Dardenne et al.[279]) and of phthalonitriles (Kiessling and Moll[280]) have been achieved. Di Lorenzo and Russo[281] used two columns to analyze mixtures of acrylonitrile with acetonitrile and water, one (1.5 m × 4 mm) filled with 8% of Carbowax on

Teflon and another (0.5 m × 4 mm) filled with 20% of 3,3'-oxydipropionitrile on Chromosorb.

For the gas-chromatographic separation of aliphatic isocyanides, Kelso and Lacey[282] observed that glass tubing was a satisfactory column container, but metal was not. It was found that the effect is due to reversible adsorption. The performance of metal tubes was improved with use of a silanized support.

Whereas the vast majority of heterocyclic nitrogen compounds are basic (see Chapter 6), there are some that belong to the neutral category. Thus neutral benzo-1,4-diazepine derivatives can be directly separated by gas chromatography (Lafargue et al.[283]). Nonbasic nitrogen compounds in petroleum (e.g., indoles, carbazoles) are analyzed by elution chromatography (Snyder and Buell[284]). Tetrazolium salts and their formazans have been separated on silica gel thin layers (Tyrer et al.,[285] Seidler and Thieme[286]).

7.4 SEPARATION OF NEUTRAL SULFUR COMPOUNDS

7.4.1 Mercaptans, Sulfides

Mixtures of aliphatic and aromatic mercaptans (thiols) can be separated by direct gas chromatography.[287] Carson and Wong[288] converted aliphatic mercaptans into alkyl dinitrophenyl sulfides, which they separated by liquid chromatography, while Obara et al.[289] recommended the separation of these derivatives on silica gel thin layers.

Petronek[290] studied the gas-chromatographic separation of C_1 to C_4 dialkyl sulfides and alkyl benzyl sulfides; selective separation of branched-chain from normal alkyl sulfides was achieved with benzoquinoline as the stationary phase, whereas separation was in order of boiling points when silicone oil was used. Franc et al.[291] reduced the sulfides by Raney nickel into the corresponding hydrocarbons, which were separated by gas chromatography.

Albro and Fishbein[292] separated sulfur mustard and its analogues on glass columns (1.5 m × 2 mm) containing 3% of cyclohexylenedimethanol succinate on Gas-Chrom; the column temperature was programmed from 110 to 230°C. Stanford[293] performed the separation on silica gel thin layers using $CHCl_3$–acetone as solvent.

Orr[294] described the separation of C_1 to C_{18} alkyl sulfides by liquid chromatography on a column (48 × 0.81 cm) containing silica gel (100 to 200 mesh); the stationary phase was 70% aqueous acetic acid saturated with Hg(II) acetate, and heptane (equilibrated with 50% acetic acid) was the mobile phase. For paper-chromatographic separation of C_1 to C_6 dialkyl sulfides, Kronrod and Panek[295] prepared the corresponding dialkylsulfonium iodides by treatment of the sulfides with methanolic methyl iodide; the stationary phase was butanol–acetic acid–H_2O (4 :1 :5), and the mobile phase was a 4% solution of acetic acid in CCl_4.

7.4.2 Sulfoxides, Sulfones, Sulfonates

According to Cates and Meloan,[296] gas-chromatographic separation of sulfoxides was achieved with Carbowax on Gas-Chrom Z at 200 to 250°C; all the compounds tested, except benzyl sulfoxide, had adequate thermal stability. This column was also used for separating sulfones (Cates Meloan [297]). Karaulova et al.[298] separated dialkyl, alkyl aryl, diaryl, heterocyclic, and alicyclic sulfoxides by thin-layer chromatography on alumina without binder, using acetone–CCl_4 (1 :4) as solvent.

Imaida et al.[299] described the gas-chromatographic separation of alkylbenzenesulfonates via the methyl esters. Other methods for the separation of aromatic sulfonates (Setzkorn and Carel,[300] Nishi[301]) and alkane sulfonates (Pollerberg[302]) involve desulfuration by heating with H_3PO_4; the resulting hydrocarbons are then submitted to gas chromatography.

Barnard et al.[303] separated alkylthiosulfonates in glass U-tubes (5 ft × 4 mm) packed with 5% of silicone oil on Celite; the column was maintained at 138°C and the alkylthiosulfonates were measured by polarographic reduction. Fishbein[304] separated isomeric mono-, di-, and trichlorophenyl esters of *m*-fluorosulfonic acid on silica gel layers; either acetone–benzene (1 :39) or isoamyl alcohol–aq. NH_3–H_2O (6 :3 :1) was used as solvent.

7.4.3 Thiocyanates, Thiocarbonyls, Thiocarbonates, Thiophenes

A mixture of phenyl-, mono-, and dibromophenyl isothiocyanates was separated by gas chromatography (Hrivnak and Kalamar[305]); the column was packed with polyoxyethylene supported on Chromosorb and operated at 130°C. Thin-layer chromatography was utilized to separate isothiocyanates (Komanova and Antos[306]) and thiocarbonyls (Compaigne[307]).

Hrivnak and Konecny[308] described the separation of 10 *S*-alkyl *N*,*N*-dialkyldithiocarbamates on a glass column (80 cm × 3.5 mm) packed with 5% of stationary phase on Chromosorb, operated at temperature from 140 to 160°C.

Thiophenic compounds in petroleum samples were separated according to the number of rings by programmed-temperature gas chromatography on a column (8 ft × 0.125 in.) of 3% of diethylene glycol sebacate polyester supported on Chromosorb (Martin and Grant[309]); each component was determined by microcoulometric titration. Chlorothiophenes (Profft and Solf[310]) and bithienyls (Berk[311]) were also separated by direct gas chromatography. Staszewski et al.[312] hydrogenated thiophenes over Raney nickel and separated the hydrocarbons by gas chromatography. Sulfur-containing heterocyclic compounds have been separated on Kieselgel thin layers with ligroin–benzene, CS_2, or benzene–ethyl acetate as solvent (Mayer et al.[313]), or on alumina or silica gel layers with methanol, ligroin, or benzene–$CHCl_3$ as solvent (Curtis and Phillips[314]).

7.5 SEPARATION OF COMPOUNDS CONTAINING UNSATURATED FUNCTIONS

7.5.1 Ethylenic Hydrocarbons

Mixtures of ethylenic hydrocarbons used in the petrochemical industry are generally separated by gas chromatography. Examples are shown in Table 7.8. It is noted that in some cases, $AgNO_3$ is added in the packing. Banthorpe et al.[341] reported that thallium(I) nitrate could replace $AgNO_3$ with advantage. Szczepaniak and Nawrocki[342] recently reviewed the use of several metals as charge-transfer complexes applied either as liquid superselective phases (e.g., $AgNO_3$ in ethanediol) or as solid phases (e.g., porous salts, molecular sieves containing metal ions; and organometallic polymers).

Eppert et al.[343] converted mono- and dialkenes into the corresponding epoxides by reaction with peroxybenzoic acid; the epoxides were then separated by liquid chromatography on a silica gel column with dimethyl sulfoxide as solvent. McDonough and George[344] also recommended the epoxide derivatives for separating *cis–trans*-isomers of olefins, but the epoxides were further acetylated and the acetates were submitted to gas chromatography.

Many naturally occurring compounds possess the ethylenic function, and analogous compounds are frequently found in one source. The terpene hydrocarbons may be cited as examples. Table 7.9 gives the methods for separating terpene hydrocarbons. Valkanas and Iconomou-Petrovitch[356] conducted systematic analysis of terpenic hydrocarbons by gas chromatography and found that the logarithms of the relative retention volumes obtained with one column, when plotted against the logarithms of the values obtained with a column of the same stationary phase, but different operating conditions, gave rectilinear graphs. These results suggest that, provided that equilibrium of flow rate and temperature is attended and columns of the same stationary phase are used, it is not necessary to reproduce working conditions to compare retention values.

Table 7.8 Gas-Chromatographic Separation of Ethylenic Hydrocarbons

Mixture	Column	Reference
Ethylene, propylene		Golbert[315]
C_2 to C_4	Double, 1.8 m × 6 mm, with $AgNO_3$	Jarzynska[316]
C_2 to C_4	12 m	Rennhak[317]
C_2 to C_4	Double, 6 m × 5 mm, one with $AgNO_3$, 40°C	Inst. of Petroleum[318]
6 pentenes	20 ft × 0.25 in., 25°C	Ottmers[319]
C_5 to C_6		Knight[320]
12 hexenes	100 m × 0.01 in., 0°C	Dunning[321]
Hexenes	With $PdCl_2$	Kraitr[322]
n-Octenes	2 m, with $AgNO_3$, 42°C	Bendel[323]
Octenes	5.3 m × 64 mm, 70°C	Vigdergauz[324]
C_6 to C_{14} 1-olefins		Miwa[325]
C_8 to C_{18} 1-olefins	2 ft × 0.25 in., 75°C and higher	Poe[326]
Internal olefins	150 ft × 0.01 in., with $AgNO_3$	Chapman[327]
Isotopic isomers	With $AgNO_3$	Lee[328]
Butenes, butadiene	13 ft, 25°C	Higgins[329]
Dienes	125 cm × 5 mm, or 600 cm × 6 mm, 4 or 20°C	Gershtein[330]
Trienes	1.9 m × 3 mm, 218°C	Nambara[331]
80 stereoisomers up to trienes		Yanotovskii[332]
Long-chain mono-, diolefins	2 m × 2.5 mm, with $AgNO_3$, 45–60°C	Fauvet[333]
Stilbenes	2 m × 3 mm, 180–215°C	Hemingway[334]
Divinylbenzenes	2 m × 0.25 in., 150°C	Wiley[335]
Divinylbenzene, ethylstyrene		Blasius[336]
Octahydronaphthalenes	4 m, 120°C	Powell[336a]
Chloroethylenes	1.7 m × 6 mm, 63°C	Sorochkina[337]
Chlorobutenes		Demina[338]
Phenylperfluoro olefins	10 ft × 0.25 in., 70°C	Herkes[339]
Perfluoro alicyclic olefins	10 ft × 0.25 in., 40°C	Johnson[340]

Table 7.9 Separation of Unsaturated Terpene Hydrocarbons

Separation method	Terpenes and conditions	Reference
Gas chromatography	Terpenes, tritolyl phosphate as stationary phase	Haslam[345]
	Terpenes, on capillary column	Bernhard[346]
	Turpentine oils, with dipropyline glycol on Celite	Miltenberger[347]
	Sesquiterpenes, with polyoxyethylene glycol adipate	Lukes[348]
	Irradiated terpenes, with petroleum grease on kieselguhr	Wang[349]
	Diterpenes	Appleton[350]
	35 terpenes	Sidorov[351]
Thin-layer chromatography	Monoterpenes, on silica gel + $AgNO_3$	Lawrence[352]
	Monoterpenes, on Kieselgel + $AgNO_3$	Schantz[353]
	Terpenes, on alumina or silica gel	Attaway[354]
	Triterpenes, on alumina	Ikan[355]

7.5.2 Acetylenic Hydrocarbons

Miocque and Blanc-Guenee[357] reviewed the literature on the separation of acetylenic compounds by gas chromatography. According to Rudenko et al.,[358] conditions should be chosen to provide as short a retention time as possible to avoid chemical reaction on the column; a glass column (0.3 to 1 m) containing 0.5% of polydimethylsiloxane on NaCl with hydrogen as carrier gas and a flame ionization detector were used. Bendel et al.[359] separated hexynes on a column (18 m × 4.75 mm) packed with Chromosorb coated with 25% of a saturated solution of $AgBF_4$ in 3,3′-oxydipropionitrile and operated at 62°C.

Mixtures containing both acetylenic and ethylenic functions present no difficulty in separation by gas chromatography. For example, Vigdergauz and Andreev[360] determined acetylene in ethylene. Skarvada and Zbytek[361] analyzed mixtures of acetylene, ethylene, and propene. Melkonyan and Vartanyan[362] reported the separation of vinylacetylene compounds.

Nebbia and Pagani[363] determined acetylene and diacetylene in the presence of methyl-, ethyl-, phenyl- and vinylacetylenes by liquid chromatography using a column packed with Al_2O_3 (96%) and Cu_2Cl_2 (4%). Roomi et al.[364] separated acetylenic and ethylenic compounds by thin-layer chromatography on silica gel plates.

7.5.3 Unsaturated Compounds Containing Other Neutral Functions

Neutral compounds possessing the double or triple bond plus another function can be separated by gas chromatography without special precautions, provided that the column has been tested with compounds containing the other neutral function. For instance, Belenkii and Orestova[365] analyzed methyl- and ethyl acraldehydes on a column (3 m × 5 mm, packed with polyoxyethylene glycol on diatomite) for separating aldehydes and alcohols. Rawshaw[366] described the separation of vinyl ketones, Yanotovskii[367] studied homologous series of ketones with various numbers of double bonds; the relationship between the logarithm of retention volume and the degree of unsaturation was rectilinear using stationary phases of poly-(ethanediol adipate) or Apiezon supported on Celite. Barnes et al.[368] separated *cis*- and *trans*-crotyl alcohols; Tyihak[369] separated stereoisomeric farnesols. Viswanathan et al.[370] separated homologues, vinylogues, and geometrical isomers of methyl-substituted vinyl ethers on an aluminum column (6 ft × 0.125 in.) packed with 20% of ethyleneglycol succinate polyester plus 2% of H_3PO_4 supported on Gas-Chrom and operated at 125°C. Kikuchi et al.[371] performed quantitative analysis of ketene produced from diketene using a column (1.5 m) containing Porapak beads.

According to Viswanathan et al.,[370] thin-layer chromatography on silica gel plates impregnated with $AgNO_3$ was also effective for separating the above-mentioned vinyl ethers. Hashmi et al.[372] employed circular thin-layer chromatography to separate α-unsaturated aldehydes. Akhrem et al.[373] separated acetylenic alcohols on alumina thin layers with benzene–ethyl ether as solvent.

McCullough[374] described the separation of acrylate, methacrylate, and tiglate ions by paper chromatography; these ions are not separated by the common solvent system of phenol and water, but the addition of methyl acetate to the system allows good separation. Zhenodarova et al.[375] separated vinyl esters after converting these compounds into their methoxymercury derivatives; the developing solvent consisted of ethanol–butanol–H_2O–conc. NH_4OH (40 : 40 : 20 : 3).

7.6 SEPARATION OF MISCELLANEOUS NEUTRAL COMPOUNDS

7.6.1 Alkanes, Alicyclic Hydrocarbons, and Aromatic Hydrocarbons

Whether found in nature or produced by industrial processes, alkanes and alicyclic or aromatic hydrocarbons are always obtained as complex mixtures.

While these materials are usually used in impure form, it is desirable to know their compositions. Before 1960, however, it was very difficult to separate and determine the individual compounds in hydrocarbon mixtures owing to their chemical inertness. For example, very elaborate fractionating columns were constructed, but the separation results were still unsatisfactory for precise analysis. The development of gas chromatography has had a tremendous impact on the analysis of hydrocarbons. It should be mentioned also, that other methods continue to be investigated. For instance, Matsunaga and Yagi[376] recently described a liquid-chromatographic procedure and recommended the use of gradient elution on activated alumina to separate aromatic hydrocarbons in heavy petroleum fractions.

For the analysis of complex hydrocarbon mixtures, the first step may involve the preliminary separation of the components into hydrocarbon types. Thus Marquart et al.[377] separated *n*-alkanes in petroleum heavy distillate by forming the urea clathrates; subsequently, the latter were decomposed by water and the *n*-alkanes were then separated by gas chromatography. Molecular sieves also can be used (Griesmer and Kiyonaga,[378] Eppert,[379] Chevron Research Corp.[379a]). Schenck and Eisma[380] determined alkanes in mixtures of saturated hydrocarbons by placing a regular gas-chromatographic column in series with a thermal-conductivity cell, a 10 cm column packed with molecular sieve 5A, and then a second conductivity cell; two chromatograms (one with and one without the alkanes) were obtained from a single injected sample and the percentage of alkanes was calculated directly from a comparison of peak areas. Niedzielska and Slowik[381] separated *n*-alkanes from branched-chain isomers by dissolving the mixture in cyclohexane and shaking with molecular sieve 5A; after filtration and drying, the *n*-alkanes were recovered by desorption. Snyder[381a] separated the five aromatic hydrocarbon types in catalytically cracked gas oils by linear-elution adsorption chromatography.

In the following paragraphs the methods for separating mixtures containing one hydrocarbon type into the individual components are discussed. It is noted that all procedures mentioned below involve direct separation of the compounds. Chromatographic separations of saturated hydrocarbons as a rule leave intact the compounds originally present in the analytical sample; exceptions are methods combining gas chromatography with hydrogenation (Cough and Walker[382]) or with mass spectrometry (Studier and Hayatsu[383]) that result in fragmentation of the molecules. Turkeltaut[384] has presented a mathematical formula for comparing the efficiencies of separation of hydrocarbons by various gas-chromatographic techniques; in general, the best separation of saturated hydrocarbons and alkylbenzenes is achieved on nonpolar stationary phases, but separation of compounds with close boiling points is best accomplished on highly polar stationary phases.

Alkanes. Table 7.10 gives examples of alkane (paraffin) mixtures that have been separated by gas chromatography. Various column packings and temperatures can be used for low-molecular-weight alkanes. For the separation of solid paraffins, a liquid stationary phase that does not decompose or volatilize at high temperatures must be chosen; Hildebrand et al.[406] tested several liquids and found only a Paraflow type pour-point depressant was suitable. Karr et al.[407] recommended gas–solid chromatography in a pair of alumina-coated aluminum capillary tubes (50 ft × 0.02 in.); with temperature programmed from 190 to 400°C, separation of C_{15} to C_{35} *n*-alkanes was completed in 20 min. For the separation of isotopic methanes or ethanes, the columns must be kept at low temperature; thus Van Hook and Kelly[413] reported that the best separation of C_2H_6 from C_2H_5D occurred at −115°C.

Cycloparaffins. Willis and Engelbrecht[414] employed a capillary column (200 ft × 0.02 in.) coated with didecyl phthalate to separate alicyclic and aliphatic hydrocarbons. For the determination of carbon number and for naphthene–paraffin separations, Brunnock and Luke[415] used glass columns (1 or 3 ft × 0.125 in.) packed with molecular sieve 13X and operated at up to 450°C. Alessi and Kikic[416] recommended reversed-phase liquid chromatography to separate cyclohexane, methylcyclopentane, and methylcyclohexane, using a stainless-steel tube (1 m × 6 mm) packed with Chromosorb, with 3,3′-oxydipropionitrile as eluent.

Alkylbenzenes. The gas chromatograph was first marketed[417] in 1955 for the determination of toluene in benzene, and it has become the mainstay for the analysis of mixtures of alkylbenzenes. Examples are given in Table 7.11. A large variety of column packings have been reported for the separation of this class of hydrocarbons. Nabivach[431] studied the selectivity of 70 stationary phases. It is interesting to note that substances like picric acid (Popescu and Blidise[434]) and chlorodnitrobenzene (Malinowska[438]) were recommended, although normally they are considered thermally unstable.

Alkylnaphthalenes. Kobot and Ettre[444] analyzed naphthalene homologues by gas chromatography using a liquid phase of phenyl silicone fluid on a 300 ft column operated at 160°C. Ohta et al.[445] separated nine monomethyl- and nine dimethyl-1-phenylnaphthalenes on columns of SE-30 on Chromosorb. Beschea and Popescu[446] studied the separation of mixtures containing C_{11} to C_{12} alkylnaphthalenes, diisopropylbenzenes, and hexadecane on a copper column (3 m × 4 mm) operated at 135°C; it was necessary to use packings in series, consisting of 285 cm of Carbowax on Celite and 15 cm of hexadecylnaphthalene on kieselguhr. Kemula et al.[447] employed liquid chromatography to separate 1- and 2-methylnaphthalenes on a column of

Table 7.10 Separation of Alkanes by Gas Chromatography

Alkanes	Column conditions	Reference
C_1 to C_4	Ethyldimethyloxamate on firebrick, 0 and 25°C	Richmond[385]
C_1 to C_4	Copper, Porapak	Papic[386]
C_1 to C_4	Dibutyl maleate + oxydipropionitrile	Miyake[387]
C_1 to C_5	Glass, Al_2O_3	McToggart[388]
C_1 to C_5	200 ft × 0.01 in., dedecane + dedec-1-ene, 0°C	Beckham[389]
C_1 to C_6	Squalene on kieselguhr	Sojak[390]
C_1 to C_6	Ucon on various supports	Gawlowski[391]
C_1 to C_8	Various columns, temperatures	Csicsery[392]
C_2 to C_5, in CH_4	Al_2O_3 on Carbowax, 45°C	Zocchi[393]
C_3 to C_8, in naphthas	Capillary, hexadecane–Fluorolube Oil	Leveque[394]
C_3 to C_{12}	200 ft × 0.01 in., squalene, −5 to 105°C	Sanders[395]
C_5 to C_8	Tritolyl phosphate on Chromosorb, 40–80°C	Blaustein[396]
n-C_5 to C_{11}, in gasoline	Bis(2-cyanoethyl)formamide	Albert[397]
C_6 isomers	2 m, quinoline–brucine + 1 m, tetradecane–quinoline	Waksmundzki[398]
C_6 in cigarette smoke	Oxydipropionitrile on Al_2O_3	Philippe[399]
C_7 to C_{12}	Silicone oil on diatomite	Berezkin[400]
C_7 to C_{20}	Molecular sieve, silicone	Barrall[401]
n-C_9 to C_{26}	Molecular sieve, Apiezon on Chromosorb	Sojok[402]
C_{10} to C_{14}	H_2O on Chromosorb, 20°C	Karger[403]
C_{10} to C_{18}	Copper, H_2O on Porasil, 20–40°C	Karger[404]
C_{11} to C_{18}	Apiezon on KNO_3	Wolf[405]
n-Paraffins to $>C_{24}$	Paraflow	Hildebrand[406]
n-C_{15} to C_{35} in tobacco	Copper, 2 ft × 0.25 in., Apiezon on Diatoport	Johnston[408]
n-C_{20} to C_{32}		O'Connor[409]
Wax, to C_{41}	Silicone on glass beads, 100–300°C	Dietz[410]
Isotopic methanes	Glass, silica gel	Bruner[411]
CH_4, deuteromethanes	Copper, 100 ft × 0.125 in., Porapak, 0°C	Czubryt[412]
Deuterated ethanes	Copper, methylcyclopentane on firebrick, −78°C	Van Hook[413]

Table 7.11 Separation of Alkylbezenes by Gas Chromatography

Alkylbenzenes	Column conditions	Reference
C_6 to C_8	Stainless-steel, polyoxyethylene glycol	Inst. of Petroleum[418]
C_6 to C_{11}	Capillary, bis(*m*-phenoxyphenoxy)-benzene	Walker[419]
C_6, C_7, xylenes	1 m × 6 mm, trimethylene glycol on diatomite	Vyakhrev[420]
C_6, C_7, xylenes	3.3 m × 4 mm, silicone on Celite	Vigdergauz[421]
C_6, C_7, ethylbenzene	10 ft × 0.125 in, Apiezon on Chromosorb	Fett[422]
C_6, C_7, ethylbenzene, xylenes	16 m × 0.3 mm, naphthaldehyde, 50°C	Gudkov[422a]
Xylenes, trimethylbenzenes	5 ft × 0.4 mm, Apiezon on Celite, 100°C	UKAEA[423]
Xylenes	Copper, 35 m × 0.25 mm, triethylene glycol butyrate	Vitt[424]
m-, *p*-xylenes	Bentone aerosol	Nesvadba[425]
m-, *p*-xylenes	7:8-Benzoquinoline	Desty[426]
m-, *p*-xylenes	1,8-Diaminonaphthalene	Case[427]
m-, *p*-xylenes	Pentaerythritol tetrabenzoate	Kruglov[428]
m-, *p*-xylenes	Copper, Bentone + dinonyl phthalate on Chromosorb	Gupta[429]
Xylenes, ethylbenzene	Bentone + diisodecyl phthalate on Chromosorb	Spencer[430]
Xylenes, ethylbenzene	Various stationary phases	Nabivach[431]
Xylenes, ethyltoluenes	Bentone + dinonyl phthalate	Strnad[432]
Xylenes, alkylbenzenes	Bentone + squalane + dinonyl phthalate	Paterok[433]
Methyl-, ethylbenzenes	Copper, picric acid on Celite, 130°C	Popescu[434]
Phenyl-C_9 to C_{14}-alkanes	Capillary, 150 ft × 0.01 in., Apiezon, 120 to 170°C	Carnes[435]
1- to 6-Phenyldodecanes	12 ft, asphalt on firebrick, 240–320°C	Spencer[436]
Diisopropylbenzenes	Didecyl phthalate, 120–130°C	Vlodavets[437]
C_7 to C_{10} alkylbenzenes	1-Chloro-2,4-dinitrobenzene on Celite	Malinowska[438]
Diethylbenzenes	20 ft × 0.25 in., silicone on Celite	Blake[439]
Propyl-, butylbenzenes	1.7 m × 6 mm, dinonyl phthalate on firebrick	Ioffe[440]
Cymene isomers	6 m, Bentone on Chromosorb	Rihani[441]
Cyclohexylbenzene, dicyclohexyl	Dioctyl phthalate on firebrick, 190°C	Hendriks[442]
Deuterobenzenes, benzene	250 m, squalene, 10°C	Burner[443]

$Ni(4\text{-picoline})_4(SCN)_2$; the eluent was *M* NH_4SCN containing 1% of 4-picoline and 45% of dimethylformamide; the individual alkylnaphthalenes were determined by polarographic double-layer capacity measurements. For the analysis of catalytic cracking products containing naphthalene and its derivatives, Szynagel and Jedrychowska[448] reported that direct gas chromatography with Apiezon on Celite gave unsatisfactory results; therefore liquid chromatography on silica gel was performed as a preliminary step to isolate the aromatic fraction by eluting with benzene; the benzene solution was then evaporated and the residue was separated on an 8 m gas-chromatographic column containing silicone on Chromosorb (5 m) and Apiezon on Celite (3 m) and operated at 190°C. Ognyanov[449] described the separation of naphthalene and methylnaphthalenes by thin-layer chromatography on alumina plates.

Di- and Polyphenyls. In spite of their high boiling points, di- and polyphenyls have been successfully separated by gas chromatography. Onuska et al.[450] separated diphenyl and terphenyls on graphitized carbon black and on Chromosorb coated with cesium chloride. Protsidim and Lavroskii[451] analyzed polyphenyls on a column (3 m × 4 mm) of Apiezon on firebrick at 275°C, while Basili et al.[452] used capillary columns (18 or 30 m × 0.3 mm) coated with silicone rubber or Apiezon with temperature programming. Gasco-Sanchez and Burriel-Marti[453] studied the stability of some stationary phases and recommended mixtures of silicone oil with Carbowax supported on Chromosorb. Perez-Garcia[454] compared the responses of the thermal-conductivity and flame ionization detectors to those of diphenyl and terphenyls and the effects of the concentration of each component in a mixture. Separations of alkyldiphenyls and polyphenyls by liquid chromatography were described by West[455] and Hellmann et al.[455a] Onuska and Janak[456] performed thin-layer chromatography of diphenyl, terphenyls, and quaterphenyl on graphitized carbon black. Ritter et al.[457] used Bentone–Celite thin layers to separate polyphenyl isomers, with heptane as solvent for multiple development. Rostovtseva and Fish[458] separated polyphenyls on paper impregnated with dimethylformamide, using dichloroethane as the developing solvent.

Polycyclic Aromatic Hydrocarbons. Thin-layer chromatography is well suited for the separation of polycyclic aromatic hydrocarbons, since these compounds do not volatilize under atmospheric pressure and exhibit strong ultraviolet fluorescence. Thus Biernoth[459] employed this technique to separate 13 polycyclic hydrocarbons on alumina layers with trimethylpentane as developing solvent and determined each compound, after extraction into ethanol, by ultraviolet spectrophotometry. Hood and Winefordner[460] described the separation of 16 polycyclic hydrocarbons, including 8 carcinogens, on silica gel with hexane–pyridine as solvent and the quantitative measurement of fluorescence or phosphorescence at $-196°C$. Ikan et al.[461] studied the separation of fluorenes on thin layers of silica gel, alumina, cellulose acetate, or cellulose acetate containing $CaSO_4$ using several developing solvents. Wieland et al.[462] prepared a highly acetylated cellulose powder to be applied on glass plates to separate polycyclic hydrocarbons with CH_3OH–$(C_2H_5)_2O$–H_2O (4 :4 :1) as the developing solvent. Kelenffy[463] used thin-layers of Kieselgel and developed the chromatograms with cyclohexane–benzene. For mixtures of polycyclic hydrocarbons not easily separated on silica gel, alumina, or Florisil, Harvey and Halonen[464] recommended the use of silica gel impregnated with an acceptor of the charge-transfer type, for example, trinitrobenzene or trinitrofluorenone; either the acceptor was included in the slurry before coating the plates or the activated plates were dipped in a 2% solution of the acceptor in benzene and then dried at 60°C; the chromatograms were developed with benzene–heptane or with CCl_4. Howard and Haenni[465] described the separation of polynuclear hydrocarbons in paraffin wax by paper chromatography. Lijinsky et al.[466] surveyed the determination of polynuclear hydrocarbons in mineral oil and coal tar by paper or column chromatography and reported that gas-chromatographic separation on silicone-coated glass beads, operated at 160 to 210°C, was also successful. Konyashina et al.[467] separated mixtures containing phenanthrene and anthracene by gas chromatography at 270°C on a column (2 m × 4 mm) of 30% of $CaCl_2$ on Chromosorb. Pinchin and Pritchard[468] analyzed anthracene oil using a 17-ft column containing Apiezon on Celite, operated at 195 to 250°C. For highly condensed polycyclic hydrocarbons that cannot be separated by gas chromatography, Thomas and Zander[469] used high-pressure liquid chromatography on a column (25 cm × 4 mm) of LiChrosorb with hexane as the mobile phase and ultraviolet detection. Jentoft and Gouw[470] employed columns packed with dimethylformamide–Carbowax on Gas-Chrom with trimethylpentane as eluent. Commins[471] separated polycyclic hydrocarbons on alumina columns with cyclohexane as solvent and eluent and determined successive fractions spectrophotometrically. Tyne and Bell[472] proposed a method using columns containing 1,3,5-trinitrobenzene supported on Columpak; it is based on the formation of complexes between polycyclic aromatic hydrocarbons and the nitro compound. Similarly, Buu-Hoi and Jacquignon[473] described a procedure that depends on the affinity of polycyclic hydrocarbons towards anhydrides and imides of tetrahalogenated phthalic acids; elution is carried out at an alkaline pH.

7.6.2 Halocarbons

Gas chromatography is the method of choice for separating halocarbons, since they are stable in the vapor state. Table 7.12 gives a number of examples. Various packings and temperatures are prescribed for different mixtures. For some halogenated hydrocarbons, for example, polychlorinated diphenyls used as pesticides,[506] the column temperature can be as high as 320°C. Beckman and Bevenue[507] found that quartz tubing gave higher recoveries for DDT, DDD, DDE, heptachlor, endrin, and dichloran than copper, stainless-steel or aluminum columns. In certain cases the column is very selective. Thus, for the separation of low-molecular-weight fluorocarbons, Bright and Matula[508] obtained good results using copper tubes (10 ft × 0.25 in.) packed with Porapak and operated at 100 to 175°C with helium as carrier gas, but under these conditions octafluorocyclobutane and decafluorobutane were not separated from each other.

Some examples of liquid chromatographic separation of halogenated hydrocarbons are given below. Schunter and Schnitzerling[509] separated DDT and related compounds on a silicic acid column using gradient elution with CCl_4–hexane followed by benzene. Kemula and Krzeminska[510] separated DDT isomers by passing the aqueous dimethylformamide solution through a powdered rubber–heptane column and determined each isomer by polarography. Amell and Helt[511] analyzed mixtures of DDT and DDD by running the sample in *n*-hexane solution (saturated with nitromethane) through a column of silicic acid. Drechsler[512] separated chloromethyltoluenes on a column of silica gel treated with nitromethane, with cyclohexane as the mobile phase. Schultz and Purdy[513] described the separation of stereoisomeric bromophenylbenzenes by column fractional precipitation. Brinkman and de Kok[514] separated isomers of di-, tri-, tetra-, penta-, or hexachlorodiphenyl by high-speed liquid chromatography at 27°C on a

Table 7.12 Gas Chromatographic Separation of Halocarbons

Halocarbons	Column conditions	Reference
Cl—C_1	Paraffin on kieselguhr	Vyakhirev[474]
Cl—C_1, C_2	Porapak	Foris[475]
Cl—C_1, C_2	Copper, paraffin on silica gel	German[476]
Cl—C_1, C_2	Tritolyl phosphate	Shuman[477]
Br—C_1, C_2	3 m, oxydipropionitrile on firebrick	Takacs[478]
Cl—C_1 to C_3	40 m × 0.3 mm, didecyl phthalate	Tantsyrev[479]
F—C_1 to C_4	Dodecafluoroheptyl acrylate on Chromosorb	Greene[480]

Table 7.12 (*Contd.*)

Halocarbons	Column conditions	Reference
Cl—C_2	Dinonyl phthalate–triethanolamine on PTFE	Tischer[481]
Cl—C_2	Polyoxyethylene glycol on firebrick	Balandina[482]
F, Cl—C_2, C_3	Stainless-steel, 50 m × 0.3 mm, dibutyl phthalate, 44°C	Tesarik[483]
Cl—C_3	Polyglycol ether–urethane	Nishimura[484]
Cl, Br, I—C_4	Stainless-steel, 4 m, squalane on Chromosorb, 40°C	Chaudri[485]
Br—C_5	12 ft × 4 mm, paraffin on Anakrom	Wesselman[486]
Br, I—C_5	Stainless-steel, 4 m, squalane on Chromosorb	Chaudri[487]
F—C_6	Copper, thiourea on Chromosorb	Mailen[488]
Cl—C_7	2 m × 38 mm, Bentone–Apiezon on Celite	Petranek[489]
Cl, Br—C_7, C_8, C_{10}	200 m × 0.5 mm, Ucon	Bendel[490]
Cl—C_8		Gates[491]
Chlorocarbons	100 ft × 0.01 in., squalene	Hollis[492]
Perfluorocarbons	6 ft × 4 mm, Carbowax on Carbopack	Shields[493]
F—cyclo C_6	Dibutyl maleate	Martin[494]
Cl—cyclo C_6	Glass, 6 ft × 0.25 in., Apiezon on Gas-Chrom, 160°C	Davis[495]
Cl—cyclo C_6	1 m, high-vacuum grease on Disphorite	Fürst[496]
Cl—cyclo C_6	3 m, polyoxypropylene glycol on glass beads, 170°C	Guillemin[497]
Cl—cyclo C_6	1 m, Apiezon–polyoxyethylene glycol on kieselguhr, 180°C	Esselborn[498]
Cl—benzenes	4 m, dinonyl phthalate on diatomite, 140°C	Sobolev[499]
Cl—benzenes	On series, 75 cm × 3 mm, polyoxyethylene glycol on Shimalite, 225 cm × 3 mm, Bentone on Celite, 140°C	Abe[500]
Cl—benzenes	Stainless-steel, 10 m × 0.25 mm, diethylene glycol succinate, 50 to 150°C	Hrivnak[501]
Br—benzenes	Stainless-steel, 2.5m × 4 mm, Bentone on Chromosorb, 190°C	Larsen[502]
Cl—toluenes	Glass, 1 m × 4.5 mm, paraffin on diatomite, 150°C	Waksmundzki[503]
Cl—toluenes	Silicone, 160°C	Haring[504]
Cl—styrenes	Bentone + bis(phenoxyphenyl)ether	Raley[505]
Cl—diphenyls	Glass, 60 m × 0.33 mm, SE-30, 180 to 260°C	Schulte[506]

stainless-steel column (25 cm × 3 mm) packed with LiChrosorb (5 μm), with dry hexane as the mobile phase and detection at 205 or 210 nm.

Thin-layer chromatography of chlorine-containing pesticides can be performed on Kieselgel (Lauckner and Fürst,[515] Diemair et al.[516]), silica gel (Abbott et al.[517]), or alumina (Kovacs[518]). Kawahara et al.[519] separated 14 chlorocarbons in waste water into three fractions on alumina thin layers and then analyzed each fraction by gas chromatography at 175 or 200°C on columns containing 5% of DC-200 on Chromosorb, with electron-capture or microcoulometric titrimetric detection. Dreshsler[520] separated chloromethyl compounds by paper chromatography; the paper was impregnated with 0.7% potassium formate solution and the mobile phase consisted of butanol–85% formic acid–water (5 :1 :5).

7.6.3 Neutral Organophosphorus Compounds

With a few exceptions, neutral organosphosphous compounds are esters. As shown in Table 7.13, thin-layer and paper chromatography are convenient methods for separating these compounds. When gas chromatography is employed, high-molecular-weight esters are preferably converted into the more volatile methyl esters (Hardy,[522] Nebbia and Bellotti[523]); this reaction can be easily carried out by means of diazomethane (Roper and Ma[543]). Quantitation after paper chromatography is best achieved (Moule and Greenfield[538]) by cutting out the spots and determining the phosphorus contents by digestion and spectrophotometry (Ma and Rittner[544]).

7.6.4 Organosilicon Compounds

Organosilicon compounds are known for their volatility; hence gas chromatography is generally used for their separation. Conversion of certain functional groups into volatile silyl derivatives is discussed previously in various sections. Table 7.14 gives the conditions recommended for separating some silicon compounds, and two examples of trimethylsilyl derivatives.

Liquid chromatography is applicable to the separation of certain types of organosilicon compounds. For instance, Nagy and Brandt-Petrik[559] performed adsorption analysis of alkyl and alkyl aryl siloxanes by placing the sample in methanolic solution in a column of activated carbon; the straight-chain compounds were eluted with methanol, followed by the cyclic species, which were eluted wish benzene.

Franc and Senkyrova[560] described thin-layer chromatography of organosilicon compounds containing phenyl groups; thin layers of silica gel bonded with starch were employed, with heptane–CCl_4–toluene (6 :1 :1) as developing solvent for compounds of molecular weight <1000, and CCl_4–toluene

Table 7.13 Separation of Neutral Organophosphorus Compounds

Method	Mixture	Conditions	Reference
Gas chromatography	Trialkyl phosphines	Reoplex, 206°C	Feinland[521]
	Mono-, dialkyl phosphates	Silicone on firebrick, 188°C	Hardy[522]
	Dialkyl phosphorodithioates	PTFE column, 115°C	Nebbia[523]
	10 pesticides	Glass, 160°C	McCaulley[524]
	20 insecticides	Cs thermionic detection	Krasue[525]
	Insecticides Dibrom, Dichlorvos	Coulometric detector	Boone[526]
Ion-exchange	Alkyl alkanephosphonates	Dowex cation exchanger, 50°C	Varon[527]
	Mono-, dialkyl phosphates	Dowex 50X4	Lew[528]
Thin-layer chromatography	Monoesters of H_3PO_4	Cellulose, C_4H_9OH–CH_3COOH–H_2O	Rabinowitz[529]
	Esters of H_3PO_4, H_3PO_3, H_3PO_2	Kieselgel, $(CH_3)_2CO$–H_2O–CH_3OH	Lamotte[530]
	Phosphites, phosphates, phosphonates	Silica gel, $CHCl_3$–$(CH_3)_2CO$	Neubert[531]
	Neutral acyl-, aryl phosphorus compounds	Silica gel, C_6H_{14}–$(CH_3)_2CO$	Lamotte[532]
	Carbohydrate phosphites, phosphonates	Alumina, CH_3NO_2–$HCON(CH_3)_2$	Nifantev[533]
	Alkyl phosphorothioates	Alumina, C_6H_{14}–$(CH_3)_2CO$	Petschik[534]
	10 pesticides	Silica gel	Salame[535]
	10 phosphorus esters	Silica gel	Giang[536]
Paper chromatography	Mono-, di-, tributyl phosphates	3 solvents successively	Cvjeticanin[537]
	Esters of H_3PO_4, H_3PO_3	Several solvents	Moule[538]
	Alkyl phosphates, phosphites	$HCOOH$–$C_5H_{11}OH$–C_3H_7OH–NH_4OH	Gabov[539]
	Pesticides	2 immobile phases	Cortes[540]
	Pesticides	Hexane–toluene	Batara[541]
Paper electrophoresis	Mono-, dialkyl phosphates	CH_3COOH–pyridine–H_2O, pH 4.4, 750 V	Luh[542]

Table 7.14 Separation of Some Organosilicon Compounds by Gas Chromatography

Silicon compounds	Column conditions	Reference
Ethoxypropoxysilanes	4 ft × 0.125 in., Triton on Chromosorb, 80 to 160°C	Taylor[545]
Methylchlorosilanes	2.5 m × 4 mm, 31 stationary phases, 25°C	Sivtsova[546]
Methylchlorosilanes	3 m each, tritolyl phosphate, dioctyl phosphate on kieselguhr, 58°C	Oiwa[547]
Methylchlorosilanes	4 m each, paraffin, H-132 on firebrick	Kawazumi[548]
Methylchlorosilanes	1.6 m, nitrobenzene on Celite	Garzo[549]
Ethylchlorosilanes	Silicone on unglazed tile, 90°C	Joklik[550]
Dimethylsiloxanes	Silicone on Rysorb, 150–195°C	Wurst[551]
Chlorosilanes	270–350 cm, benzylbenzoate, etc. on Celite, 30°C	Palamarchuk[552]
Methylcyclopolysiloxanes	Stainless-steel, polydimethylsiloxane on Chromosorb, 160–190°C	Preisler[553]
Vinylethoxysilanes	Silicone on Chromosorb	Wurst[554]
Chloromethylvinylsilanes	1.6 m × 5 mm, silicone on Celite, 40°C	Wurst[555]
Methylphenylpolysilanes	1 m, polymethylsiloxane on firebrick	Luskina[556]
Trimethylsilyl monosaccharides	1.7 m × 6 mm, poly(butane-1,4-diol succinate) on kieselguhr	Bilik[557]
Isotopic trimethylsilyl-glucose	Glass, 84 × 0.25 in., Apiezon on Gas-Chrom, 180°C	Waller[558]

(10 : 1) was used for compounds of molecular weight > 1000. Separation of trimethylsilyl derivatives on silica gel thin layers is exemplified by those of glycols (Leibman and Ortiz[561]), carbohydrates (Lehrfeld[562]), and terpenes (Lindgren and Srahn[563]). Franc and Ceeova[564] separated methylphenyl siloxanes by paper chromatography, with H_2O–dimethylformamide (1 : 8) as the mobile phase on Whatman No. 1 paper impregnated with silicone or Apiezon as the stationary phase.

7.6.5 Organometallic Compounds

Selected examples of mixtures of organometallic compounds that have been successfully separated by differential migration are shown in Table 7.15. When the organometallic compounds are prepared as chelates and found to be volatile, gas chromatography is generally employed since this technique can separate the organometallics from all metal ions that do not form

Table 7.15 Examples of Separation Methods for Organometallic Compounds

Separation method	Organometallics	Conditions	Reference
Gas chromatography	Alkyl lead	Stainless-steel, 10 ft × 0.125 in., tris(2-cyanoethoxy) propane on Chromosorb, 72°C	Bonnelli[565]
	Alkyl mercury	Silicone, 190–220°C	Broderson[566]
	Alkyl selenium	5 ft × 0.125 in., bis(3-phenoxyphenoxy)benzene on Chromosorb, 150°C	Evans[567]
	Trimethylgermanyl ethylenes	1 m × 4 mm, carbon black, 100–200°C	Vyazankin[568]
	Arene tricarbonyl chromium	Glass, 4 ft × 4 mm, SE on Gas-Chrom, 100 to 200°C	Veening[569]
	Hexafluoroacetyl-acetone chelates of aluminum, beryllium, chromium, copper, gallium, and rhodium	1.5 m × 4 mm, SE or QF on Chromosorb	Arakawa[570]
	Aluminum, boron, beryllium, and antimony compounds	1 m, paraffin–triphenylamine on Chromosorb, 73–165°C	Longi[571]
	Germanium, tin compounds	Apiezon, *o*-nitrotoluene	Garzo[571a]
Liquid chromatography	Tris(2,2-bipyridyl) Ru(II) compounds	Ion-paired with $CH_3SO_3^-$ or $C_7H_{15}SO_3^-$	Valenty[572]
Thin-layer chromatography	Alkyl and alkoxylmercury	Silica gel, C_4H_9OH–CH_3COOH–H_2O	Rissanen[573]
	Mercurials	Silica gel, cuclohexane–acetone	Johnson[574]
	Bicyclic mercury	Alumina, ligroin, CCl_4, etc.	Ptitsyna[575]
	Alkyl tin	Silica gel, several solvents	Figge[576]
	Trisglycinatocobalt	Cellulose, ethanol–H_2O	Jursik[577]
Paper chromatography	Alkyl lead		Barbierri[578]
	Titanium compounds	Decane, then heptane–acetone	Kozina[579]

complexes. As long as the organometallic compounds do not decompose upon volatilization, no special precaution is necessary for their gas-chromatographic analysis. Thus Veening et al.[569] separated and determined five arene tricarbonylchromium complexes using a glass column (4 ft × 4 mm) packed with 3.6% of SE-30 on Gas-Chrom Q (100 to 200 mesh). For flame ionization detection, nitrogen was the carrier gas (60 ml per min) and the temperature was programmed from 100 to 200°C at 7.5°C per min, while argon–methane (19 : 1) was used for electron-capture detection with an isothermal column temperature of 135 to 145°C; the precision of the method was ±3% in the concentration range of 0.1 to 1 mg per ml. Arakawa and Tanikawa[570] separated metal chelates of hexafluoroacetylacetone on a column (1.5 m × 4 mm) containing 10% of QF-1 on Chromosorb W (60 to 80 mesh) at 60°C for berylium, aluminum gellium, and chromium compounds, and 10% of SE-52 at 90°C for iron and copper compounds; symmetrical elution curves were obtained in all cases.

For organometallic nonvolatile compounds, thin-layer or paper chromatography is suitable. For instance, Barbieri et al.[578] performed quantitative analysis of mixtures of organolead compounds of the types R_3PbCl, R_2PbCl_2, and R_4Pb (R = alkyl or phenyl) by paper-chromatographic separation followed by conversion into $(PbI_4)^{2-}$ and spectrophotometric measurement at 357 nm. Kozina and Subbotina[579] separated organotitanium compounds of cyclopentadiene by dissolving the sample (up to 50 μg) in dichloroethane and developing successively with decane and heptane–acetone; the spots were detected with iodine vapor, cut out, extracted with acetone and then treated with $K_4Fe(CN)_6$; absorbance was measured at 533 nm.

REFERENCES

1. N. D. Cheronis and T. S. Ma, *Organic Functional Group Analysis*, Wiley, New York, 1964.
2. N. D. Cheronis and T. S. Ma, *Organic Functional Group Analysis*, Wiley, New York, p. 9.
3. M. Singliar and J. Dykyj, *Collect. Czech. Chem. Commun.*, **34**, 767 (1969).
4. N. N. Kosenko, G. F. Baranova and S. M. Turvich, *Zh. Prikl. Khim. Leningr.*, **41**, 324 (1968).
5. M. Wilcox, *J. Pharm. Sci.*, **56**, 642 (1967).
6. A. A. Andersons, A. Y. Karmilchik, M. V. Shimanskaya, and S. A. Giller, *Izv. Akad. Nauk Latv. SSR, Ser. Khim.*, **1963**, (2), 168.
7. N. I. Shuikin, B. L. Lebedev and V. V. An, *Izv. Akad. Nauk SSSR, Otd. Khim. Nauk*, **1962**, 1868.
8. T. A. Polyakova, T. S. Sokolova, and Y. A. Tsarfin, *Zavod. Lab.*, **29**, 664 (1963).

9. R. H. Nealey, *J. Chromatogr.*, **14**, 120 (1964).
10. R. Hänsel and H. Rimpler, *Z. Anal. Chem.*, **207**, 270 (1965).
11. R. Giebelmann, *Pharmazie*, **19**, 703 (1964).
12. M. M. Schachter and T. S. Ma, *Mikrochim. Acta*, **1966**, 55.
13. T. Mitsu and Y. Kitamura, *Microchem. J.*, **7**, 141 (1963).
13a. W. H. Gutenmann and D. J. Lisk, *J. Agric. Food Chem.*, **11**, 470 (1963).
14. P. W. Albro and J. C. Dittmer, *J. Chromatogr.*, **38**, 230 (1968).
15. R. Wood and F. Snyder, *Lipids*, **1**, 62 (1966).
16. A. Lucchesi and M. Giorgini, *Chim. Ind. (Milano)*, **51**, 289 (1969).
17. United Kingdom Atomic Energy Authority, Rep. PG 690(w), 1966.
18. L. V. Andreev, *Zavod. Lab.*, **32**, 1059 (1966).
19. E. McMeans and J. R. Chipault, *Anal. Chem.*, **37**, 151 (1965).
20. N. J. Berridge and J. D. Watts, *J. Sci. Food Agric.*, **5**, 417 (1954).
21. R. V. Golovnya and V. P. Uralets, *Dokl. Akad. Nauk SSSR*, **177**, 350 (1967).
22. R. V. Golovnya and V. P. Uralets, *Zh. Anal. Khim.*, **24**, 449 (1969).
23. R. Wood and R. D. Harlow, *J. Lipid Res.*, **10**, 463 (1969).
24. V. I. Yakerson, L. A. Gorskaya, and A. M. Rubinshtein, *Zh. Anal. Khim.*, **23**, 293 (1968).
25. S. P. Fore and H. P. Dupuy, *J. Gas Chromatogr.*, **6**, 522 (1968).
26. R. Komers and K. Kochloefl, *Collect. Czech. Chem. Commun.*, **29**, 1803 (1964).
27. A. N. Bagaev and Y. V. Vodzinski, *Tr. Khim.*, **1967**, 186; *Zh. Anal. Khim.*, **3**, 1553 (1968).
28. A. L. Prabucki and F. Lenz, *Helv. Chim. Acta*, **45**, 2012 (1962).
29. G. E. Martin, F. J. Feeny, and F. P. Scaringelli, *J. Assoc. Off. Agric. Chem.*, **47**, 561 (1964).
30. J. M. Pepper, M. Manolopoulo, and R. Burton, *Can. J. Chem.*, **40**, 1976 (1962).
31. M. M. Buzlanova and V. F. Stepanovskaya, *Zh. Anal. Khim.*, **20**, 859 (1965).
32. J. Franc, *Collect. Czech. Chem. Commun.*, **21**, 581 (1956).
33. E. Gudrinietse and D. Kreitsberga, *Izv. Akad. Nauk Latv. SSSR, Ser. Khim.*, 1963, 515.
34. J. H. Dhont and G. J. C. Dijkman, Report of the Central Institute for Nutrition and Food Research Ultrecht, 1965.
35. M. H. Klouwen, R. ter Heide, and J. G. J. Kok, *Fette, Seifen Anstrichm.*, **65**, 414 (1963).
36. P. Blanc, P. Bertrand, G. de Saqui-Sannes, and R. Lescure, *Chim. Anal. (Paris)*, **47**, 354 (1965).
36a. L. Durisnova and D. Bellus, *J. Chromatogr.*, **32**, 584 (1968).
37. K. E. Schulte and C. B. Storp, *Fette, Seifen, Anstrichm.*, **57**, 600 (1955).
38. J. Fitelson, *J. Assoc. Off. Agric. Chem.*, **47**, 1161 (1964).
39. E. Sawicki, T. W. Stanley, W. C. Elbert, and M. Morgan, *Talanta*, **12**, 605 (1965).

40. A. J. Fatiadi, *J. Chromatogr.*, **20**, 319 (1965).
41. K. Christofferson, *Svensk. Pappersidn.*, **70**, 540 (1967).
42. R. P. Sandman and O. N. Miller, *J. Biol. Chem.*, 230 (1958).
43. J. Sherma, D. A. Goldstein, and R. Lucek, *J. Chromatogr.*, **38**, 54 (1968).
44. L. M. Andronov and Y. D. Norikov., *Zh. Anal. Khim.*, **20**, 1007 (1965).
45. P. Ronkainen and S. Brummer, *J. Chromatogr.*, **45**, 341 (1969).
46. E. Fedeli and M. Cirimele, *J. Chromatogr.*, **15**, 435 (1964).
47. G. M. Nano and P. Sancin, *Experientia*, **19**, 323 (1963).
48. T. Takeuchi, Y. Suzuki, and Y. Yamazaki, *Jap. Analyst*, **18**, 948 (1969).
49. L. M. Libbey and E. A. Day, *J. Chromatogr.*, **14**, 273 (1964).
50. M. Severin, *Bull. Inst. Agron. Bembloux*, **32**, 122 (1964).
51. E. Bloem, *J. Chromatogr.*, **35**, 108 (1968).
52. W. Y. Cobb, *J. Chromatogr.*, **14**, 512 (1964).
53. R. Ellis, A. M. Gaddis, and G. T. Currie, *Anal. Chem.*, **30**, 475 (1958).
54. S. Maruta and Y. Suzuki, *J. Chem. Soc. Jap., Ind. Chem. Sect.*, **61**, 1147 (1958).
55. K. J. Monty, *Anal. Chem.*, **30**, 1350 (1958).
56. A. M. Parsons, *Analyst*, **91**, 297 (1966).
57. E. A. Corbin, *Anal. Chem.*, **34**, 1244 (1962).
58. J. Franc and G. Celikovska, *Collect. Czech. Chem. Commun.*, **26**, 667 (1961).
59. L. A. Jones and R. J. Monroe, *Anal. Chem.*, **37**, 935 (1965).
60. J. W. Ralls, *Anal. Chem.*, **36**, 946 (1964).
61. S. Nishi, *Jap. Analyst*, **11**, 415 (1962).
62. L. M. Kaliberdo and A. S. Vaabel, *Zh. Anal. Khim.*, **22**, 1590 (1967).
63. R. L. Vanetten and A. T. Bottini, *J. Chromatogr.*, **21**, 408 (1966).
64. C. R. Glowacki, P. J. Menardi, and W. E. Link, *J. Am. Oil Chem. Soc.*, **47**, 225 (1970).
65. J. A. Fioriti, M. J. Kanuk, and R. J. Sims, *J. Chromatogr., Sci.*, **7**, 448 (1969).
66. H. Puschmann, *Tenside*, **5**, 207 (1968).
67. R. Wickbold, *Fette, Seifen, Anstr.*, **70**, 688 (1968).
68. N. E. Skelly and W. B. Crummett, *J. Chromatogr.*, **21**, 257 (1966).
69. H. Hadorn and K. Zürcher, *Mitt. Geb. Lebensmittunters Hyg.*, **58**, 209 (1967).
70. E. R. Adlard, M. J. Smith, and B. T. Whitnam, *Journées Hellènes d'étude des Méthods de Separation Immédiate et de Chromatographie*, Athens, Greece, 1965, p. 125.
71. M. Jernejcic and L. Premru, *J. Oil Colour Chem. Assoc.*, **52**, 623 (1969).
72. S. Ito and K. Fukuzumi, *J. Chem. Soc. Jap., Ind. Chem. Sec.*, **65**, 1963 (1962).
73. F. Pupin and R. Vuillaume, *Ann. Falsif. Expert. Chim.*, **61**, 13 (1968).
74. L. D. Metcalfe, A. A. Schmitz, and J. R. Pelka, *Anal. Chem.*, **38**, 514 (1966).
75. H. M. Kellog, E. Brochmann-Hanssen, and A. B. Sevendsen, *J. Pharm. Sci.*, **53**, 420 (1964).

76. C. W. J. Chang and S. W. Pelletier, *Anal. Chem.*, **38**, 1247 (1966).
77. G. Brus, R. Bentejac, and F. Prevot, *Ann. Falsif. Expert. Chim.*, **61**, 233 (1968).
78. W. W. Christie, *J. Chromatogr.*, **37**, 27 (1968).
79. G. R. Allen and M. J. Saxby, *J. Chromatogr.*, **37**, 312 (1968).
80. B. Fell, E. Bendel, M. Lauscher, and H. Hübner, *J. Chromatogr.*, **24**, 161 (1966).
81. C. Litchfield, A. F. Isbell, and R. Reiser, *J. Am. Oil Chem. Soc.*, **39**, 330 (1962).
82. J. B. Tepe and H. J. Wesselman, *J. Am. Pharm. Assoc. Sci. Ed.*, **47**, 457 (1958).
83. V. Prochazkova, J. Benes, and K. Veres, *J. Chromatogr.*, **21**, 402 (1966).
84. S. Mori and T. Takeuchi, *J. Chromatogr.*, **46**, 137 (1970).
85. B. A. Rudenko and V. F. Kucherov, *Izv. Akad. Nauk SSSR, Otd. Khim. Nauk*, **1963**, 220.
86. A. N. Genkin and B. I. Boguslavskaya, *Tr. Komiss. Anal. Khim., Akad. Nauk SSSR*, **13**, 263 (1963).
87. E. M. Kazinik, V. V. Platonov, L. M. Lvovich, and V. N. Aleksandrov, *Tr. Vses. Nauchno-Issled. Proekt. Inst. Monomerov*, **1**, 123 (1969).
88. N. Ohashi and T. Hara, *J. Chem. Soc. Jap., Pure Chem. Sec.*, **88**, 195 (1967).
89. P. G. Olafsson and A. Gambino, *J. Chromatogr.*, **24**, 222 (1966).
90. M. Covello and O. Schettino, *Riv. Ital. Sostanze Grasse*, **41**, 75 (1964).
91. J. Simal-Lozano, A. Charro-Arias, and F. Gonzalez-Pena, *An. Bromatol.*, **18**, 401 (1966).
92. E. Vioque and R. T. Holman, *J. Am. Chem. Soc.*, **39**, 63 (1962).
93. V. V. S. Mani and G. Lakshminarayana, *J. Chromatogr.*, **39**, 182 (1969).
94. A. Dolev, W. K. Rohwedder, and H. J. Dutton, *Lipids*, **1**, 231 (1966).
95. J. Sherma, *Chemist-Analyst*, **52**, 114 (1963).
96. C. R. Scholfield, E. P. Jones, R. O. Butterfield, and H. J. Dutton, *Anal. Chem.*, **35**, 386 (1963).
97. J. K. Haken and I. D. Allen, *J. Chromatogr.*, **23**, 375 (1966); **51**, 415 (1970).
98. J. K. Haken and P. Souter, *J. Gas Chromatogr.*, **4**, 295 (1966).
99. R. W. Germaine and J. K. Haken, *J. Chromatogr.*, **43**, 33 (1969).
100. D. W. Connell, *J. Chromatogr.*, **14**, 104 (1964).
101. R. L. Stern, B. L. Karger, W. J. Keanne, and H. C. Rose, *J. Chromatogr.*, **39**, 17 (1969).
102. A. J. Sheppard, S. A. Meeks, and L. W. Elliot, *J. Gas Chromatogr.*, **6**, 28 (1968).
103. W. O. Caster, P. Ahn, and R. Pogue, *Chem. Phys. Lipids*, **1**, 393 (1967).
104. F. A. Vandenheuvel, *Anal. Chem.*, **35**, 1186 (1963).
105. M. A. Volodina and I. V. Konkova, *Vestn. Mosk. God. Univ., Ser. Khim.*, **1970**, 119.
106. R. E. Leitch, H. L. Rothbart, and W. Rieman, III, *Talanta*, **15**, 213 (1968).
107. H. D. Spitz, H. L. Rothbart, and W. Rieman, III, *J. Chromatogr.*, **29**, 94 (1967).
108. J. R. Watson and P. Crescuolo, *J. Chromatogr.*, **52**, 63 (1970).

109. F. K. Kawahara, *Anal. Chem.*, **40**, 2073 (1968).
110. K. T. Wang and I. S. Y. Wang, *Nature*, **210**, 1039 (1966).
111. D. Nurok, G. L. Taylor, and A. M. Stephen, *J. Chem. Soc.*, **B**, **1968**, 291.
112. M. R. Sahasrabudhe and J. J. Legari, *J. Am. Oil Chem. Soc.*, **45** (3), 148 (1968).
113. H. Wessels and N. S. Rajagopal, *Fette, Seifen, Anstrichm.*, **71**, 543 (1969).
114. G. Jurriens, B. de Vries, and L. Schouten, *J. Lipid Res.*, **5**, 267 (1964); *Riv. Ital. Sostanze Grasse*, **41**, 4 (1964).
115. J. Pokorny and O. Prochazkova, *Sb. Vys. Sk. Chem. Technol. Praze, Potravin. Technol.*, **8**, 93 (1964).
116. C. Litchfield, M. Farquhar, and R. Reiser, *J. Am. Oil Chem. Soc.*, **41**, 588 (1964).
117. J. Sliwick and Z. Kwapniewski, *Mikrochim. Acta*, **1965**, 1.
118. E. Distler and F. J. Baur, *J. Assoc. Off. Agric. Chem.*, **48**, 444 (1965).
119. B. C. Black and E. G. Hammond, *J. Am. Oil Chem. Soc.*, **40**, 575 (1963).
120. S. Smith, R. Watts, and R. Dils, *J. Lipid Res.*, **9**, 40, 52 (1968).
121. C. Dumazert, C. Ghiglione, and T. Pugnet, *Bull. Soc. Chim. Fr.*, **1963**, 475.
122. M. R. Sahasrabudhe and R. K. Chada, *J. Am. Oil Chem. Soc.*, **46**, 8 (1969).
123. T. L. Mounts and H. J. Dutton, *Anal. Chem.*, **37**, 641 (1965).
124. P. E. Brandt, N. Krog, J. B. Lauridsen, and O. Tolboe, *Acta Chem. Scand.*, **22**, 1691 (1968).
125. R. D. Wood, P. K. Raju, and R. Reiser, *J. Am. Oil Chem. Soc.*, **42**, 161 (1965).
126. R. Watts and R. Dils, *J. Lipid Res.*, **10**, 33 (1969).
127. B. Freedman, *J. Am. Oil Chem. Soc.*, **44**, 113 (1967).
128. A. Kuksis and B. A. Gordon, *Can J. Biochem. Physiol.*, **41**, 1355 (1963).
129. A. Kamibayashi, M. Miki, and H. Ono, *J. Agric. Chem. Soc. Jap.*, **35**, 968 (1961).
130. C. Bighi, A. Betti, F. Dondi, and R. Francesconi, *J. Chromatogr.*, **42**, 176 (1969).
131. C. Bluestein and H. N. Posmanter, *Anal. Chem.*, **38**, 1865 (1966).
132. V. A. Sokolov and L. P. Kolesnikova, *Neftekhimiya*, **1**, 564 (1961).
133. L. P. Kolesnikova, V. T. Gurevich, and L. L. Starobinets, *Neftekhimiya*, **4**, 340 (1964).
134. P. Urone and R. L. Pecsok, *Anal. Chem.*, **35**, (1963).
135. M. Rogozinski, L. M. Shorr, and A. Warshawsky, *J. Chromatogr.*, **8**, 429 (1962).
136. H. W. Doelle, *J. Chromatogr.*, **42**, 541 (1969).
137. H. G. Struppe, *Z. Chemie, Leipz.*, **6**, 272 (1966).
138. J. H. Kahn, F. M. Trent, P. A. Shipley, and R. A. Vordenberg, *J. Assoc. Off. Anal. Chem.*, **51**, 1330 (1968).
139. A. Maurel, *Bull. Soc. Chim. Fr.*, **1963**, 316.
140. A. G. Ionescu, L. Stanescu, V. Bozgan, S. Trestianu, and D. Sandulescu, *Rev. Chim. (Buchar.)*, **14**, 347 (1963).
141. J. W. Robinson, *Anal. Chim. Acta*, **27**, 377 (1962).
141a. S. Raccagni, *Riv. Ital. Sostanze Grasse*, **41**, 459 (1964).

142. R. Wood, *J. Gas Chromatogr.*, **6**, 94 (1968).
143. B. Drews, H. Specht, and G. Offer, *Z. Anal. Chem.*, **189**, 325 (1962).
144. U. P. Schlunegger, *J. Chromatogr.*, **24**, 165 (1966).
145. P. J. Porcaro and V. D. Johnston, *Anal. Chem.*, **33**, 361 (1961).
146. L. P. Kolesnikova, V. V. Kamzolkin, and M. I. Khotimskaya, *Neftekhimiya*, **2**, 355 (1962).
147. E. Bendel, G. Mahr, B. Fell, and M. F. El Daoushy, *J. Chromatogr.*, **16**, 216 (1964).
148. E. Gil-Av, R. Charles-Sigler, G. Fischer, and D. Nurok, *J. Gas Chromatogr.*, **4**, 51 (1966).
149. G. P. Cartoni, A. Liberti, and A. Pela, *Anal. Chem.*, **39**, 1618 (1967).
150. V. Arkima, *Mischr. Brau.*, **21** (2), 25 (1968).
151. A. A. Casselman and R. A. B. Bannard, *J. Chromatogr.*, **20**, 424 (1965).
152. R. Komers, K. Kochloefl, and V. Bazant, *Chem. Ind.* (*Lond.*), **1958**, 1405; *Collect. Czech. Commun.*, **28**, 46 (1963).
153. L. Fishbein and W. L. Zielinski, Jr., *J. Chromatogr.*, **28**, 418 (1967).
154. L. G. Gomes, *An. Inst. Vinho Porto*, **25**, 131 (1974).
155. H. Terada, S. Tsuda, and T. Shono, *J. Chem. Soc. Jap., Ind. Chem. Sect.*, **65**, 1569 (1962).
156. P. Capella, E. Fedeli, and M. Cirimele, *Chem. Ind. (Lond.)*, **1963**, 1590.
157. J. Pollerberg, *Fette, Seifen, Anstrichm.*, **68**, 561 (1966).
158. E. J. Singh and L. L. Gershbein, *J. Chromatogr.*, **23**, 180 (1966).
159. C. Bandyopadhyay and M. M. Chakrabarty, *J. Chromatogr.*, **32**, 297 (1968).
160. G. P. McSweeney, *J. Chromatogr.*, **17**, 183 (1965).
161. E. Stahl and H. Vollmann, *Talanta*, **12**, 525 (1965).
162. L. Syper, *Dissnes. Pharm. Warsz.*, **17**, 229 (1965).
163. B. Ruzicka, *Chem. Anal. (Warsaw)*, **10**, 1165 (1965).
164. D. Locke, J. Sherma, and D. E. Thompson, Jr., *Anal. Chim. Acta*, **5**, 312 (1961), **32**, 181 (1965).
165. A. I. Novoselov, A. M. Afanasev, E. V. Kalyazin, and V. E. Zakharov, *Zh. Anal. Chim.*, **25**, 386 (1970).
165a. H. G. Nadeau and D. M. Oaks, *Anal. Chem.*, **32**, 1760 (1960).
166. O. A. Maksimenko, L. A. Zyukova, E. V. Ignateva, and R. M. Fedorovich, *Zh. Anal. Khim.*, **28**, 1588 (1973).
167. J. F. Dorlini, *J. Chromatogr. Sci.*, **7**, 319 (1969).
168. S. Spencer and H. G. Nadeau, *Anal. Chem.*, **33**, 1626 (1961).
169. H. Kuschmann, *Fette, Seifen, Anstrichm.*, **65**, 1 (1963).
170. F. C. Gross, *J. Assoc. Off. Agric. Chem.*, **48**, 647 (1965).
171. V. N. Balakhontseva and R. M. Poltinina, *Zh. Anal. Khim.*, **19**, 757 (1964).
172. V. A. Vaver, A. N. Ushakov, and L. D. Bergelson, *Izv. Akad. Nauk SSSR, Ser. Khim.*, **1968**, 400.

173. K. Assmann, O. Serfas, and G. Geppert, *J. Chromatogr.*, **26**, 495 (1967).
174. L. Dooms, D. Declarck, and H. Verachtert, *J. Chromatogr.*, **42**, 349 (1969).
175. F. H. de la Court, N. J. P. Van Cassel, and J. A. M. Van der Valk, *Farbe Lack*, **75**, 218 (1969).
176. N. Sen, M. Keating, and C. B. Barrett, *J. Gas Chromatogr.*, **5**, 269 (1967).
177. K. Konishi and S. Yamaguchi, *Anal. Chem.*, **40**, 1720 (1968).
178. M. B. Naff, A. S. Naff, and J. A. Strite, *J. Chromatogr.*, **11**, 496 (1963).
179. G. Haeseler and K. Misselhorn, *Z. Lebensm. Forsch.*, **129**, 71 (1966).
180. V. Castagnola, *Boll. Chim. Farm.*, **102**, 784 (1963).
181. D. Waldi, *J. Chromatogr.*, **18**, 417 (1965).
182. H. Grasshof, *J. Chromatogr.*, **14**, 513 (1964).
183. I. G. Borisovich, L. N. Orobinskaya, and N. A. Vasyunina, *Izv. Akad. Nauk SSSR, Ser. Khim.*, **1970**, 2361.
184. E. V. Dyatlovitskaya, V. V. Voronkova, and L. D. Bergelson, *Dokl. Akad. Nauk SSSR*, **145**, 325 (1962).
185. V. de Simone and M. Vicedomini, *J. Chromatogr.*, **37**, 538 (1968).
186. F. Fischer and H. Koch, *J. Chromatogr.*, **16**, 246 (1964).
187. E. A. Adelberg, *Anal. Chem.*, **25**, 1553 (1953).
188. J. Zajic, *Sb. Vys. Sk. Chem.-Technol. Praze Oddil Fak. Potravin. Technol.*, **6**, 179 (1962).
189. R. Sargent and W. Rieman, III, *Anal. Chim. Acta*, **16**, 144 (1957).
190. N. S. Dabagov and A. A. Balandin, *Izv. Akad. Nauk SSSR, Ser. Khim.*, **1966**, 1308.
191. O. Samuelson and H. Strömberg, *Acad. Chem. Scand.*, **22**, 1252 (1968).
192. O. Samuelson and H. Strömberg, *J. Food Sci.*, **33**, 308 (1968).
193. H. Matsui, E. Paart, and O. Samuelson, *Chem. Scrip.*, **1**, 45 (1971).
194. S. A. Barker, M. J. How, P. V. Peplow, and P. J. Somers, *Anal. Biochem.*, **26**, 219 (1968).
195. T. S. Ma and V. Horak, *Microscale Manipulations in Chemistry*, Wiley, New York, 1976, p. 265.
196. V. G. Berezkin and V. S. Kruglikova, *Zh. Anal. Khim.*, **24**, 455 (1969).
197. R. Wood, W. J. Baumann, F. Snyder, and H. K. Mangold, *J. Lipid Res.*, **10**, 128 (1969).
198. B. Smith and L. Tullberg, *Acta Chem. Scand.*, **19**, 605 (1965).
199. W. Seidenstücker, *Z. Anal. Chem.*, **237**, 280 (1968).
200. A. Prevot and C. Barbati, *Rev. Fr. Corps Gras.*, **15**, 157 (1968).
201. B. Smith and O. Carlsson, *Acta Chem. Scand.*, **17**, 455 (1963).
202. M. Sato, H. Miyake, M. Mitooka, and K. Asano, *Bull. Chem. Soc. Jap.*, **38**, 884 (1965).
203. G. Jung, H. Pauschmann, W. Voelter, E. Breitmaier, and E. Bayer, *Chromatographia*, **3**, 26 (1970).

204. M. Sanz-Burata, S. Julia-Arechaga, and J. Irrure-Perez, *Afinidad*, **25**, 371 (1968).
205. R. Celades and C. Paquot, *Rev. Fr. Corps Gras*, **9**, 145 (1962).
206. F. W. Hefendehl, *Naturwissenschaften*, **51** (6), 138 (1964).
207. J. C. Wekell, C. R. Houle, and D. C. Malins, *J. Chromatogr.*, **14**, 529 (1964).
208. C. D. Hurd and H. T. Miles, *Anal. Chem.*, **36**, 1375 (1964).
209. J. O. Deferrari, R. M. de Lederkremer, B. Matsuhiro, and J. F. Sproviero, *J. Chromatogr.*, **9**, 283 (1962).
210. M. von Schantz, S. Juvonen, A. Oksanen, and I. Hakama, *J. Chromatogr.*, **38**, 364 (1968).
211. G. Lefebvre, J. Berthelin, M. Maugras, R. Gay, and E. Urion, *Bull. Soc. Chim. Fr.*, **1966**, 266.
212. J. Churacek, K. Komarek, V. Vanasek, and M. Jurecek, *Collect. Czech. Chem. Commun.*, **33**, 3876 (1968).
213. M. R. Tulus and A. Güran, *Arch. Pharm.*, **296**, 623 (1963).
214. V. G. Berezkin, A. E. Mysak, and L. S. Polak, *Zavod. Lab.*, **31**, 282 (1965).
215. I. Katz and M. Keeney, *Anal. Chem.*, **36**, 231 (1964).
216. W. L. Zielinski, Jr., and L. Fishbein, *Anal. Chem.*, **38**, 41 (1966).
217. S. W. Bukata, L. L. Zabrocki, and M. F. McLaughlin, *Anal. Chem.*, **35**, 885 (1963).
218. M. L. Vlodavets, K. A. Golbert, E. Y. Chirvinskaya, N. V. Perovskaya, and L. P. Ternovskaya, *Neftekhimiya*, **5**, 613 (1965).
219. R. V. Kucher, M. A. Kovbuz, M. E. Teodorovich, and R. N. Rud., *Zavod. Lab.*, **27**, 1331 (1961), **29**, 19 (1963).
220. P. F. Ivanenko and A. S. Volga, *Zavod. Lab.*, **30**, 797 (1964).
221. D. Bernhardt, *Z. Chem.*, **8**, 237 (1968).
222. P. Fijolka and R. Gnauck, *Plaste Kautsch.*, **13**, 343 (1966).
223. S. Hayano, T. Ota, and Y. Fukushima, *Jap. Analyst*, **15**, 365 (1966).
224. E. Knappe and D. Peteri, *Z. Anal. Chem.*, **190**, 386 (1962).
225. M. M. Buzlanova, V. F. Stepanovskaya, A. F. Nesterov, and V. L. Antonovskii, *Zh. Anal. Khim.*, **21**, 506 (1966).
226. K. Paluch, *Chem. Anal. (Warsaw)*, **9**, 1129 (1964).
227. T. Furuya, S. Shibata, and H. Iizuka, *J. Chromatogr.*, **21**, 116 (1966).
228. P. Dugan and D. Lundgren, *Anal. Biochem.*, **8**, 312 (1964).
229. M. T. Yanotovskii, E. I. Kozlov, E. A. Obolnikova, O. I. Volkova, and G. I. Samokhvalov, *Dokl. Akad. Nauk SSSR*, **179**, 733 (1968).
230. E. Nyström and J. Sjovall, *J. Chromatogr.*, **24**, 212 (1966).
231. G. Pettersson, *J. Chromatogr.*, **12**, 352 (1963).
232. M. Kotakemori and K. Okada, *Agric. Biol. Chem.*, **30**, 935 (1966).
233. M. Danilovic and O. Naumovic-Stevanoyic, *J. Chromatogr.*, **19**, 613 (1965).
234. W. Poethke and H. Behrendt, *Pharm. Zentralhalle*, **104**, 549 (1965).
235. R. P. Labadie, *Pharm. Weekbl. Ned.*, **104**, 257 (1969).

236. M. V. Nikolaeva, L. N. Vertyulina, N. I. Malyugina, and D. A. Vyakhirev, *Tr. Khim. Gorkii*, **1967**, 127.
237. I. M. Starshov and F. Z. Rayanov, *Tr. Kazan. Khim. Tekhnol. Inst.*, **1967**, 531.
238. W. Biernacki and T. Urbanski, *Bull. Acad. Pol. Sci.*, **10**, 601 (1962).
239. J. C. Courtier, L. Etienne, J. Tranchant, and S. Vertalier, *Bull. Soc. Chim. Fr.*, **1965**, 3181.
240. L. Etienne and J. Tranchant, *Mem. Poud.*, **46**, 167 (1966).
241. D. G. Gehring and J. E. Shirk, *Anal. Chem.*, **39**, 1315 (1967).
242. W. Kemula and D. Sybilska, *Chem. Anal. (Warsaw)*, **4**, 123 (1959); *Rocz. Chem.*, **38**, 861 (1964).
243. W. Kemula, K. Butkiewicz, D. Sybilska, A. Kwiecinska, and A. Kurjan, *Rocz. Chem.*, **39**, 73, 1101 (1965); *Chem. Anal.* (*Warsaw*), **12**, 869 (1967).
244. W. Kemula, D. Sybilska, and K. Duszcyk, *Microchem. J.*, **11**, 296 (1966).
245. I. Pejkovic-Tadic, M. Hranisavlkevic-Jakovkjevic, and M. Maric, *J. Chromatogr.*, **21**, 123 (1966).
246. T. B. Beider, T. P. Bochkareva, and E. S. Mikhaleva, *Zavod. Lab.*, **34**, 540 (1968).
247. E. Trachman, A. Fono, and T. S. Ma, *Mikrochim. Acta*, **1968**, 1185.
248. J. C. Hoffsommer and J. F. McCullough, *J. Chromatogr.*, **38**, 508 (1968).
249. B. Bortoletti and T. Perlotto, *Farmaco, Ed. Prat.*, **23**, 371 (1968).
250. D. M. Colman, *J. Chromatogr.*, **8**, 399 (1962); *Anal. Chem.*, **35***m* 652 (1963).
251. S. Kitahara, A. Ito, and H. Hiyama, *Sci. Ind., Jap.*, **37**, 31 (1963).
252. A. A. Cherkasskii and L. V. Merzloukhova, *Zavod. Lab.*, **28**, 1177 (1962).
253. C. G. Crawforth and D. J. Waddington, *J. Gas Chromatogr.*, **6**, 103 (1968).
254. R. M. Harrison and F. J. Stevenson, *J. Gas Chromatogr.*, **3**, 240 (1965).
255. G. Paraskevopoulos and R. J. Cvetanovic, *J. Chromatogr.*, **25**, 479 (1966).
256. V. G. Pimkin, G. I. Bass, and S. L. Dobychin, *Zavod. Lab.*, **34**, 802 (1968).
257. M. H. Litchfield, *Analyst*, **93**, 653 (1968).
258. E. Camera and D. Pravisani, *Anal. Chem.*, **36**, 2108 (1964).
259. K. R. K. Rao, A. K. Bhalla, and S. K. Sinha, *Curr. Sci.*, **33**, 12 (1964).
260. E. Fedeli, A. Lanzani, and A. F. Valentini, *Chim. Ind. (Milan)*, **47**, 989 (1965).
261. W. Freytag, *Fette, Seifen, Anstrichm.*, **65**, 603 (1963).
262. D. P. Schwartz, *J. Chromatogr.*, **9**, 187 (1962).
263. K. de Jong, K. Mostert, and D. Sloot, *Rec. Trav. Chim. Pays-Bas*, **82**, 837 (1963).
264. E. F. L. J. Anet, *J. Chromatogr.*, **9**, 291 (1962).
265. H. H. Stroh and W. Schüler, *Z. Chem.*, **4**, 188 (1964).
266. H. J. Haas and A. Seeliger, *J. Chromatogr.*, **13**, 573 (1964).
267. F. C. Hunt, *J. Chromatogr.*, **35**, 111 (1968).
268. H. Rasmussen, *J. Chromatogr.*, **27**, 142 (1967).
269. G. E. Secor and L. M. White, *J. Chromatogr.*, **15**, 111 (1964).
270. B. J. Camp and F. O'Brien, *J. Chromatogr.*, **20**, 178 (1965).

271. D. W. Russell, *J. Chromatogr.*, **19**, 199 (1965).
272. R. J. Gritter, *J. Chromatogr.*, **20**, 416 (1965).
273. D. W. Bowins, R. C. DeGeiso, and L. G. Donaruma, *Anal. Chem.*, **34**, 1321 (1962).
274. I. Pejkovic-Tadic, M. Hranisavljevic-Jakovljevic, and S. Nesic, *J. Chromatogr.*, **21**, 239 (1966).
275. M. Hranisavklevic-Jakovljevic, I. Pejkovic-Tadic, and A. Stojijkovic, *J. Chromatogr.*, **12**, 70 (1963).
276. J. Toul and A. Okac, *Chem. Listy*, **59**, 1468 (1965).
277. A. I. Parimskii and I. K. Shelomov, *Maslob-Zhir. Prom.*, **1964**, (6), 28.
278. Y. Arad-Talmi, M. Levy, and D. Vofsi, *J. Chromatogr.*, **10**, 417 (1963).
279. G. A. Dardenne, M. Severin, and M. Marlier, *J. Chromatogr.*, **47**, 176 (1970).
280. W. Kiessling and K. K. Moll, *J. Prakt. Chem.*, **311**, 876 (1969).
281. A. Di Lorenzo and G. Russo, *Chim. Ind. (Milan)*, **51**, 170 (1969).
282. A. G. Kelso and A. B. Lacey, *J. Chromatogr.*, **18**, 156 (1965).
283. P. Lafargue, P. Pont, and J. Meunier, *Ann. Pharm. Fr.*, **28**, 477 (1970).
284. L. R. Snyder and B. E. Buell, *Anal. Chem.*, **36**, 767 (1964); *Anal. Chim. Acta*, **33**, 285 (1965).
285. J. H. Tyrer, M. J. Eadie, and W. D. Hooper, *J. Chromatogr.*, **39**, 312 (1969).
286. E. Seidler and H. Thieme, *Z. Med. Labortech.*, **9**, 252 (1968).
287. T. S. Ma and A. S. Ladas, *Organic Functional Group Analysis by Gas Chromatography*, Academic, London, 1976, p. 100.
288. J. F. Carson and F. F. Wong, *J. Organ. Chem.*, **22**, 1725 (1957).
289. Y. Obara, Y. Ishikawa, and C. Nishi, *Agric. Biol. Chem.*, **30**, 164 (1966).
290. J. Petranek, *J. Chromatogr.*, **5**, 254 (1961).
291. J. Franc, J. Dvoracek, and V. Kolouskova, *Mikrochim. Acta*, **1965**, 4.
292. P. W. Albro and L. Fishbein, *J. Chromatogr.*, **46**, 202 (1970).
293. F. G. Stanford, *Analyst*, **92**, 64 (1967).
294. W. L. Orr, *Anal. Chem.*, **38**, 1558 (1966).
295. L. Kronrad and K. Panek, *Z. Anal. Chem.*, **191**, 199 (1962).
296. V. E. Cates and C. E. Meloan, *Anal. Chem.*, **35**, 658 (1963).
297. V. E. Cates and C. E. Meloan, *J. Chromatogr.*, **11**, 472 (1963).
298. E. N. Karaulova, T. S. Bobruiskaya, and G. D. Galpern, *Zh. Anal. Khim.*, **21**, 893 (1966).
299. M. Imaida, T. Sumimoto, M. Yada, M. Yoshida, K. Koyama, and N. Kunita, *J. Food Hyg. Soc. Jap.*, **16**, 218 (1975).
300. E. A. Setzkorn and A. B. Carel, *J. Am. Oil Chem. Soc.*, **40**, (2) 57 (1963).
301. S. Nishi, *Jap. Analyst*, **14**, 912 (1965).
302. J. Pollerberg, *Fette, Seifen, Anstrichm.*, **67**, 927 (1965).
303. D. Barnard, M. B. Evans, G. M. Higgins, and J. F. Smith, *Chem. Ind. (Lond.)*, **1961**, 20.

304. L. Fishbein, *J. Chromatogr.*, **32**, 596 (1968).
305. J. Hrivnak and J. Kalamar, *Chem. Zvesti*, **20**, 462 (1966).
306. E. Komanova and K. Antos, *Chem. Zvesti*, **20**, 85 (1966).
307. E. Campaigne and M. Georgiadis, *J. Organ. Chem.*, **28**, 1044 (1963).
308. J. Hrivnak and V. Konecny, *Collect. Czech. Chem. Commun.*, **32**, 4139 (1967).
309. R. L. Martin and J. A. Grant, *Anal. Chem.*, **37**, 649 (1965).
310. E. Profft and G. Solf, *Annalen*, **649**, 100 (1961).
311. S. Berk, *J. Chromatogr.*, **15**, 540 (1964).
312. R. Staszewski, J. Janak, and T. Wojdala, *J. Chromatogr.*, **36**, 429 (1968).
313. R. Mayer, P. Rosmus, and J. Fabian, *J. Chromatogr.*, **15**, 153 (1964).
314. R. F. Curtis and G. T. Phillips, *J. Chromatogr.*, **9**, 366 (1962).
315. K. A. Golbert and A. V. Alekseeva, *Zavod. Lab.*, **24**, 688 (1958).
316. M. Jarzynska and T. Mirecka, *Chem. Anal. (Warsaw)*, **8**, 91 (1963).
317. S. Rennhak, C. E. Döring, G. Schmid, D. Schneller, H. Stürtz, and E. Werner, *Chem. Tech. (Berl.)*, **17**, 688 (1965).
318. Institute of Petroleum, *J. Inst. Petroleum*, **55**, 48 (1969).
319. D. M. Ottners, G. R. Say, and H. F. Rase, *Anal. Chem.*, **38**, 148 (1966).
320. H. S. Knight, *Anal. Chem.*, **30**, 9 (1958).
321. R. W. Dunning and J. A. Leonard, *Chromatographia*, **2**, 293 (1969).
322. M. Kraitr, R. Komers, and F. Cuta, *Collect. Czech. Chem. Commun.*, **39**, 1440 (1974).
323. E. Bendel and M. Kern, *Angew. Chem., Int. Ed.*, **1**, 599 (1962).
324. M. S. Vigdergauz and M. I. Afanasev, *Neftekhimiya*, **3**, 425 (1963).
325. S. Miwa and A. Tatemastsu, *Jap. Analyst*, **17**, 816 (1968).
326. R. W. Poe and E. F. Kaelble, *J. Am. Oil Chem. Soc.*, **40**, 347 (1963).
327. L. R. Chapman and D. F. Kuemmel, *Anal. Chem.*, **37**, 1598 (1965).
328. E. K. C. Lee and F. S. Rowland, *Anal. Chem.*, **36**, 2181 (1964).
329. C. E. Higgins and W. H. Baldwin, *Anal. Chem.*, **36**, 473 (1964).
330. N. A. Gershtein, *Trudy Komiss. Anal. Khim., Akad. Nauk SSSR*, **13**, 231 (1963).
331. T. Nambara, T. Iwata, and S. Honma, *J. Chromatogr.*, **50**, 400 (1970).
332. M. T. Yanotovskii and B. A. Rudenko, *Zh. Anal. Khom.*, **20**, 848 (1965).
333. J. E. Fauvet, A. Pazdzerski, and B. Blouri, *Bull. Soc. Chim. Fr.*, **1967**, 4732.
334. R. W. Hemingway, W. E. Hillis and K. Bruerton, *J. Chromatogr.*, **50**, 391 (1970).
335. R. H. Wiley and R. M. Dyer, *J. Polym. Sci., A*, **2**, 3153 (1964).
336. E. Blasius and H. Lohde, *Talanta*, **13**, 701 (1966).
336a. J. W. Powell and M. C. Whiting, *Tetrahedron*, **12**, 163 (1961).
337. R. M. Sorochkina and A. Y. Lazaris, *Zh. Anal. Khim.*, **21**, 248 (1966).
338. N. D. Demina, Z. S. Smolyan, and I. V. Bodrikov, *Zavod. Lab.*, **32**, 929 (1966).
339. F. E. Herkes and D. J. Burton, *J. Chromatogr.*, **28**, 396 (1967).

340. R. L. Johnson and D. J. Burton, *J. Chromatogr.*, **20**, 138 (1965).
341. D. V. Banthorpe, C. Gatford, and B. R. Hollebone, *J. Gas Chromatogr.*, **6**, 61 (1968).
342. W. Szczepaniak and J. Nawrocki, *Chem. Anal. (Warsaw)*, **20**, 3 (1975).
343. G. Eppert, H. Prinzler, and K. Deutrich, *Z. Chem.*, **14**, 318 (1974).
344. L. M. McDonough and D. A. George, *J. Chromatogr. Sci.*, **8**, 158 (1970).
345. J. Haslam and A. R. Jeffs, *Analyst*, **87**, 658 (1962).
346. R. A. Bernhard, *Anal. Chem.*, **34**, 1576 (1962).
347. K. H. Miltenberger and G. Keicher, *Farbe Lack*, **69**, 677 (1963).
348. V. Lukes and R. Komers, *Collect. Czech. Chem. Commun.*, **29**, 1598 (1964).
349. L. S. Wang, A. Bekker, C. C. Yang, and A. N. Nesmeyanov, *Izv. Vyssh. Uchebn. Zaved. Khim.*, **6**, 597 (1963).
350. R. A. Appleton and A. McCormick, *Tetrahedron*, **24**, 633 (1968).
351. R. I. Sidorov, and G. A. Rudakov, *Gidroliz. Lesokhim. Prom.*, **1976**, (5), 20.
352. B. M. Lawrence, *J. Chromatogr.*, **38**, 535 (1968).
353. M. V. Schantz, S. Juvonen, and R. Hemming, *J. Chromatogr.*, **20**, 618 (1965).
354. J. A. Attaway, L. J. Barabas, and R. W. Wolford, *Anal. Chem.*, **37**, 1289 (1965).
355. R. Ikan, J. Kashman, and E. D. Bergmann, *J. Chromatogr.*, **14**, 275 (1964).
356. G. Valkanas and N. Iconomou-Petrovitch, *J. Chromatogr.*, **1** , 536 (1963).
357. M. Miocque and J. Blanc-Guéné, *Ann. Pharm. Fr.*, **24**, 377 (1966).
358. B. A. Rudenko, M. V. Mavrov, A. R. Derzhinskii, and V. F. Kucherov, *Izv. Akad. Nauk SSSR, Ser. Khim.*, **1968**, 1174.
359. E. Bendel, B. Fell, W. Gartzen, and G. Kruse, *J. Chromatogr.*, **31**, 531 (1967).
360. M. S. Vigdergauz and L. V. Andreev., *Zavod. Lab.*, **31**, 550 (1965).
361. A. Skarvada and P. Zbytek, *Ropa Uhlie*, **10**, 78 (1968).
362. S. A. Melkonyan and S. A. Vartanyan, *Zh. Anal. Khim.*, **22**, 930 (1967).
363. L. Nebbia and B. Pagani, *Chim. Ind. (Milan)*, **37**, 200 (1955).
364. M. W. Roomi, M. R. Subbaram, and K. T. Achaya, *J. Chromatogr.*, **16**, 106 (1964).
365. B. G. Belenkii and V. A. Orestova, *Zavod. Lab.*, **31**, 1328 (1965).
366. E. H. Ramshaw, *J. Chromatogr.*, **10**, 303 (1963).
367. M. T. Yanotovskii, *Zh. Anal. Khim.*, **24**, 1247 (1969).
368. D. Barnes, P. C. Uden, and P. Zuman, *Anal. Lett.*, **3**, 633 (1970).
369. E. Tyihak, D. Vagujalvi, and P. L. Hagony, *J. Chromatogr.*, **11**, 45 (1963).
370. C. V. Viswanathan, F. Phillips, and V. Mahadevan, *J. Chromatogr.*, **30**, 405 (1967).
371. Y. Kikuchi, T. Kikikawa, and R. Kato, *J. Gas Chromatogr.*, **5**, 261 (1967).
372. M. H. Hashmi and M. A. Shahid, *Mikrochim. Acta*, **1968**, 1045.
373. A. A. Akhrem, A. I. Kuznetsova, Y. A. Titov, and I. S. Levina, *Izv. Akad. Nauk SSSR, Otd., Khim.*, **1962**, 657.
374. T. McCullough, *J. Chromatogr.*, **44**, 188 (1969).

375. S. M. Zhenodarova, M. N. Adamova, and L. A. Lyapota, *Zh. Anal. Khim.*, **18**, 285 (1963).
376. A. Matsunaga and M. Yagi, *Anal. Chem.*, **50**, 753 (1978).
377. J. R. Marquart, G. B. Dellow, and E. R. Freitas, *Anal. Chem.*, **40**, 1968.
378. G. J. Griesmer and K. Kiyonaga, British Patent 954,851 (1960).
379. G. Eppert, British Patent, 1,123,718 (1966).
379a. Chevron Research Co., British Patent, 1,074,481 (1964).
380. P. A. Schenck and E. Eisma, *Nature*, **199**, 170 (1963).
381. K. Niedzielska and A. Slowik, *Chem. Anal. (Warsaw)*, **20**, 1163 (1975).
381a. L. R. Snyder, *Anal. Chem.*, **36**, 774 (1964).
382. T. A. Gough and E. A. Walker, *J. Chromatogr. Sci.*, **8**, 134 (1970).
383. M. H. Studier and R. Hayatsu, *Anal. Chem.*, **40**, 1011 (1968).
384. N. M. Turkeltaub, *Trudy Komiss, Anal Khim., Akad. Nauk SSSR*, **13**, 225 (1963).
385. A. B. Richmond, *J. Chromatogr. Sci.*, **7**, 321 (1969).
386. M. Papic, *J. Gas Chromatogr.*, **6**, 493 (1968).
387. H. Miyake and M. Mitooka, *J. Chem. Soc. Jap., Pure Chem. Sect.*, **84**, 593 (1963).
388. N. G. McTaggert, C. A. Miller, and B. Pearce, *J. Inst. Petroleum*, **54**, 265 (1968).
389. R. D. Beckham and R. Libers, *J. Gas Chromatogr.*, **6**, 188 (1968).
390. L. Sojuk, M. Gregorik, and J. Kurcova, *Ropa Uhlie*, **5**, 289 (1963).
391. J. Gawlowski, J. Niedzielski, and A. Bierzynski, *Chem. Anal. (Warsaw)*, **15**, 721 (1970).
392. S. M. Csicsery and H. Pines, *J. Chromatogr.*, **9**, 34 (1962).
393. F. Zocchi, *J. Gas Chromatogr.*, **6**, 100 (1968).
394. R. E. Leveque, *Anal. Chem.*, **39**, 1811 (1967).
395. W. N. Sanders and J. B. Maynard, *Anal. Chem.*, **40**, 527 (1968).
396. B. D. Blaustein, C. Zahn, and G. Pantages, *J. Chromatogr.*, **12**, 104 (1963).
397. D. K. Albert, *Anal. Chem.*, **35**, 1918 (1963).
398. A. Waksmundzki, Z. Suprynowicz, and T. Pietrusinska, *Chem. Anal. (Warsaw)*, **10**, 377 (1965).
399. R. J. Philippe, R. G. Honeycutt, and J. M. Ruth, *J. Chromatogr.*, **20**, 250 (1965).
400. V. G. Berezkin and L. S. Polak, *Tr. Komiss. Anal. Khim., Akad. Nauk SSSR*, **13**, 205 (1963).
401. E. M. Barrall, Jr., and F. Baumann, *J. Gas Chromatogr.*, **2**, 256 (1964).
402. L. Sojak and B. Bucinska, *Ropa Uhlie*, **10**, 572 (1968).
403. B. L. Karger and A. Hartkopf, *Anal. Chem.*, **40**, 215 (1968).
404. B. L. Karger, A. Hartkopf, and H. Posmanter, *J. Chromatogr. Sci.*, **7**, 315 (1969).
405. F. Wolf, A. Losse, and K. Franke, *J. Chromatogr.*, **30**, 378 (1967).
406. G. Hildebrand, C. Peper, and B. Dahlke, *Chem. Tech. (Leipz.)*, **15**, 147 (1963).
407. C. Karr, Jr. and J. R. Comberiati, *J. Chromatogr.*, **18**, 394 (1965).
408. R. L. Johnston and L. A. Jones, *Anal. Chem.*, **40**, 1728 (1968).

409. J. G. O'Connor, F. H. Burow, and M. S. Norris, *Anal. Chem.*, **34**, 82 (1962).
410. W. A. Dietz, P. K. Starnes, and R. A. Brown, *Tappi Spec. Tech. Assoc. Publ.* **2**, 33 (1963).
411. F. Bruner, G. P. Cartoni, and M. Possanzini, *Anal. Chem.*, **41**, 1122 (1969).
412. J. J. Czubryt and H. D. Gesser, *J. Chromatogr.*, **6**, 41 (1968).
413. W. A. Van Hook and M. E. Kelly, *Anal. Chem.*, **37**, 508 (1965).
414. D. E. Willis and R. M. Engelbrecht, *J. Gas Chromatogr.*, **5**, 536 (1967).
415. J. V. Brunnock and L. A. Luke, *Anal. Chem.*, **40**, 2158 (1968); *J. Chromatogr.*, **39**, 502 (1969).
416. P. Alessi and I. Kikic, *Gass. Chim. Ital.*, **103**, 445 (1973).
417. Vapor Fractometer, demonstrated in British Chemical Exposition, Earls Court, London, 1955.
418. Institute of Petroleum, *J. Inst. Petroleum*, **55**, 51 (1969).
419. J. Q. Walker and D. L. Ahlberg, *Anal. Chem.*, **35**, 2022 (1963).
420. D. A. Vyakhirev and M. I. Ostasheva, *Tr. Khim. Khim. Tekhnol.*, **1959**, 128.
421. M. S. Vigdergauz and K. A. Golbert, *Tr. Komiss. Anal. Khim. Akad. Nauk SSSR*, **13**, 257 (1963).
422. E. R. Fett, D. J. Christoffersen, and L. R. Snyder, *J. Gas Chromatogr.*, **6**, 572 (1968).
422a. S. F. Gudkov and N. A. Teterina, *Tr. Vses. Nauchno-Issled. Inst. Prir. Gazov*, **1969**, 40.
423. United Kingdom Atomic Energy Authority Report PG 695(W), 1 (1966).
424. S. V. Vitt, V. B. Bondarec, V. L. Polinin, and M. I. Rosengart, *Izv. Akad. Nauk SSSR, Ser. Khim.*, **1963**, 2043.
425. K. Nesvadba, J. Matena, L. Odstrcil, and M. Slavik, *Chem. Prum.*, **16**, 392 (1966).
426. D. H. Desty, A. Goldup, and W. T. Swanton, *Nature*, **183**, 107 (1959).
427. L. C. Case, *J. Chromatogr.*, **6**, 381 (1961).
428. E. A. Kruglov, K. M. Vaisberg, and Z. I. Abramovich, *Tr. Nauchno-Issled. Inst. Neftekhim. Proizvod.*, **1969**, 186.
429. P. L. Gupta and P. Kumar, *Anal. Chem.*, **40**, 992 (1968).
430. S. F. Spencer, *Anal. Chem.*, **35**, 592 (1963).
431. V. M. Nabivach, *Neftekhimiya*, **1962**, 906; *Neftepererabotka Nauch-Tekh. Sb.*, **1963** (12), 25.
432. P. Strnad, *Collect. Czech. Chem. Commun.*, **30**, 2132 (1965).
433. N. Paterok and H. Pilarczyk, *Chem. Anal. (Warsaw)*, **10**, 1227 (1965).
434. R. Popescu and I. Blidisel, *Petrol Gaze*, **17**, 433 (1966).
435. W. J. Carnes, *Anal. Chem.*, **36**, 1197 (1964).
436. C. F. Spencer and J. F. Johnson, *J. Chromatogr.*, **4**, 244 (1960).
437. M. L. Vlodavets, K. A. Golbert, F. A. Tokareva, E. N. Skuryat, and A. A. Rodionov, *Neftekhimiya*, **5**, 445 (1965).

438. K. Malinowska, *Chem. Anal. (Warsaw)*, **9**, 353 (1964).
439. C. A. Blake, Jr., *Anal. Chem.*, **35**, 1759 (1963).
440. B. V. Ioffe and B. V. Stolyzrov, *Neftekhimiya*, **2**, 911 (1962).
441. D. N. Rihani and G. F. Froment, *J. Chromatogr.*, **18**, 150 (1965).
442. W. J. Hendriks, R. M. Soemantri, and H. I. Waterman, *J. Inst. Petrol.*, **43**, 288 (1957).
443. F. Brunner and G. P. Cartoni, *J. Chromatogr.*, **10**, 396 (1963).
444. F. J. Kobot and L. S. Ettre, *Anal. Chem.*, **36**, 250 (1964).
445. A. Ohta, Y. Ogihara, K. Nei, N. Ikekawa, and S. Shibata, *Chem. Pharm. Bull. Jap.*, **11**, 1078 (1963).
446. C. Beschea and R. Popescu, *Revta Chim.*, **19**, 121 (1968).
447. W. Kemula, B. Behr, K. Chlebicka, and D. Sybilska, *Rocz. Chem.*, **39**, 1315 (1965).
448. P. Szynagel and M. Jedrychowska, *Chem. Anal. (Warsaw)*, **12**, 773 (1967).
449. I. Ognyanov, *C. R. Acad. Bulg. Sci.*, **16**, 265 (1963).
450. F. Onuska, J. Janak, K. Tesarik, and A. V. Kiselev, *J. Chromatogr.*, **34**, 81 (1968).
451. P. S. Protsidim and K. P. Lavrovskii, *Neftekhimiyz*, **8**, 302 (1968).
452. N. Basili, C. Bordonali, and C. Patimo, *Ann. Chim.*, **54**, 1081 (1964).
453. L. Gasco-Sanchez and F. Burriel-Marti, *Anal. Chim. Acta*, **36**, 460 (1966); *An. R. Soc. Esp. Fis. Quim.*, **B, 61**, 771 (1965).
454. M. Perea-Garcia, *An. R. Soc. Esp. Fis. Quim.*, **B**, **61**, 999 (1965).
455. W. W. West, U.S. Atomic Energy Comm. Report AECY-4699 (1960).
455a. M. Hellmann, R. L. Alexander, Jr., and C. F. Coyle, *Anal. Chem.*, **30**, 1206 (1958).
456. F. Onuska and J. Janak, *J. Chromatogr.*, **32**, 403 (1968).
457. F. J. Ritter, G. M. Meyer, and F. Geiss, *J. Chromatogr.*, **19**, 304 (1965).
458. L. I. Rostovtseva and Y. L. Fish, *Neftekhimiyz*, **5**, 275 (1965).
459. G. Biernoth, *J. Chromatogr.*, **36**, 325 (1968).
460. L. V. S. Hood and J. D. Winefordner, *Anal. Chim. Acta*, **42**, 199 (1968).
461. R. Ikan, I. Kirson, and E. D. Bergmann, *J. Chromatogr.*, **18**, 526 (1965).
462. T. Wieland, G. Lüben, and H. Determann, *Experientia*, **18**, 432 (1962).
463. S. Kelenffy, *Egeszsegtudomany*, **12**, 125 (1968).
464. R. G. Harvey and M. Halonen, *J. Chromatogr.*, **25**, 294 (1966).
465. J. W. Howard and E. O. Haenni, *J. Assoc. Off. Agric. Chem.*, **46**, 933 (1963).
466. W. Lijinsky, I. Domsky, G. Mason, H. Y. Ramahi, and T. Savavi, *Anal. Chem.*, **35**, 952 (1963).
467. R. A. Konyashina, T. S. Nikiforova, and V. P. Pakhomov, *Khim. Tverd. Topl.*, **1967**, (3), 60.
468. E. J. Pinchin and E. Pritchard, *Chem. Ind. (Lond.)*, **1962**, 1753.
469. R. Thomas and M. Zander, *Z. Anal. Chem.*, **282**, 443 (1976).
470. R. E. Jentoft and T. H. Gouw, *Anal. Chem.*, **40**, 1787 (1968).

471. B. T. Commins, *Analyst*, **83**, 386 (1958).
472. R. Tye and Z. Bell, *Anal. Chem.*, **36**, 1612 (1964).
473. N. P. Buu-Hoi and P. Jacquignon, *Experientia*, **13**, 375 (1057).
474. D. A. Vyakhirev and L. E. Reshetnikova, *Zh. Prikl. Khim.*, **31**, 802 (1958).
475. A. Foris and J. G. Lehman, *Sep. Sci.*, **4**, 225 (1969).
476. I. A. German and N. Ciot, *Revta Chim.*, **16**, 382 (1965).
477. H. Shuman, *J. Assoc. Off. Agric. Chem.*, **47**, 607 (1964).
478. J. Takacs, J. Inczedy, and L. Erdey, *J. Chromatogr.*, **9**, 247 (1962).
479. G. D. Tantsyrev and S. T. Kozlov, *Zh. Anal. Khim.*, **23**, 1881 (1968).
480. S. A. Greene and F. M. Wachi, *Anal. Chem.*, **35**, 928 (1963).
481. G. Tischer, *Chem. Tech. (Berl.)*, **19**, 691 (1967).
482. L. A. Balandina and A. I. Subbotina, *Zavod. Lab.*, **34**, 154 (1968).
483. K. Tesarik, A. Post, and O. Paleta, *Collect. Czech. Chem. Commun.*, **33**, 596 (1968).
484. S. Nishimura and I. Ichizuka, *J. Chem. Soc. Jap., Ind. Chem. Sect.*, **64**, 780 (1961).
485. B. A. Chaudri and H. R. Hudson, *J. Chromatogr.*, **27**, 240 (1967).
486. H. J. Wesselman and G. W. Mills, *J. Gas Chromatogr.*, **2**, 344 (1964).
487. B. A. Chaudri, H. R. Hudson, and W. S. Murphy, *J. Chromatogr.*, **29**, 218 (1967).
488. J. C. Mailen, T. M. Reed, III, and J. A. Young, *Anal. Chem.*, **36**, 1883 (1964).
489. J. Petranek, M. Kolimsky, and D. Lim, *Nature*, **207**, 1290 (1965).
490. E. Bendel, W. Meltzow, and L. H. Kung, *Angew. Chem., Int. Ed.*, **3**, 750 (1964).
491. P. N. Gates, W. Gerrard, and E. F. Mooney, *J. Chem. Soc.*, **1964**, 3480.
492. O. L. Hollis and W. V. Hayes, *Anal. Chem.*, **34**, 1223 (1962).
493. R. R. Shields and J. A. Nieman, *Anal. Chem.*, **50**, 661 (1978).
494. H. Martin and K. Abraham, *Z. Anal. Chem.*, **197**, 221 (1963).
495. A. Davis and H. M. Joseph, *Anal. Chem.*, **39**, 1016 (1967).
496. H. Fürst, H. Köhler, and J. Lauckner, *Chem. Tech. (Berl.)*, **16**, 105 (1964).
497. C. L. Guillemin, *Anal. Chim. Acta*, **27**, 213 (1962).
498. W. Esselborn and K. G. Krebs, *Pharm. Ztg.*, **107**, 464 (1962).
499. A. S. Sobolev and I. N. Kaluhskaya, *Dokl. Neftekhim. Sekts. Bashk. Resp. Pravl. Vse. Khim. Obshch. Mendeleeva*, **1967** (3), 91.
500. S. Abe, A. Hongo, and E. Shirakawa, *Jap. Analyst*, **16**, 399 (1967).
501. J. Hrivnak and M. Michalek, *Chromatographis*, **3**, 123 (1970).
502. E. Larsen, K. E. Siekierska, and J. Fenger, *J. Gas Chromatogr.*, **6**, 171 (1968).
503. A. Waksmundzki, Z. Suprynowicz, and T. Piertrusinska, *Chem. Anal. (Warsaw)*, **9**, 731 (1964).
504. H. G. Haring and J. Kroon, *J. Chromatogr.*, **16**, 285 (1964).
505. C. F. Raley and J. W. Kaufman, *Anal. Chem.*, **40**, 1371 (1968).
506. E. Schulte and L. Acker, *Z. Anal. Chem.*, **268**, 260 (1974).

507. H. F. Beckman and A. Bevenue, *J. Chromatogr.*, **10**, 231 (1963).
508. R. N. Bright and R. A. Matula, *J. Chromatogr.*, **37**, 217 (1968).
509. C. A. Schuntner and H. J. Schnitzerling, *J. Chromatogr.*, **21**, 483 (1966).
510. W. Kemula and A. Krzeminska, *Chem. Anal. (Warsaw)*, **5**, 611 (1960).
511. A. R. Amell and R. Helt, *J. Agric. Food Chem.*, **3**, 53 (1955).
512. G. Drechsler, *Z. Chem.*, **3**, 104 (1963).
513. W. W. Schulz and W. C. Purdy, *Anal. Chem.*, **35**, 2044 (1963).
514. U. A. T. Brinkman and J. J. de Kok, *Z. Anal. Chem.*, **283**, 203 (1977).
515. J. Lauckner and H. Fürst, *Chem. Tech. (Berl.)*, **20**, 236 (1968).
516. W. Diemair, G. Maier, and K. Schloegel, *Z. Lebensmitt-Forsch.*, **139**, 294 (1969).
517. D. C. Abbott, H. Egan, E. W. Hammond, and J. Thomson, *Analyst*, **89**, 480 (1964).
518. M. F. Kovacs, Jr., *J. Assoc. Off. Agric. Chem.*, **46**, 884 (1963).
519. F. K. Kawahara, R. L. Moore, and R. W. Gorman, *J. Gas Chromatogr.*, **6**, 24 (1968).
520. G. Drechsler, *J. Prakt. Chem.*, **22**, 282 (1963).
521. R. Feinland, J. Sass, and S. A. Buckler, *Anal. Chem.*, **35**, 920 (1963).
522. C. J. Hardy, *J. Chromatogr.*, **13**, 372 (1964).
523. L. Nebbia and V. Bellotti, *Chim. Ind. (Milan)*, **52**, 369 (1970).
524. D. F. McCaulley, *J. Assoc. Off. Agric. Chem.*, **48**, 659 (1965).
525. C. Krause and J. Kirchhoff, *Dtsch. Lebensmitt-Rundsch.*, **66**, 194 (1970).
526. G. H. Boone, *J. Assoc. Off. Agric. Chem.*, **48**, 748 (1965).
527. A. Varon, F. Jakob, K. C. Ciric, and W. Rieman, III, *Talanta*, **9**, 573 (1962).
528. R. B. Lew, H. Gard, and F. Jakob, *Talanta*, **10**, 911 (1963).
529. J. Rabinowitz, B. Baehler, and G. Weber, *Helv. Chim. Acta*, **49**, 590 (1966).
530. A. Lamotte and J. C. Merlin, *J. Chromatogr.*, **38**, 296 (1968).
531. G. Neubert, *J. Chromatogr.*, **20**, 342 (1965).
532. A. Lamotte, A. Francina, and J. Merlin, *J. Chromatogr.*, **44**, 75 (1969).
533. E. E. Nifantev, *Zh. Obsch. Khim.*, **35**, 1980 (1965).
534. H. Petschik and E. Steger, *J. Chromatogr.*, **9**, 307 (1962).
535. M. Salame, *J. Chromatogr.*, **16**, 476 (1964).
536. B. Y. Giang and H. F. Beckman, *J. Agric. Food Chem.*, **17**, 63 (1969).
537. N. Cvjeticanin and J. Cvoric, *Bull. Inst. Nucl. Sci. Belgrade*, **13**, 35 (1962).
538. H. A. Moule and S. Greenfield, *J. Chromatogr.*, **11**, 77 (1963).
539. N. I. Gabov and A. I. Shafiev, *Zh. Anal. Khim.*, **21**, 1107 (1966).
540. A. Cortes and D. R. Gilmore, *J. Chromatogr.*, **19**, 450 (1965).
541. V. Batara, J. Kovac, M. Behusova, and J. Kovacicova, *J. Chromatogr.*, **35**, 277 (1968).
542. H. Y. Luh, K. X. Ye, and C. Y. Yuen, *Acta Chim. Sin.*, **30**, 471 (1964).
543. R. Roper and T. S. Ma, *Microchem. J.*, **1**, 246 (1957).

544. T. S. Ma and R. C. Rittner, *Modern Organic Elemental Analysis*, Dekker, New York, 1978.
545. J. H. Taylor, *J. Gas Chromatogr.*, **6**, 557 (1968).
546. E. V. Sivtsova, V. B. Kogan, and S. K. Ogorodnikov, *Zh. Prikl. Khim.*, **38**, 2609 (1965).
547. T. Oiwa, M. Sato, Y. Miyakawa, and I. Miyazaki, *J. Chem. Soc. Jap., Pure Chem. Sect.*, **84**, 409 (1963).
548. K. Kawazumi, S. Kataoka, and K. Maruyama, *J. Chem. Soc. Jap., Ind. Chem. Sect.*, **64**, 784 (1961).
549. T. Garzo, F. Rill, and I. Till, *Magy Kem. Foly.*, **68**, 327 (1962).
550. J. Joklik, *Collect. Czech. Chem. Commun.*, **26**, 2079 (1961).
551. M. Wurst, *Collect. Czech. Chem. Commun.*, **29**, 1458 (1964).
552. N. A. Palamarchuk, S. V. Syavtsillo, N. M. Turkeltaub, and V. T. Shemyatenkova, *Tr. Komiss. Anal. Khim., Akad. Nauk SSSR*, **13**, 277 (1963).
553. L. Preisler, *Z. Anal. Chem.*, **240**, 389 (1968).
554. M. Wurst and R. Dusek, *Collect. Czech. Commun.*, **27**, 2391 (1962).
555. M. Wurst, *Collect Czech. Chem. Commun.*, **30**, 2038 (1965).
556. B. M. Luskina, *Zavod. Lab.*, **33**, 1496 (1967).
557. V. Bilik, S. Bauer, I. Jezl, and M. Furdik, *Chem. Zvesti*, **19**, 28 (1965).
558. G. R. Waller, S. D. Sastry, and K. Kinneberg, *J. Chromatogr. Sci.*, **7**, 577 (1969).
559. J. Nagy and E. Brandt-Petrik, *Period. Polytech.*, **10**, 443 (1966).
560. J. Franc and J. Senkyrova, *J. Chromatogr.*, **36**, 512 (1968).
561. K. C. Leibman and E. Ortiz, *J. Chromatogr.*, **32**, 757 (1968).
562. J. Lehrfeld, *J. Chromatogr.*, **32**, 685 (1968).
563. B. O. Lindgren and C. M. Svahn, *Acta Chem. Scand.*, **20**, 1763 (1966).
564. J. Franc and J. Ceeova, *Collect. Czech. Chem. Commun.*, **33**, 1570 (1968).
565. E. J. Bonnelli and H. Hartmann, *Anal. Chem.*, **35**, 1980 (1963).
566. K. Broderson and U. Schlenker, *Z. Anal. Chem.*, **182**, 421 (1961).
567. C. S. Evans and C. M. Johnson, *J. Chromatogr.*, **21**, 202 (1966).
568. N. S. Vyazankin, G. N. Bortnikov, I. A. Migunova, A. V. Kiselev, Y. I. Yashin, A. N. Egorochkin, and V. F. Mironov, *Izv. Akad. Nauk SSSR, Ser. Khim.*, **1969**, 186.
569. H. Veening, N. J. Graver, D. B. Clark, and B. R. Willeford, *Anal. Chem.*, **41**, 1655 (1969).
570. K. Arakawa and K. Tanikawa, *Jap. Analyst*, **16**, 812 (1967).
571. P. Longi and R. Mazzocchi, *Chim. Ind. (Milan)*, **48**, 718 (1966).
571a. G. Garzo, J. Fekete, and M. Blazso, *Acta Chim. Hung.*, **51**, 359 (1967).
572. S. J. Valenty and P. E. Behnken, *Anal. Chem.*, **50**, 834 (1978).
573. K. Rissanen and J. K. Miettinen, *Ann. Agric. Fenn. Suppl.*, **7**, 22 (1968).
574. G. W. Johnson and C. Vickers, *Analyst*, **95**, 356 (1970).

575. O. A. Ptitsyna, S. I. Orlov, and O. A. Reutov, *Vestn. Mosk. God. Univ., Ser. Khim.*, **1966**, (5), 50.

576. K. Figge, *J. Chromatogr.*, **39**, 84 (1969).

577. F. Jursik, *J. Chromatogr.*, **35**, 126 (1968).

578. R. Barbieri, U. Belluco, and G. Tagliavini, *Ric. Sci.*, **30**, 1671 (1960).

579. I. Z. Kozina and A. I. Subbotina, *Zavod. Lab.*, **36**, 670 (1970); *Tr. Khim. Borkii*, **1968**, 153.

Author Index

Numbers indicate pages where references are given in full.

Subject Index

Abbreviations: detn. = determination; sepn. = separation.